A New Approach to Scientific Computation

Notes and Reports in Computer Science and Applied Mathematics

Editor
Werner Rheinboldt
University of Pittsburgh

1. JOHN R. BOURNE. Laboratory Minicomputing

2. CARL TROPPER. Local Computer Network Technologies

3. KENDALL PRESTON, JR. AND LEONARD UHR (Editors). Multicomputers and Image Processing: Algorithms and Programs

4. STEPHEN S. LAVENBERG (Editor). Computer Performance Modeling Handbook

5. R. MICHAEL HORD. Digital Image Processing of Remotely Sensed Data

6. SAKTI P. GHOSH, Y. KAMBAYASHI, AND W. LIPSKI (Editors). Data Base File Organization: Theory and Applications of the Consecutive Retrieval Property

7. ULRICH W. KULISCH AND WILLARD L. MIRANKER (Editors). A New Approach to Scientific Computation

A New Approach to Scientific Computation

Edited by

Ulrich W. Kulisch

Institute for Applied Mathematics
University of Karlsruhe
Karlsruhe, West Germany

Willard L. Miranker

Department of Mathematical Sciences
IBM Thomas J. Watson Research Center
Yorktown Heights, New York

ACADEMIC PRESS, INC.

(Harcourt Brace Jovanovich, Publishers)

Orlando San Diego New York London
Toronto Montreal Sydney Tokyo

Academic Press Rapid Manuscript Reproduction

Proceedings of the Symposium on A New Approach
to Scientific Computation
Sponsored by IBM and Held at the IBM Thomas J. Watson
Research Center
Yorktown Heights, New York, August 3, 1982

ACADEMIC PRESS, INC.
Orlando, Florida 32887

United Kingdom Edition published by
ACADEMIC PRESS, INC. (LONDON) LTD.
24/28 Oval Road, London NW1 7DX

Library of Congress Cataloging in Publication Data

Main entry under title:

A new approach to scientific computation.

(Notes and reports in computer science and applied mathematics)
Includes bibliographical references.
1. Numerical analysis--Data processing. 2. Algebra--Data processing. 3. Computer arithmetic. I. Kulisch, Ulrich W. II. Miranker, Willard L. III. Series.
QA297. N47 1983 519.4 83-15595
ISBN 0-12-428660-7 (alk. paper)

PRINTED IN THE UNITED STATES OF AMERICA

85 86 87 88 9 8 7 6 5 4 3 2

This volume is dedicated to J. Hartmut Bleher,
a friend indeed to scientific computation

Contents

Contributors

Numbers in parentheses indicate the pages on which the authors' contributions begin.

Gerd Bohlender (247, 269, 311), Institut für Angewandte Mathematik, Universität Karlsruhe (TH), D-7500 Karlsruhe 1, West Germany

Harald Böhm (121, 311), Institut für Angewandte Mathematik, Universität Karlsruhe (TH), D-7500 Karlsruhe 1, West Germany

Kurt Grüner (247, 311), Institut für Angewandte Mathematik, Universität Karlsruhe (TH), D-7500 Karlsruhe 1, West Germany

Jürgen Wolff von Gudenberg (225, 311), Institut für Angewandte Mathematik, Universität Karlsruhe (TH), D-7500 Karlsruhe 1, West Germany

Edgar Kaucher (139, 311), Institut für Angewandte Mathematik, Universität Karlsruhe (TH), D-7500 Karlsruhe 1, West Germany

Rudi Klatte (311), Institut für Angewandte Mathematik, Universität Karlsruhe (TH), D-7500 Karlsruhe 1, West Germany

Walter Krämer (311), Institut für Angewandte Mathematik, Universität Karlsruhe (TH), D-7500 Karlsruhe 1, West Germany

Ulrich W. Kulisch (1, 269, 311), Institut für Angewandte Mathematik, Universität Karlsruhe (TH), D-7500 Karlsruhe 1, West Germany

Willard L. Miranker (165, 311), IBM Thomas J. Watson Research Center, Yorktown Heights, New York 10598

Louis B. Rall (291), Mathematics Research Center, University of Wisconsin, Madison, Wisconsin 53706

Siegfried M. Rump (27, 51, 311), IBM Deutschland, Entwicklung und Forschung, 7030 Boeblingen, West Germany

Christian P. Ullrich (199, 311), Institut für Angewandte Mathematik, Universität Karlsruhe (TH), D-7500 Karlsruhe 1, West Germany

Preface

The contributions in this volume are based on papers delivered at a symposium at the IBM Thomas J. Watson Research Center on August 3, 1982. They treat various aspects of a new approach to scientific computation, which derives from a new and systematic theory of computer arithmetic. For its implementation on computers, the new arithmetic requires new techniques in both hardware and software and affects higher programming languages as well. The new arithmetic not only leads to better results in traditional numerical analysis, but also suggests promising new approaches. It changes the interplay between computation and numerical analysis in a qualitative way. Very general methods have been developed that solve algebraic problems of many types with maximum accuracy. Other algorithms compute approximations to the continuous solution of initial or boundary value problems for ordinary differential equations and provide upper and lower bounds for the solution with full machine accuracy. The new ideas are applicable to integral equations, functional equations, and partial differential equations as well. They do more than provide sharp bounds for the solution, also permitting the computer to verify simultaneously the existence and uniqueness of the solution within the computed bounds. If there is no unique solution, for instance in the case of a singular matrix, the computer delivers this information to the user. The new methods have been called E-methods, referring to the three German words *Existenz* (existence), *Eindeutigkeit* (uniqueness), and *Einschliessung* (inclusion).

The first contribution reviews concepts and results of a new theory of computer arithmetic. A survey of a computer demonstration that shows the effectiveness of the new methodology follows.

The next three contributions deal with the principal ideas of E-methods, the first examining algebraic problems, the second arbitrary expression evaluation, and the third functional equations. In each of these contributions, algorithms are developed that lead to computable bounds of high accuracy for the solution. To obtain these results for functional equations, the principal ideas that led to the new theory of computer arithmetic must now be extended to function spaces. These developments and considerations are discussed in the following contribution on ultra-arithmetic.

The next four contributions examine the question of effective computer implementation of the new arithmetic. The implications of the new methods for higher programming languages are given in the two contributions that deal with appropriate extensions of FORTRAN and PASCAL, respectively (the latter called MATRIX PASCAL). The third contribution deals with the principal technical algorithms for the implementation of the new arithmetic on computers, and the fourth deals with questions of hardware implementation of the new arithmetic. Algorithmic and flowchart descriptions of an appropriate hardware unit are given.

The volume concludes with two contributions: an application and an extension of PASCAL-SC. The application concerns symbolic differentiation, and the extension, MATRIX PASCAL, provides an effective PASCAL language extension for the requirements of the new arithmetic, including, in particular, the evaluation of expressions with maximum accuracy.

Acknowledgments

There have been many sources of support for the original work, the symposium itself, and this volume. Among them are Bundesministerium für Forschung und Technologie, Deutsche Forschungsgemeinschaft, Alexander von Humboldt Stiftung, IBM Deutschland, IBM Thomas J. Watson Research Center, Mathematics Research Center, University of Wisconsin, and Nixdorf Computer AG. We are grateful to all of these institutions and to the people associated with them who supported the work presented here.

The manuscript of the volume was prepared at the IBM Thomas J. Watson Research Center. We are grateful to the Graphics Department for the preparation of the many complicated figures. We are also most grateful to Jo Genzano, who helped with the organization and logistics of the symposium, prepared the entire text on the word processor, and coordinated the long and complicated process of interaction with the authors.

A NEW ARITHMETIC FOR SCIENTIFIC COMPUTATION

Ulrich Kulisch

Institute for Applied Mathematics
University of Karlsruhe
Karlsruhe, West Germany

The paper summarizes an extensive research activity in scientific computation that went on during the last fifteen years as well as the experience gained through various implementations of a new approach to arithmetic on diverse processors including even microprocessors.

We begin with a complete listing of the spaces that occur in numerical computations. This leads to a new and general definition of computer arithmetic.

Then we discuss aspects of traditional computer arithmetic such as the definition of the basic arithmetic operations, the definition of the operations in product spaces and some consequences of these definitions for error analysis of numerical algorithms.

In contrast to this we then give the new definition of computer arithmetic. The arithmetic operations are defined by a general mapping principle which is called a semimorphism. We discuss the properties of semimorphisms, show briefly how they can be obtained and mention the most important features of their implementation on computers.

Then we show that the new operations can not be properly addressed by existing programming languages. Correcting this limitation led to extensions of PASCAL and FORTRAN.

A demonstration of a computer that has been systematically equipped with the new arithmetic will follow the talk. The new arithmetic turns out to be a key property for an automatic error control in numerical analysis. By means of a large number of examples we show that guaranteed bounds for the solution with maximum accuracy can be obtained. The computer even proves the existence and uniqueness of the solution within the calculated bounds. If there is no unique solution (e.g., in case of a singular matrix) the computer recognizes it.

Further details will be discussed in subsequent papers.

1. INTRODUCTION

Considerable progress was made in computer arithmetic during the last 15 years. A summarizing representation of this progress in the form of lecture notes was first published in [12]. This manuscript gives for the first time a complete formal mathematical description of the spaces that occur if numerical computations are done by computers. Moreover, it suggests a new definition of computer arithmetic by so-called semimorphisms. Fast algorithms for the implementation of the newly defined operations are also given. Between 1976 and 1979 the first computer, which is completely equipped with the new arithmetic, was built at the author's

ISBN 0-12-428660-7

institute. It allows tremendous improvements in numerical analysis. In 1979 a new edition of the material was worked out at the IBM Research Center and was published in [15]. This book also deals with the question how existing programming languages have to be extended in order to address the new arithmetic in an optimal sense.

The new developments show that a great deal of mathematics is necessary for a complete description of computer arithmetic and the derivation of fast algorithms for its implementation. Ideas and motivation are easily lost among so much mathematics and details. Therefore in this treatise, we try to avoid too much formalism and mathematical proofs, so that comparison of the new developments with traditional techniques, and some motivation are made clear.

The question itself is certainly one of the most fundamental questions that occurred in mathematics during the last 25 years. How far can the numerical properties and behavior of computers be described and controlled in terms of mathematics..

Arithmetic in a certain range of the integers may be performed exactly on computers. In this paper, therefore, we consider the cases of the real numbers as well as product spaces over the real numbers.

2. THE SPACES OF NUMERICAL COMPUTATIONS

We begin with a complete listing of the spaces that occur when numerical computations are done on computers.

In addition to the integers, numerical algorithms are usually defined in the space R of real numbers and vectors VR or matrices MR over the real numbers. Additionally, the corresponding complex spaces C, VC and MC also occur occasionally. All these spaces are ordered with respect to the order relation $\leq$. (In all product sets the order relation is defined componentwise.). Using the order relation $\leq$ the concept of intervals can be defined in all these spaces which is of interest to numerical analysts who also define and study algorithms for intervals. If we denote the set of intervals over an ordered set $\{M,\leq\}$ by IM, we obtain

		R	$\supset$	D	$\supset$	S
		VR	$\supset$	VD	$\supset$	VS
		MR	$\supset$	MD	$\supset$	MS
PR	$\supset$	IR	$\supset$	ID	$\supset$	IS
PVR	$\supset$	IVR	$\supset$	IVD	$\supset$	IVS
PMR	$\supset$	IMR	$\supset$	IMD	$\supset$	IMS
		C	$\supset$	CD	$\supset$	CS
		VC	$\supset$	VCD	$\supset$	VCS
		MC	$\supset$	MCD	$\supset$	MCS
PC	$\supset$	IC	$\supset$	ICD	$\supset$	ICS
PVC	$\supset$	IVC	$\supset$	$IVCD$	$\supset$	$IVCS$
PMC	$\supset$	IMC	$\supset$	$IMCD$	$\supset$	$IMCS$

Figure 1

the spaces IR, IVR, IMR and IC, IVC, IMC. See the second column of Figure 1.

Every algorithm in numerical analysis is defined in one or several of these spaces. But in general these algorithms cannot be executed within them. For an execution we have to use computers. On a computer there is only a system S (i.e., single precision floating-point numbers) available. If a prescribed accuracy for a computation cannot be achieved by operating within S, we use a larger subset D (double precision floating-point numbers) of R with the property $R \supset D \supset S$. Vectors (n-tuples), matrices ($n \times n$-tuples), complexifications (pairs), vectors and matrices of such pairs as well as the corresponding sets of intervals can now be defined over S and D. Doing so, we obtain the spaces VS, MS, IS, IVS, IMS CS, VCS, MCS, ICS, $IVCS$, $IMCS$ and the corresponding spaces over D. These two collections of spaces are listed in the third and fourth columns of Figure 1. By computer arithmetic, we understand all operations that have to be defined in all of these sets listed in the third and fourth columns of Figure 1 as well as in certain combinations of these sets. In a good programming system, these operations should be available as operators for all admissible combinations of data types.

The number of operations just mentioned is larger than might be expected. A complex matrix, for instance, can be multiplied by another complex matrix, but also by a complex vector or a complex number as well as by a real matrix, a real vector or a real number or an integer matrix, an integer vector or an integer number. In the sense of the type concept of programming languages these multiplications are to be understood as being different. A count of the numbers of inner and outer operations that occur in the fourth column of Figure 1 shows them to be several hundred in number. Before defining these operations by the newly developed mathematical concept of semimorphism we first discuss a few features of the definition of the arithmetic operations on traditional computers. We expect that this helps to clarify the differences of the two methods.

3. TRADITIONAL DEFINITION OF COMPUTER ARITHMETIC: THE VERTICAL METHOD

From the several hundred operations that occur in the column under S in Figure 1, traditional computers in general provide only four, the addition, subtraction, multiplication and division of floating-point numbers in S. All the others have to be defined by the user himself in case he requires them. He can only do so by means of subroutines. Each occurrence of an operation in an algorithm then causes a procedure call. This is a complicated, time consuming and numerically unprecise approach.

When the traditional programming languages like FORTRAN and ALGOL were defined in the 50's it was a common consensus that these should not include the properties of the arithmetic operations. The definition and realization of the latter were left to the computer manufacturers. As a consequence the user nowadays does not know precisely what happens if he writes $+, -, \times$ or $/$ in an algorithm. Moreover, two computers of different manufacturers in general differ in the properties of their arithmetic operations. This is an uncertain situation and numerical analysis, therefore, is based on unsafe ground. All that can be done in numerical analysis is to accept the basic assumptions about the error analysis of numerical algorithms as a substitute for a precise definition of the mathematical properties of the arithmetic operations of the computer. We begin with a review of these properties.

3.1 The Basic Operations

Error analyses of numerical algorithms make use of the following assumptions about the arithmetic. Provided that no underflow and no overflow occurs for the rounded image $\square a$ of a real number a, the following property holds:

$$\square a = a(1-\varepsilon) \text{ with } |\varepsilon| \leq \varepsilon^* \tag{1}$$

For the computer operations $\boxast$, $* \in \{+,-,\times,/\}$, we have

$$a \boxast b = (a*b)(1-\varepsilon) \text{ with } |\varepsilon| < \varepsilon^*, \tag{2}$$

i.e., the relative error of a rounding respectively of each operation is in absolute value less than a machine constant ε^*:

$$|\square a - a| < \varepsilon^* |a| \tag{3}$$

$$|a \boxast b - a*b| < \varepsilon^* |a*b| \tag{4}$$

The typical value of the constant ε^* implies that the error of a rounding or of an operation is confined to the last digit of the mantissa only.

Numerical analysis shows that with these error formulas an error analysis of simple algorithms can be given. So far it is a matter of fact, however, that error estimates obtained with these formulas, cannot be used to derive reliable error bounds for the solution of the problem on all computers. This would only be possible if (1) and (2) would hold for all $a \in R$ respectively for all $a,b \in S$ provided that no under- and no overflow occurs. For many computers on the market (1) and (2) do not hold strictly in this sense.

To illustrate this deficiency we cite an actual example on a real and commonly used computer. We subtract the two numbers $a = 134217728.0$ and $b = 134217727.0$. We obtain the result 2 and not 1 as expected [25]. In this example the relative error is 1 and not 10^{-8} as was expected by (2). What happens on the computer? In the binary system the two numbers are $a = 1.00...0 \cdot 2^{27}$ and $b = 0.11...1 \cdot 2^{27}$. They are representable on the computer without rounding. Before entering the addition routine the computer transfers a into a normalized floating-point number. Then the exponent parts of the numbers are equalized. This is achieved by shifting the second operand b one digit to the right. Then the result is obtained as follows:

$$\begin{array}{r|l} 0.1\;0\;0\;\ldots\;0 & *2^{28} \\ 0.0\;1\;1\;\ldots\;1 & 1*2^{28} \\ \hline 0.0\;0\;0\;\ldots\;0 & 1*2^{28} \end{array} \qquad (5)$$

On the real computer the last digit 1 is cut off. This leads to a difference of unity one place further to the left. The computed difference is thus 2.

Similar examples can be shown for other computers. The effect is obviously independent of the length of the mantissa. This is indicated by the dots in (5).

Now let b denote the base of the number system in use and ℓ the number of digits in the mantissa of the floating-point numbers. If a computer executes subtraction with an accumulator of length ℓ, then there exist approximately b^{ℓ} subtractions for which (2) does not hold. An error analysis based on (2) is then dubious, because of the many holes in these basic assumptions. Modern computers are able to execute 100 million floating-point operations per second. A computation of this speed can easily run into one such hole. Then the result of an error analysis based on (2) is not valid for the entire computation. The way out of this situation is a clear mathematical definition of the arithmetic properties of the computer (possibly in the context of programming languages), and the requirement that they be realized and fulfilled strictly by the manufacturers.

3.2 Higher Arithmetic Operations

For the sake of clarity let us restrict our consideration to a subset of the spaces displayed in Figure 1. Let us consider the spaces R, C and MC as well as their computer representable subsets S, CS and MCS. See Figure 2. Complex matrices occur in many applications such as the Fast Fourier Transform. In analysis the operations for complex numbers in C are defined by those of the real numbers in R through well known formulas. Similarly, the operations for complex matrices of MC are defined by the operations on the complex numbers (the componentwise addition and the scalar product in case of multiplication). On traditional computers a floating-point arithmetic is usually available. We indicate this by an arrow drawn between the sets R and S in Figure 2. The question mark along this arrow reminds us to the uncertainties of this arithmetic mentioned in the previous section. Floating-point operations in CS and MCS are now defined by taking the formulas which define the operations in C and MC and executing them by the given floating-point operations in S. Figure 2 shows this connection. It is clear that proceeding like this the uncertainties indicated by the question mark reoccur in the arithmetic in CS and MCS also.

$$\begin{array}{ccc} R & \xrightarrow{\ ?\ } & S \\ \downarrow & & \downarrow ? \\ C & & CS \\ \downarrow & & \downarrow ? \\ MC & & MCS \end{array}$$

Figure 2

We now discuss yet another dilemma which occurs if we define the arithmetic in CS and MCS as indicated. Let $\alpha = \alpha_1 + i\alpha_2$ and $\beta = \beta_1 + i\beta_2$ be complex numbers and $a = (a_{ij})$ and $b = (b_{ij})$ be two complex matrices. The following product formulas are well known:

$$\alpha \bullet \beta = (\alpha_1\beta_1 - \alpha_2\beta_2,\ \alpha_1\beta_2 + \alpha_2\beta_1),$$
$$a \bullet b = \left(\sum_{k=1}^{n} a_{ik} \bullet b_{kj} \right).$$

Four multiplications and two additions occur in $\alpha \bullet \beta$. For a description of the associated computational error in the sense of (2), we need six inaccuracy factors ε_i. In $a \bullet b$ products of complex floating-point numbers appear under the summation sign. For an error analysis of one such product we need six inaccuracy factors ε_i. If the matrices a and b are comparatively small with perhaps 100 rows and columns, we have to sum over 100 such products. For an error description of these products we need 600 inaccuracy factors ε_i. The real- and imaginary parts of these products then have to be added. This gives 200 inaccuracy factors ε_i. Thus, in order to describe the error of one single component of the product matrix $a \bullet b$ we need about 800 inaccuracy factors ε_i. Since we have a 100×100-matrix this means that eight million inaccuracy factors are necessary to describe the error of only one matrix product. If we take a 200×200-matrix we obtain 64 million inaccuracy factors. Matrices, that occur in practice, sometimes have thousands of rows of columns.

It is in general impossible to study the error propagation of several million ε_i in a complicated algorithm. Because of the very rapidly increasing number of inaccuracy factors ε_i, an error analysis of the type in question for numerical algorithms is hardly possible. In order to come through one has to globalize, that is, estimate and take norms early on. If one reaches a final estimation at all, this in general leads to crude and unrealistic bounds.

In this section we are discussing the traditional way of defining computer arithmetic. The arithmetic operations in the sets of the third and fourth column of Figure 1 are defined by making use of operations in S respectively in D that are given somehow. We call this the *vertical method* of defining the arithmetic operations in the sets of the third and fourth columns. It is clear that imprecision in the definition and properties of the arithmetic in S resp. D reoccurs in all sets in the columns under S resp. D.

3.3 Error Analysis of Numerical Algorithms

Traditionally the error analysis of numerical algorithms takes one of the two forms: "forward analysis" or "backward analysis". The more natural way of looking at round off errors is the forward analysis. It simply asks how wrong the computed answers are compared to the correct solution of the problem. It is virtually axiomatic that the forward analysis of numerical algorithms is too difficult and usually impossible to carry out.

Instead of it, often a backward analysis is carried out. In general, it is characterized by asking how little change in the data of the problem would be necessary to cause the computed answer to be the exact solution of the altered problem. Backward analyses have been carried out successfully for many problems. One should remember, however, that the backward analysis is based on formulas like (1) and (2) also. And we saw already that such estimations are not strictly realized on many computers.

Now let M denote any set of Figure 1 in which the operations are known and N the subset on its right in the same row. For each operation $*$ in M we define an operation $\boxed{*}$ in N as follows:

(RG) $$\bigwedge_{a,b\in N} a \boxed{*} b := \square(a*b) \quad \text{for all } *.$$

Here $\square$: $M \rightarrow N$ denotes a mapping with the following properties

(R1) $$\bigwedge_{a\in N} \square a = a \qquad \text{(rounding)}$$

(R2) $$\bigwedge_{a,b\in M} (a \le b \Rightarrow \square a \le \square b) \qquad \text{(monotone)}$$

(R4) $$\bigwedge_{a\in M} \square(-a) = -\square a \qquad \text{(antisymmetric)}$$

In the case of the interval sets in Figure 1 we also require that the rounding $\square$ has the property

(R3) $$\bigwedge_{a\in M} a \le \square a \qquad \text{(upwardly directed)}$$

In this case the order relation $\le$ means set inclusion.

In mathematical context it sometimes occurs that a set with certain operations is considered, but the operations are in fact not implementable. Mathematicians then usually look for another set with implementable operations and try to arrange an isomorphism between the two sets and corresponding operations. The isomorphism is the strongest relevant mathematical mapping principle. It has the property that the inverse image of the result of a computation in the image set is the result that would have been obtained if the computation could have been executed in the original set.

Since the operations in the leftmost element of each row of Figure 1 are not computer implementable we have a similar situation. However, in Figure 1 set-subset pairs occur which are of different cardinality, and between such sets, isomorphisms cannot be established.

The new definition of computer arithmetic, which we shall discuss in the next section, provides tools that allow computation of sharp bounds for the solution of problems by the computer itself. In general, this is possible even for problems with imprecise data. The need for an error analysis which for special problems is done by mathematical methods applying the forward or backward analysis then seems to be secondary. Since a mathematical theory in general, considers a whole class of problems the computed bounds making use of the individual data of the problem will in general be sharper anyhow.

4. THE NEW DEFINITION OF COMPUTER ARITHMETIC: THE HORIZONTAL METHOD

Mathematicians often tried to specify and formalize the concept of computer arithmetic. All these attempts considered the mapping of the real numbers onto the floating-point numbers only. The real numbers, however, have so many special properties that it certainly is hard to seize the essential mathematical properties out of this model. In general, it is difficult to develop a mathematical theory just using one model. Therefore, we have to look for further models. The several rows of Figure 1 provide us with a means for doing this.

4.1 Properties of a Semimorphism

We are now going to define the operations in the sets of the third and fourth columns of Figure 1 by means of a general mapping principle.

In several sets of the second column of Figure 1 we know the operations as well as the resulting structure very well. These are the spaces R, VR, MR and C, VC and MC. Now let M be one of these sets and $*$ be one of the operations defined in M. Then in the powerset PM, which is the set of all subsets of M, an operation $*$ can be defined by

$$\bigwedge_{A,B \in PM} A*B := \{a*b \mid a \in A \wedge b \in B\}. \tag{6}$$

With (6), for each operation in M a corresponding operation in PM is defined. Summarizing we can say that the operations in the sets listed in the left most element of every row in Figure 1 are always known. We now use these operations in order to define operations in the subsets on the right hand side of the same row by a general method.

The next weaker mathematical mapping principle is that of a homomorphism. This still leaves groups invariant. However, it can also be shown by simple examples [11] or a mathematical theorem [24] that even homorphisms cannot be established. Now we can try to weaken the mapping properties of a homomorphism still further. Then we reach our mapping properties (RG), (R1), (R2), (R3), (R4). They can be derived as necessary conditions for a homomorphism between ordered algebraic structures [12], [15]. Therefore, we call the mapping which they characterize a semimorphism. With the properties of a semimorphism we go as far as possible to a homomorphism. We also define all outer operations that occur in Figure 1 (scalar times vector, matrix times vector, etc.) by corresponding semimorphims.

We call this method of defining the operations in a subset N of a set M via semimorphism the *horizontal method* of defining computer arithmetic. See Figure 3. It is important to understand that the operations defined by semimorphism in general are different from those defined by the vertical method which we discussed previously.

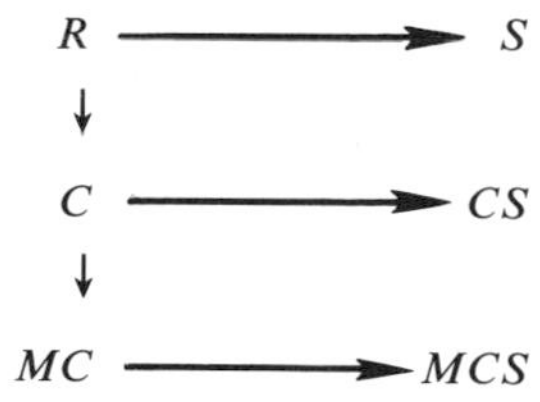

Figure 3

The horizontal method defines the operations in a subset N of a set M directly by making use of the operations in M. So a semimorphism directly links an operation and its approximation in the subset N. The operations in MCS (see Figure 3), for instance, are directly defined by the operations in MC and not in a round about way via C, R, S, CS to MCS as in the case of the vertical method (see Figure 2).

We have already seen above that the operations in the leftmost element of every row in Figure 1 are well known. It is easy to see now that a semimorphism of a semimorphism is a semimorphism again. With this discovery operations by semimorphisms in all sets of Figure 1 are directly defined. We define the outer operations in Figure 1 also by semimorphisms.

The new operations defined in all sets of Figure 1 deliver maximum accuracy in the sense that between the correct result of an operation and its approximation in the subset is no further element of the subset [12], [15]. This fundamental result is guaranteed by (RG), (R1) and (R2).

Further for the mapping of R into S (see the first row of Figure 1) the error formulas (1) and (2), that we assumed above, can now be proved to follow as a consequence [12], [15] in particular, with the for all quantifiers as in the formulas below.

These formulas now are not only valid for row 1 of Figure 1 but in the same simple form for all the other rows of Figure 1 as well:

$$\bigwedge_{a \in M} |\square a - a| < \varepsilon^* |a|,$$

$$\bigwedge_{a,b \in N} |a \boxast b - a * b| < \varepsilon^* |a * b|.$$

These formulas hold in all cases with just one single scalar factor ε^* which is in fact the same as the one in (3) and (4). Compared with the eight million ε_i that occurred in the case of just one single multiplication of two complex 100×100-matrices the simplification now is striking. The new error bounds for the operations in all sets of Figure 1 are much more accurate and of a simpler form. Thus, they allow a simpler error analysis of numerical algorithms and lead to more accurate error estimations and bounds [6].

4.2 Comments on the Derivation of Semimorphisms

We have already noted that certain essential properties of a semimorphism can be obtained as necessary conditions for a homomorphism between ordered algebraic structures. There are still other possibilities for deriving the properties of a semimorphism. They can be derived directly by considering special models of sets in Figure 1. Let us take, for instance, the mapping of the powerset of the complex numbers PC into the intervals of the complex numbers IC. An interval $[a,b]$ of two comparable complex numbers $a \leq b$ is a rectangle in the complex plane (Figure 4a). If we multiply two complex intervals A and B in the sense of (6) we generally do not obtain an interval once again, but a more arbitrary element $A \bullet B$ of the powerset PC (Figure 4b).

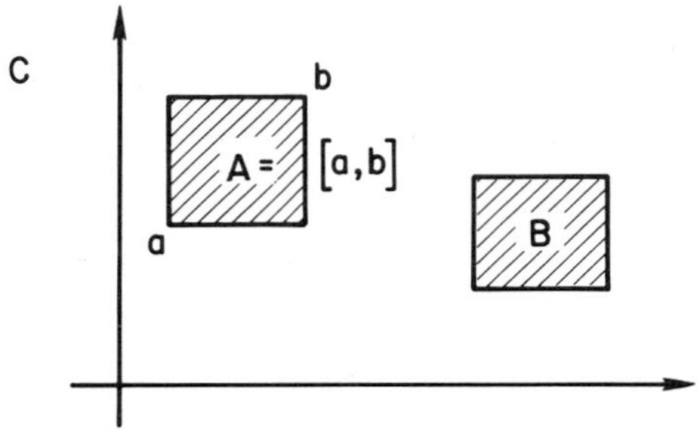

Figure 4a

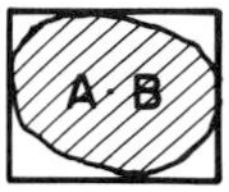

Figure 4b

The result of an interval operation, however, must be an interval. The best we can do to obtain an interval is to map the powerset product $A \bullet B$ on the least interval that contains it (Figure 4b). It is now easy to see that this mapping $\square$: $PC \rightarrow IC$ has all properties (R1, 2, 3, 4) and (RG) of a semimorphism: In our case the order relation is set inclusion $\subseteq$. If the set $A \bullet B$ is already an interval, the rounding has no effect, i.e., (R1) holds. If we enlarge the set $A \bullet B$ somewhat, this enlarges the least interval that includes it also, i.e., (R2) holds. (R3) is obvious and (R4) holds by reasons of symmetry. The result of our operation fulfills (RG) $A \boxed{*} B = \square(A * B)$ by construction.

If the properties of a semimorphism are once obtained they can easily be geometrically verified in case of other models of Figure 1. For instance, if we add two floating-point numbers a and b (row 1), then the correct sum $c = a + b$ is not in general a floating-point number (Figure 5). In order to obtain a floating-point number we round the result into the floating-point screen.

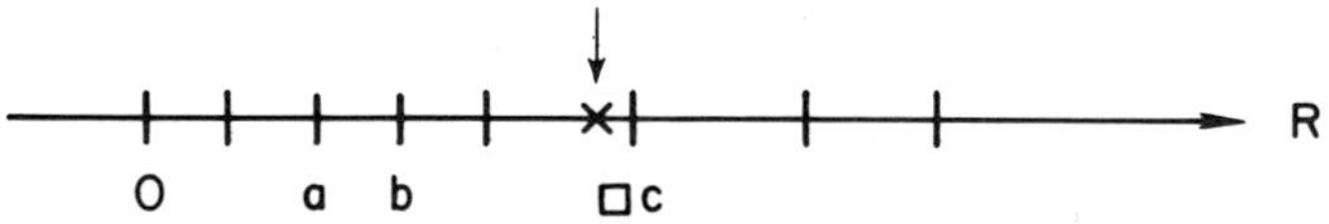

Figure 5

It is easy to see now that the process of rounding fulfills the properties (R1), (R2) and (R4) again. There the order relation is $\leq$. The floating-point operation is defined by (RG) $a \boxplus b = \square(a + b)$.

It is only natural that the properties of a semimorphism are not obvious in all set-subset pairs of Figure 1. This is not at all necessary. If they once are discovered they can be applied in all other cases in an abstract manner.

Finally, we note that the mapping principle of semimorphism acquires deeper significance since in all cases of Figure 1 it leaves certain reasonable mathematical structures, the so-called ordered or weakly ordered ringoids and vectoids, invariant. These properties cannot be discussed here. See [12], [15]. They are necessary in order to derive implementable algorithms for the diverse operations.

4.3 Implementation of Semimorphisms on Computers

We now discuss the question whether in all cases of Figure 1 the operations defined by semimorphism can be implemented on computers by means of fast algorithms. A method that defines computer arithmetic is a reasonable one if it can be implemented by algorithms that are comparable in speed to those of competitive definitions of computer arithmetic. At first sight it seems doubtful that formula (RG), in particular, can be implemented on computers at all. In order to determine the approximation $a \boxast b$, the result $a*b$ seems to be necessary. For instance, if in the case of addition in a decimal floating-point system, a is of magnitude 10^{50} and b of magnitude 10^{-50}, about 100 decimal digits in the mantissa would be necessary in order to represent $a + b$. Not even the largest computers have such long accumulators. Still more difficult situations for implementation of (RG) arise in the case of floating-point matrix multiplication or in the case of the division of two complex floating-point numbers.

Nevertheless, one can prove by means of special algorithms [12], [15] that the formula of a semimorphism, especially (RG) can be realized in all cases of Figure 1. These algorithms show that whenever $a*b$ is not reasonably representable on the computer, it is sufficient to replace it by an appropriate and representable value $\widetilde{a*b}$ with the property

$\square(a*b) = \square(a\tilde{*}b)$. Then $a\tilde{*}b$ can be used to define $a \boxed{*} b := \square(a*b) := \square(a\tilde{*}b)$. These algorithms are comparable in speed with the corresponding algorithms of the vertical definition of the arithmetic operations. A careful implementation of floating-point arithmetic easily delivers the two directed roundings $\triangledown$ and $\triangle$ also as well as the corresponding operations $\overset{*}{\triangledown}$ and $\overset{*}{\triangle}$, $* \in \{+, -, \times, /\}$. The roundings $\triangledown$ and $\triangle$ are defined by (R1), (R2) and

(R3) $\bigwedge_{a \in R} \triangledown a \leq a \qquad \bigwedge_{a \in R} a \leq \triangle a$

and the operations by

(RG) $\bigwedge_{a,b \in S} a \overset{*}{\triangledown} b := \triangledown(a*b) \qquad \bigwedge_{a,b \in S} a \overset{*}{\triangle} b := \triangle(a*b), \quad * \in \{+, -, \times, /\}.$

These roundings and operations are necessary to obtain guarantees in computation. They are equivalent to the operations on intervals which are defined over S. Algorithms for an implementation of monotone and antisymmetric roundings $\square$, the two directed roundings $\triangledown$ and $\triangle$ and the corresponding operations $\boxed{*}$, $\overset{*}{\triangledown}$, $\overset{*}{\triangle}$, $* \in \{+, -, \times, /\}$, defined by (RG) are given in [12], [15]. A precise analysis shows that with these algorithms, the implementation of semimorphisms is possible for the rows, 1, 2, 4, and 5 of Figure 1.

We now turn to the implementation of the arithmetic operations for row 3 of Figure 1. Here difficulties occur particularly in the case of the multiplication of two floating-point matrices $a = (a_{ij})$ and $b = (b_{ij})$. (RG) demands an implementation of the matrix multiplication by

$$a \boxdot b := \square(a \cdot b) = \square\Big(\sum_{k=1}^{n} a_{ik} \cdot b_{kj}\Big).$$

Here the sum and multiplication signs mean the correct operations in the sense of real numbers. The rounding may only be executed once (componentwise) at the end of the computation. Since the a_{ij} and b_{ij} are floating-point numbers the products $a_{ik} \cdot b_{kj}$ can be exactly executed in a double length accumulator. Thus, the problem is reduced to a computation of the sum

$$c = \square\Big(\sum_{k=1}^{n} c_k\Big), \tag{7}$$

where the c_i, $i = 1(1)n$, are double length floating-point numbers. c is a single precision floating-point number. The summation sign in (7) means the correct addition in the sense of real numbers. Since we need to be able to multiply matrices of arbitrary dimension, (7) requires algorithms for an implementation of all sums of floating-point numbers of double length with maximum accuracy. Such an algorithm was first given in [13]. Another very elegant algorithm, which solves the same problem by a different technique, was then given in [2]. Both algorithms are described in [15] also. In the meantime additional algorithms which solve the same problem were discovered. It is easy to see that with these algorithms the problem of implementing semimorphisms is solved for the rows 3, 7, 8 and 9 of Figure 1. We turn now to the implementation of semimorphisms for the remaining rows 6, 10, 11 and 12 of Figure 1.

In these cases we are confronted with a new situation. The operations in IMR (row 6 of Figure 1) are defined by

$$\text{(RG)} \qquad \bigwedge_{A,B \in IMR} A \boxast B := \square(A * B), \text{ for all } * \in \{+,-,\times\}.$$

Here $*$ denotes the operation in the powerset PMR, which is defined by (6). The powerset operations, however, are not executable in principle.

However, independently of the set IMR we can consider the set MIR. The elements of this set are matrices whose components are intervals. In MIR operations may be defined by

$$\bigwedge_{A = (A_{ij}),\ B = (B_{ij}) \in MIR} \begin{cases} A \oplus B := (A_{ij} \boxplus B_{ij}) \\[2ex] A \odot B := \left(\boxed{\sum_{k=1}^{n}} A_{ik} \boxdot B_{kj} \right). \end{cases}$$

This definition makes use of the executable operations in IR and the vertical method. It was already shown in [10] that these operations in MIR and those defined in IMR by (RG) are isomorphic with respect to the mapping

$$([a_{ij}^1, a_{ij}^2]) \longleftrightarrow [(a_{ij}^1), (a_{ij}^2)].$$

Thereafter it was shown in [21] and [22], that corresponding isomorphisms also hold if in the last three rows of Figure 1, the letter I is shifted to the right in the letter sequence denoting the spaces:

$$IC \longleftrightarrow CIR,$$

$$IVC \longleftrightarrow VCIR,$$

$$IMC \longleftrightarrow MCIR.$$

We omit displaying the mappings which establish these isomorphisms. Of course the situation for these spaces is more complicated.

Finally, it was shown in [14] that these isomorphisms are preserved upon passage from both sides of the above isomorphisms into the subsets of the third and fourth column of Figure 1 by semimorphisms. Doing so we obtain, for instance, the following isomorphisms:

$$IMS \longleftrightarrow MIS,$$

$$ICS \longleftrightarrow CIS,$$

$$IVCS \longleftrightarrow VCIS,$$

$$IMCS \longleftrightarrow MCIS.$$

For the verification of these isomorphisms, the weak structures that hold in the corresponding subsets are used. These subsets must be analyzed very carefully for this purpose. For the outer operations defined between different sets of Figure 1 the existence of corresponding isomorphisms can be demonstrated [14]. With these isomorphisms it can be shown then that an algorithm for an optimal computation of scalar products can be used for an optimal implementation of the operations in the sets of the rows 6, 10, 11 and 12 in Figure 1 via semimorphisms. The algorithms for an optimal computation of scalar products were originally developed for implementation of the matrix operations in row 3 by semimorphisms. It turned out that these algorithms are an essential tool for an implementation of semimorphic operations in rows 7, 8, 9, 10, 11 and 12 of Figure 1 also. See especially Chapter 6 of [15]. This shows the fundamental role played by optimal scalar product algorithms in the implementation of the arithmetic operations with maximum accuracy in the most commonly used linear spaces and their interval sets.

In [12] and [15] it is shown that all (semimorphic) operations in the sets of the third and fourth columns of Figure 1 can be realized by a modular technique in terms of a higher programming language if

1. an operator concept or an operator notation is available for all operations in that higher level language and
2. the following 15 fundamental operations for floating-point numbers are available:

$$
\begin{array}{ccccc}
\boxplus & \boxminus & \boxtimes & \boxslash & \boxdot \\
\overset{\triangledown}{+} & \overset{\triangledown}{-} & \overset{\triangledown}{\times} & \overset{\triangledown}{/} & \overset{\triangledown}{\cdot} \\
\overset{\vartriangle}{+} & \overset{\vartriangle}{-} & \overset{\vartriangle}{\times} & \overset{\vartriangle}{/} & \overset{\vartriangle}{\cdot}
\end{array}
$$

Here $\boxast$, $* \in \{+, -, \times, /\}$, denotes the semimorphic operations defined by (RG) using one monotone and antisymmetric rounding (R1,2,4) as, for instance, a rounding to the nearest number of the screen. $\overset{\triangledown}{*}$ and $\overset{\vartriangle}{*}$, $* \in \{+, -, \times, /\}$, denote the operations defined by (RG) and the monotone downwardly respectively upwardly directed rounding. $\boxdot$, $\overset{\triangledown}{\cdot}$ and $\overset{\vartriangle}{\cdot}$ denote the optimal scalar products with maximum accuracy

$$\bigcirc\left(\sum_{i=1}^{n} a_i \cdot b_i\right), \quad \bigcirc \in \{\square, \triangledown, \vartriangle\}.$$

Of these 15 fundamental operations traditional numerical analysis only makes use of the four operations $\boxplus$, $\boxminus$, $\boxtimes$, $\boxslash$. Traditional interval arithmetic uses the eight operations $\overset{\triangledown}{+}$, $\overset{\triangledown}{-}$, $\overset{\triangledown}{\times}$, $\overset{\triangledown}{/}$ and $\overset{\vartriangle}{+}$, $\overset{\vartriangle}{-}$, $\overset{\vartriangle}{\times}$, $\overset{\vartriangle}{/}$. The newly proposed IEEE Computer Society arithmetic [29], [30] offers all twelve of these operations $\boxast$, $\overset{\triangledown}{*}$, $\overset{\vartriangle}{*}$, $* \in \{+, -, \times, /\}$. These twelve operations were already implemented in software on a computer Zuse Z23 in 1967 and in hardware on an ELEKTROLOGIKA X8 in 1968 at the Institute for Applied Mathematics at the University of Karlsruhe. Both implementations were supported by a high level language "TRIPLEX ALGOL-60", published in Numerische Mathematik in 1968 [31]. Many difficulties that traditionally occur in numerical analysis as well as in interval analysis cannot be avoided unless the three scalar products $\boxdot$, $\overset{\triangledown}{\cdot}$, $\overset{\vartriangle}{\cdot}$

are available on the computer for all relevant values of n. We shall comment on the significance of these optimal scalar products in numerical analysis towards the end of this paper.

Algorithms providing the three optimal scalar products can and should be made available on every computer by very fast hardware routines. We note in passing, that optimal scalar products can be provided on a computer by a black box technique, where the vector components a_i and b_i, $i = 1(1)n$, are the input and the scalar products with maximum accuracy $\boxdot$, $\triangledown$, $\triangle$ the output.

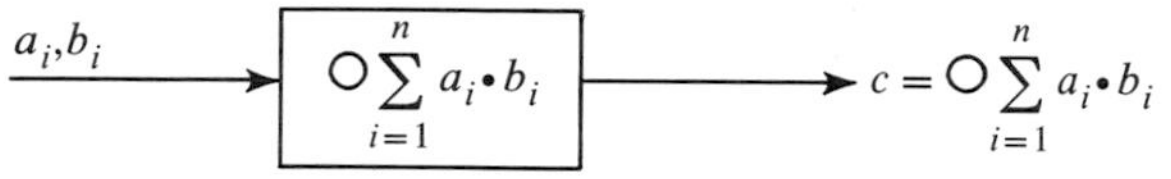

$$a_i, b_i \longrightarrow \boxed{\bigcirc \sum_{i=1}^{n} a_i \bullet b_i} \longrightarrow c = \bigcirc \sum_{i=1}^{n} a_i \bullet b_i$$

$$\bigcirc \in \{\square, \triangledown, \triangle\}.$$

The black box only needs some local storage and works independently of the main store of the computer. The size of the local storage of this optimal scalar product unit depends only on the data formats in use (base of the number system, length of the mantissa and range of the exponents). In particular, it is independent of the dimension n of the two vectors $a = (a_i)$ and $b = (b_i)$ to be multiplied. The access to this local storage is in general, much faster than access to the main storage of the computer. Additionally, portions of the sums and multiplications that occur in the scalar product can be executed simultaneously in a combined manner. This leads to a considerable gain in speed whenever scalar products occur in a computation.

5. COMPUTER ARITHMETIC AND PROGRAMMING LANGUAGES

Traditional programming languages like ALGOL, FORTRAN, PL/1, PASCAL, and so on use only two elementary data types and correspondingly provide the operations associated with these data types. These are the types integer and real. Even when complex operations are made available they are based on the real operations. All the other operations that we mentioned in the context of Figure 1 have to be defined and implemented by the user himself in form of procedures. The use of these operations for higher numerical units like vectors, matrices, complex numbers, complex vectors and matrices and intervals over all of these sets is a complicated procedure itself. Each occurrence of an operation in an algorithm causes a procedure call. This is notationally inconvenient, complicated and very time consuming. Above all, however, this method of defining the higher operations establishes the vertical definition of computer arithmetic.

In order to make this clear let us consider two real $n \times n$-matrices $a = (a_{ij})$ and $b = (b_{ij})$. In all languages, referred to above, a multiplication of a and b is programmed by using three loops like:

$$
\begin{array}{l}
\text{for } i := 1 \text{ to } n \text{ do} \\
\text{for } j := 1 \text{ to } n \text{ do} \\
\text{for } k := 1 \text{ to } n \text{ do} \\
\quad s := s + a[i,k] * b[k,j].
\end{array}
\tag{8}
$$

This clearly shows that the matrix multiplication is reduced to the real operations by way of the vertical method. We have a similar situation if we program all the other operations that occur in the third and fourth columns of Figure 1 in a traditional programming language.

This shows that in order to eliminate the vertical definition of the arithmetic operations in all product spaces of the third and fourth column of Figure 1 we have to eliminate the way these operations are programmed in traditional languages. A simple way of reaching this goal is to allow an operator notation for the operations that are newly defined by way of semimorphism. Traditional programming languages allow such a notation only for operands of the types integer, real and sometimes complex. The operational signs in expressions of these types are written with the familiar mathematical symbols $+, -, \times, /$.

In Section 1 we suggested that the arithmetic operations in the most frequently used linear spaces like the real and complex numbers, the vectors and matrices over the real and complex numbers as well as the spaces of intervals over all these sets should be defined by semimorphisms which among others provide maximum accuracy. This can easily be achieved if the programming language in use accepts elements of these spaces as special data types and allows such concepts as variables, operators, expressions, functions, comparisons and statements for all of these types. If in such a language, for instance, a, b and x are variables of the type "real matrix" the value of the product of the matrices a and b can be computed and assigned to the matrix x by the single statement

$$x := a*b. \tag{9}$$

The following statement

$$z := (a*x + b)*y + c \tag{10}$$

for example, computes the value of an arithmetic expression of the type complex, real or complex matrix, real or complex interval, or even real or complex interval matrix and assigns it to another variable z of the same type. A general operator concept was already part of ALGOL-68. Corresponding extensions of PASCAL and FORTRAN are discussed in [3], [4], [15] and [28].

An operator notation for all operations in the spaces displayed in Figure 1 greatly simplifies programming. This becomes evident if, for instance, the simple statement (10) in case of complex matrices is substituted by a corresponding FORTRAN program. A programming language equipped with such an operator notation naturally supports vector and matrix computations. So it is well prepared and suited for the use on array processors. In a programming language with an operator concept programs not only become shorter and clearer: they are easier to read and to write, easier to debug and therefore more reliable. Beyond this the operator notation is still accompanied by another essential advantage: In expressions or statements like (9) or (10) components or indices of the arrays a,b,c,x,y,z which occur in a traditional notation, have disappeared. This may save considerable computer time. Each run through the inner loop of (8) causes some address calculations, checking of index ranges or even an optimization of the for-statements by a special translation pass. All this is avoided if an operator notation is used. Whenever an operation like a multiplication sign occurs in expressions like (9) or (10) the computer calls an internal scalar product routine that does the index calculations automatically in a compact and highly efficient manner.

6. REALIZATION AND APPLICATIONS

A complete implementation of the arithmetic operations by semimorphism in all spaces of Figure 1 was first executed at the author's institute between 1977 and 1979. This implementation uses a Z80 microprocessor. Other implementations on more powerful processors have been finished in the meantime as well. PASCAL is used as the basic programming language. In order to avoid the disadvantages of the vertical definition of the arithmetic operations the programming language PASCAL was extended by a general operator concept. It is very similar to the one of ALGOL-68. The extended language was called PASCAL-SC (PASCAL for Scientific Computation) [3], [15,] and [28]. A new PASCAL-SC compiler was developed at the Institute of Prof. H.-W. Wippermann at the University of Kaiserslautern.

For implementation of the several hundred arithmetic operations mentioned in the context of Figure 1, only the 15 fundamental operations displayed and discussed toward the end of Section 4 had to be written in assembly language. All the other operations, in particular, those in the product spaces of Figure 1, were built up by a modular technique using the operator concept of the extended language PASCAL-SC. Some surprising effects should be mentioned. Although all arithmetic routines had to be implemented very carefully in order to achieve all properties of a semimorphism the new operations from row 3 on are faster as if they were implemented in basic PASCAL in the traditional manner. For example, the matrix multiplication via semimorphism implemented by use of the operator concept is twice as fast as implemented and executed in the traditional way in standard PASCAL.

The new arithmetic seems to have a tremendous impact on numerical analysis. The computer is no longer an experimental tool but it becomes an instrument of mathematics. Within a short time it was possible to develop program packages for all kinds of standard problems of numerical analysis such as linear systems of equations, inversion of matrices, eigenvalue and eigenvector problems, zeros of polynomials, nonlinear systems of equations, evaluation of arithmetic expressions, linear and quadratic optimization problems, numerical quadrature, initial and boundary value problems of ordinary differential equations, iterative solution of large systems of equations and so on. The special computer in use was only equipped with a twelve decimal digit arithmetic. For all problems just cited results can generally be guaranteed with at least 11 digits. The computation time is in all cases comparable with corresponding programs using ordinary floating-point arithmetic. Additionally, the new algorithms prove the existence and uniqueness of the solution within the computed bounds. If this is not possible (e.g., in case of a singular matrix) the computer recognizes it and informs the user by a corresponding message.

Many of these results were not obtained and could not be obtained by traditional implementation of computer arithmetic using the vertical definition of the operations. Many of these results are not achievable either, at least not in this generality, if the arithmetic operations only in the first row of Figure 1 are implemented very carefully via semimorphism, but the operations in the other rows of Figure 1 are defined by the vertical method.

It is the opinion of the author that the tacit agreement of the past, to let the manufacturers define computer arithmetic was erroneous. It is certainly a legitimate desire to control the accuracy of numerical algorithms. Then a necessary condition is to keep the accuracy of the simplest algorithms under control and to define them most precisely. These are the algorithms for the arithmetic operations in the most commonly used linear spaces and their interval sets (Fig. 1). This necessary condition turns out to be sufficient also to control the accuracy of numerical algorithms by the computer itself for many problems of numerical analysis.

A simple way out of the present situation is to define as a minimum the 15 most fundamental operations (which we explicitly listed and discussed in Section 4.3) axiomatically by the properties (RG), (R1), (R2) and (R4) respectively by (RG), (R1), (R2), (R4) and (R3) within the programming language. Then all other operations that occur in the spaces displayed in Figure 1 can be built up with maximum accuracy by a modular technique as shown in [12] and [15].

The following sentence is a golden rule in mathematics:

"*If a computation is correct one gets a correct result*".

In numerical analysis in general we cannot execute correct computations. Nevertheless, we are interested in the correct result. In the past mathematicians tried to close the gap between the result of a computation and the correct result of the problem by concepts like: stability of the process, the condition number of an algorithm or an error analysis made by hand. Computers, in the meantime, have reached speeds of several hundred million floating-point operations per second. An analysis of these concepts then often is of the same computational complexity as the computation of the solution of the original problem. Applications of the new theory of computer arithmetic show that the gap between the incorrect result of a computation and the correct result of the problem can be closed by the computer itself in a great number of cases if maximally accurate scalar products can be executed by the computer for all relevant vector dimensions n. Fast algorithms for the computation of scalar products with maximum accuracy turned out to be a key tool for an automatic error analysis and the computation of small bounds for the solution of the problem by the computer itself. It cannot be the place here to develop these ideas into further details. For the specialists we just mention that with optimal scalar products the step from the execution of the single arithmetic operations $+, -, *$ and $/$ with maximum accuracy to the computation of the value of arbitrary arithmetic expressions with maximum accuracy has been made.

REFERENCES

[1] Bohlender, G. (1977). Floating-point computation of functions with maximum accuracy. IEEE Trans. Comput. C-26, No. 7, 621-632.

[2] Bohlender, G. (1977). Genaue Summation von Gleitkommazahlen, Computing Suppl. 1, 21-32.

[3] Bohlender, G. (1978). Genaue Berechnung mehrfacher Summen, Produkte und Wurzeln von Gleitkommazahlen und Arithmetik in Höheren Programmiersprachen. Dissertation, Universität Karlsruhe.

[4] Bohlender, G., Kaucher, E., Klatte, R., Kulisch, U., Miranker, W. L., Ullrich, Ch. and Wolff von Gudenberg, J. FORTRAN for contemporary numerical computation, Report RC 8348, IBM Thomas J. Watson Research Center, 1980 and Computing 26, 277-314 (1981).

[5] Grüner, K. (1977). Fehlerschranken für lineare Gleichungssysteme, Computing Suppl. 1, 47-55.

[6] Grüner, K. (1979). Allgemeine Rechnerarithmetik und deren Implementierung. Dissertation, Universität Karlsruhe.

[7] Kaucher, E., Rump, S. M. (1980). Generalized iteration methods for bounds of the solution of fixed point operator equations, Computing 24, 131-137.

[8] Knuth, D. (1971). "The Art of Computer Programming", Vol. 2, Addison-Wesley. Reading, Massachusetts.

[9] Kulisch, U. An Axiomatic approach to rounded computations, TS Report No. 1020, Mathematics Research Center, University of Wisconsin, Madison, Wisconsin, 1969 and Numer. Math. 19, 1-17 (1971).

[10] Kulisch, U. (1970). Interval arithmetic over completely ordered ringoids, TS Report No. 1105, Mathematics Research Center, University of Wisconsin, Madison, Wisconsin.

[11] Kulisch, U. Implementation and formalization of floating-point arithmetic. Report RC 4608, IBM Thomas J. Watson Research Center, 1973 and C. Caratheodory Symp., 1973, 328-369.

[12] Kulisch, U. (1976). Grundlagen des numerischen Rechnens. Bibliographisches Institut Mannheim.

[13] Kulisch, U., Bohlender, G. (1976). Formalization and Implementation of Floating-point Matrix Operations. *Computing* 16, 239-261.

[14] Kulisch, U., Miranker, W. L. (1980). Arithmetic operations in interval spaces, Report RC 7681, IBM Thomas J. Watson Research Center, 1979; Computing Suppl. 2, 51-67.

[15] Kulisch, U., Miranker, W. L. (1981). Computer Arithmetic in Theory and Practice. Academic Press, New York.

[16] Kulisch, U. (1981). Numerisches Rechnen wie es ist und wie es sein könnte, Jahrbuch Überblicke Mathematik. Bibliographisches Institut.

[17] Lortz, B. (1971). Eine Langzahlarithmetik mit optimaler einseitiger Rundung, Dissertation, Universität Karlsruhe.

[18] Moore, R. E. (1966). "Interval Analysis". Prentice Hall, Englewood Cliffs, New Jersey.

[19] Reinsch, Ch. (1979). Die Behandlung von Rundungsfehlern in der Numerischen Analysis, Jahrbuch Überblicke Mathematik 1979, Wissenschaftsverlag des Bibliographischen Institus Mannheim, 43-62.

[20] Rutishauser, H. (1976). Eine Axiomatik des numerischen Rechnens und ihre Anwendung auf den Quotienten-Differenzen-Algorithmus, Vorlesungen über Numerische Mathematik, Band 2, 179-221, Birkhäuser-Verlag, Basel.

[21] Ullrich, Ch. (1972). Rundungsinvariante Strukturen mit äusseren Verknüpfungen, Dissertation, Universität Karlsruhe.

[22] Ullrich, Ch.: (1975). Über die beim numerischen Rechnen mit komplexen Zahlen und Intervallen vorliegenden mathematischen Strukturen, Computing 14, 51-65.

[23] Rump, S. M., Kaucher, E. (1980). Small bounds for the solution of systems of linear equations, Computing Suppl. 2, 157-164.

[24] Rump, S. M. (1980). Kleine Fehlerschranken bei Matrixproblemen, Dissertation, Universität Karlsruhe.

[25] Rump, S. M. (1980). Notiz zur Genauigkeit der Arithmetik in Rechenanlagen, Elektronische Rechenanlagen 22 H.5, S. 243-244.

[26] Wippermann, H.-W. (1968). Implementierung eines ALGOL-60 Systems mit Schrankenzahlen, Electron Datenverarb. 10, 189-191.

[27] Wolff von Gudenberg, J. Evaluation of the standard functions in generalized computer arithmetic.

[28] Wolff von Gudenberg, J. (1980). Einbettung allgemeiner Rechnerarithmetik in

PASCAL mittels eines Operatorkonzeptes und Implementierung der Standardfunktionen mit optimaler Genauigkeit. Dissertation, Universität Karlsruhe.

[29] Coonan, J., et al. (1979). A proposed standard for floating-point arithmetic, SIGNUM Newsletter.

[30] INTEL 12 1586-001. (1980). The 8086 Family User's Manual, Numeric Supplement.

[31] Apostolatos, N., Kulisch, U., Krawczyk, R., Lortz, R., Nickel, K., Wippermann, H.-W. (1968). The Algorithmic Language TRIPLEX ALGOL-60, Num. Math. 11, 175-180.

Additional References are given in [15].

COMPUTER DEMONSTRATION PACKAGES FOR STANDARD PROBLEMS OF NUMERICAL MATHEMATICS

Siegfried M. Rump*

Institute for Applied Mathematics
University of Karlsruhe
Karlsruhe, West Germany

This is a short demonstration of potential uses of programming packages for standard problems in numerical analysis implemented on a Z80 based minicomputer. The programming language is a PASCAL extension called PASCAL-SC. The mantissa length is 12 decimal digits with an exponent range -99 up to 99.

The minicomputer is programmed with a precise floating-point arithmetic and a precise dot product. This allows not only the computation of approximations to the solution of a given problem but also the computation of bounds within which there is one and only one solution to the given problem. The correctness of the computed bounds as well as the existence and the uniqueness of the solution within these bounds is verified automatically by the computer without any effort on the part of the user.

There are packages programmed for the solution of linear systems (also over- and underdetermined), eigenproblems, zeros of polynomials, evaluation of arbitrary arithmetic expressions, differential equations etc. The computed bounds are of high accuracy. In many cases the left and right bounds of the resulting intervals are adjacent in the floating-point screen, i.e. is the bounds are best possible. We call this least significant bit accuracy (lsba). The presented examples show that even for ill-conditioned problems such as a linear system with the Hilbert 15×15 matrix, bounds with the lsba-property are computable.

The language extension PASCAL-SC (PASCAL for Scientific Computation) allows the user to develop programs in a simple, easy to read manner. This is achieved especially by the general function concept and the possibility of user defined operators with arbitrary result type.

In the following the bounds of intervals are displayed with as many figures as is the negative of the exponent of the relative error between them. If, for instance, the first five figures of the left and right bound coincide, then for either bound six figures are displayed. Therefore the accuracy of the computed result can be recognized optically.

---

*Present address: IBM Deutschland, Entwicklung und Forschung, 7030 Boeblingen, West Germany

ISBN 0-12-428660-7

```
P   recise
A   rithmetic
S   afe and
C   orrect
A   ppropriate
L   anguage
-   for
S   cientific
C   omputation
```

Language extension PASCAL-SC

Interval Newton procedure in PASCAL-SC

```
BEGIN    (* Main program *)
WRITELN ('Give initial interval, e.g. [1,1.5] ');
  IREAD( INPUT,X1 );
  REPEAT
    X0 := X1;
    XM := MIDDLE X0;
    X1 := ( XM - FCT(XM)/DER(X0) ) ** X0;
    IWRITE( OUTPUT,X1 );  WRITELN
  UNTIL X0 = X1
END.
```

Interval Newton procedure in ordinary PASCAL

```
BEGIN    (* Main program *)
    WRITELN ('Give initial interval, e.g. [1,1.5] ');
    READI(INPUT,X1);
    REPEAT X0:=X1;
      MIDDLE(X0,XM);
      FCT(XM,YM);
      DER(X0,Y0);
      DIVI(YM,Y0,Y);
      SUBI(XM,Y,Z);
      IS(Z,X0,X1);
      WRITEI(OUTPUT,X1);
      WRITELN
    UNTIL EQ(X0,X1)
  END.
```

When programming the interval Newton procedure in ordinary PASCAL, the subprograms READI, MIDDLE, DIVI, SUBI, IS and WRITEI have to be written by the user. This requires about 180 lines of code. In the language extension PASCAL-SC all these subprograms are predefined in the system and can be called in the operator notation. This simplifies the reading and debugging of the program.

Computing inclusions - old and new

Execution of the interval Newton procedure for $F(X) = X*(X ** 9 - 1) - 1$

```
Give initial interval, e.g. [1,1.5]

* [1,1.6]

[                 1.0E+00,                 1.3E+00]
[                 1.0E+00,                 1.2E+00]
[                1.07E+00,                1.09E+00]
[              1.0755E+00,              1.0761E+00]
[           1.0757659E+00,           1.0757662E+00]
[  1.07576606608E+00,  1.07576606609E+00]
```

Execution of a combined Newton procedure for the same function as above

$F(X) = X*(X ** 9 - 1) - 1$

```
Give initial approximation, e.g. 1.0

* 1.3

 1.30000000000E+00
 1.19065781455E+00
 1.11558133052E+00
 1.08181447152E+00
 1.07592376424E+00
 1.07576617573E+00
 1.07576606608E+00
[  1.07576606608E+00,  1.07576606609E+00]
```

The interval Newton procedure requires an initial interval containing one and only one zero of the function in use. This interval has to be known by the user.

The second (combined) Newton procedure does not require any information on the part of the user. In contrast the user gives only an initial approximation, but no information is required about the quality of this approximation. However, a very bad initial guess may not result in an inclusion. In any case, either the correctness of the result including existence and uniqueness of the solution within the computed bounds is automatically verified by the computer, or a message is given that no inclusion could be computed. Experience shows that the latter case is rare as demonstrated in the succeeding examples. The important point is that only true information is passed to the user.

Precise dot product

The precise dot product computes the result of any inner product with maximum accuracy, i.e., between the computed result and the exact (real) result there is no other floating-point number. The special algorithm is, moreover, faster than an inner product programmed in ordinary PASCAL.

```
Enter length of inner product :

* 3

Enter X[i], Y[i] for i=1..length :

* 1000000.00   1000000.00
* 1.00   1.00
* 1000000.00   -1000000.00


Result of precise inner product :

Sum =  1.0000000000000E+00

Result of an inner product programmed in ordinary PASCAL :

Sum =  0.0000000000000E+00



Enter length of inner product :

* 3

Enter X[i], Y[i] for i=1..length :

*  1E99    1E99
*  1E-99    1.0
* -1E99    1E99


Result of precise inner product :

Sum =  1.0000000000000E-99

Result of an inner product programmed in ordinary PASCAL :

**********   Overflow   **********
```

If the rounded result of an inner product is a floating-point number then the precise inner product algorithm gives the correct value even if an intermediate overflow occurs.

Linear Systems

This package computes an inclusion of the solution of a system of linear equations with automatic verification of the correctness of the bounds as well as the existence and uniqueness of the solution within the computed bounds. The data may consist of interval data in which case an inclusion of the solution of every linear system contained in the interval system is computed.

```
Linear systems, real or interval data, over- and underdetermined.
Please enter one of the following characters :

S  to enter the matrix and the right hand side
R  to generate a matrix and right hand side randomly
H  to generate a Hilbert matrix
L  to see intermediate results
N  to omit intermediate results
A  to see the floating-point approximation computed by Gaussian elimination
I  to see this information panel
Q  to quit the package

* H

Enter dimension.

* 15
```

The matrix is :

1.16454478140E+12	5.82272390700E+11	3.88181593800E+11	2.91136195350E+11
2.32908956280E+11	1.94090796900E+11	1.66363540200E+11	1.45568097675E+11
1.29393864600E+11	1.16454478140E+11	1.05867707400E+11	9.70453984500E+10
8.95803678000E+10	8.31817701000E+10	7.76363187600E+10	
5.82272390700E+11	3.88181593800E+11	2.91136195350E+11	2.32908956280E+11
1.94090796900E+11	1.66363540200E+11	1.45568097675E+11	1.29393864600E+11
1.16454478140E+11	1.05867707400E+11	9.70453984500E+10	8.95803678000E+10
8.31817701000E+10	7.76363187600E+10	7.27840488375E+10	
3.88181593800E+11	2.91136195350E+11	2.32908956280E+11	1.94090796900E+11
1.66363540200E+11	1.45568097675E+11	1.29393864600E+11	1.16454478140E+11
1.05867707400E+11	9.70453984500E+10	8.95803678000E+10	8.31817701000E+10
7.76363187600E+10	7.27840488375E+10	6.85026342000E+10	
2.91136195350E+11	2.32908956280E+11	1.94090796900E+11	1.66363540200E+11
1.45568097675E+11	1.29393864600E+11	1.16454478140E+11	1.05867707400E+11
9.70453984500E+10	8.95803678000E+10	8.31817701000E+10	7.76363187600E+10
7.27840488375E+10	6.85026342000E+10	6.46969323000E+10	
2.32908956280E+11	1.94090796900E+11	1.66363540200E+11	1.45568097675E+11
1.29393864600E+11	1.16454478140E+11	1.05867707400E+11	9.70453984500E+10
8.95803678000E+10	8.31817701000E+10	7.76363187600E+10	7.27840488375E+10
6.85026342000E+10	6.46969323000E+10	6.12918306000E+10	
1.94090796900E+11	1.66363540200E+11	1.45568097675E+11	1.29393864600E+11
1.16454478140E+11	1.05867707400E+11	9.70453984500E+10	8.95803678000E+10
8.31817701000E+10	7.76363187600E+10	7.27840488375E+10	6.85026342000E+10
6.46969323000E+10	6.12918306000E+10	5.82272390700E+10	
1.66363540200E+11	1.45568097675E+11	1.29393864600E+11	1.16454478140E+11
1.05867707400E+11	9.70453984500E+10	8.95803678000E+10	8.31817701000E+10
7.76363187600E+10	7.27840488375E+10	6.85026342000E+10	6.46969323000E+10
6.12918306000E+10	5.82272390700E+10	5.54545134000E+10	
1.45568097675E+11	1.29393864600E+11	1.16454478140E+11	1.05867707400E+11
9.70453984500E+10	8.95803678000E+10	8.31817701000E+10	7.76363187600E+10
7.27840488375E+10	6.85026342000E+10	6.46969323000E+10	6.12918306000E+10
5.82272390700E+10	5.54545134000E+10	5.29338537000E+10	
1.29393864600E+11	1.16454478140E+11	1.05867707400E+11	9.70453984500E+10
8.95803678000E+10	8.31817701000E+10	7.76363187600E+10	7.27840488375E+10
6.85026342000E+10	6.46969323000E+10	6.12918306000E+10	5.82272390700E+10
5.54545134000E+10	5.29338537000E+10	5.06323818000E+10	
1.16454478140E+11	1.05867707400E+11	9.70453984500E+10	8.95803678000E+10
8.31817701000E+10	7.76363187600E+10	7.27840488375E+10	6.85026342000E+10
6.46969323000E+10	6.12918306000E+10	5.82272390700E+10	5.54545134000E+10
5.29338537000E+10	5.06323818000E+10	4.85226992250E+10	
1.05867707400E+11	9.70453984500E+10	8.95803678000E+10	8.31817701000E+10
7.76363187600E+10	7.27840488375E+10	6.85026342000E+10	6.46969323000E+10
6.12918306000E+10	5.82272390700E+10	5.54545134000E+10	5.29338537000E+10
5.06323818000E+10	4.85226992250E+10	4.65817912560E+10	
9.70453984500E+10	8.95803678000E+10	8.31817701000E+10	7.76363187600E+10
7.27840488375E+10	6.85026342000E+10	6.46969323000E+10	6.12918306000E+10
5.82272390700E+10	5.54545134000E+10	5.29338537000E+10	5.06323818000E+10
4.85226992250E+10	4.65817912560E+10	4.47901839000E+10	
8.95803678000E+10	8.31817701000E+10	7.76363187600E+10	7.27840488375E+10
6.85026342000E+10	6.46969323000E+10	6.12918306000E+10	5.82272390700E+10
5.54545134000E+10	5.29338537000E+10	5.06323818000E+10	4.85226992250E+10
4.65817912560E+10	4.47901839000E+10	4.31312882000E+10	
8.31817701000E+10	7.76363187600E+10	7.27840488375E+10	6.85026342000E+10
6.46969323000E+10	6.12918306000E+10	5.82272390700E+10	5.54545134000E+10
5.29338537000E+10	5.06323818000E+10	4.85226992250E+10	4.65817912560E+10
4.47901839000E+10	4.31312882000E+10	4.15908850500E+10	
7.76363187600E+10	7.27840488375E+10	6.85026342000E+10	6.46969323000E+10
6.12918306000E+10	5.82272390700E+10	5.54545134000E+10	5.29338537000E+10
5.06323818000E+10	4.85226992250E+10	4.65817912560E+10	4.47901839000E+10
4.31312882000E+10	4.15908850500E+10	4.01567166000E+10	

```
Enter the 15 components of the right hand side.

* 1 2 3 4 5 6 7 8 7 6 5 4 3 2 1


1.   Floating-point approximation :

 -3.70144248083E-06  3.90101713089E-04 -9.93527792450E-03  1.05104981172E-01
  5.56103731082E-01  1.52452075214E+00 -1.76642775931E+00 -9.59186415565E-01
  4.90243939872E+00 -4.09472893569E+00 -7.96632647673E-01  1.88744082265E+00
  1.12694074067E+00 -2.02683453639E+00  6.63018130949E-01


2.   Result of the new algorithm :

The algorithm verified that the matrix of the linear system is not singular and
that the unique solution is included in the following bounds :

       1      [ -5.78999254704E-03, -5.78999254703E-03]
       2      [  1.14415175302E+00,  1.14415175303E+00]
       3      [ -5.65419358902E+01, -5.65419358901E+01]
       4      [  1.22731367521E+03,  1.22731367522E+03]
       5      [ -1.46321958521E+04, -1.46321958520E+04]
       6      [  1.07653530434E+05,  1.07653530435E+05]
       7      [ -5.23134225388E+05, -5.23134225387E+05]
       8      [  1.74885780475E+06,  1.74885780476E+06]
       9      [ -4.11439182702E+06, -4.11439182701E+06]
      10      [  6.86925162911E+06,  6.86925162912E+06]
      11      [ -8.09380818008E+06, -8.09380818007E+06]
      12      [  6.57904885545E+06,  6.57904885546E+06]
      13      [ -3.51015512696E+06, -3.51015512695E+06]
      14      [  1.10616242295E+06,  1.10616242296E+06]
      15      [ -1.56024600400E+05, -1.56024600399E+05]
```

```
Enter H,S,R or Q. Enter I for general information.

* S

Enter the number of equations and the number of unknowns.

* 3 2

Enter the matrix rowwise and then the right hand side. To enter
are 3 equations in 2 unknowns.

* 665857 -941664
* 470832 -665857
* 470833 -665857
* 1
* 0
* 665858

1.   Floating-point approximation :

      1         6.65858261006E+05
      2         4.70832085775E+05


2.   Result of the new algorithm :

It has been verified that the matrix is of maximum rank and that the best
approximation is included within the following bounds :

      1      [  6.65858000000E+05,  6.65858000001E+05]
      2      [  4.70832707107E+05,  4.70832707108E+05]

Enter H,S,R or Q. Enter I for general information.

* Q
```

Obviously the floating-point approximation may be false to an arbitrary degree, even when using Gaussian elimination with optimal arithmetic.

The package for linear systems computes bounds with high accuracy for systems of linear equations as well as for over- or underdetermined systems (more equations resp. more unknowns). In case of an overdetermined linear system the solution with smallest residue, in case of an underdetermined linear system the solution with smallest norm is included.

In any case the computer verifies that the matrix of the linear system has maximum rank (if this is true) without any effort on the part of the user.

The condition number of the Hilbert 15×15 matrix is approximately 10**22. It is the Hilbert matrix with largest number of rows which is exactly storable with 12 decimal digit accuracy. Despite the ill-condition of the system the solution is included with least significant bit accuracy.

Inversion of a matrix

This package computes an inclusion of the inverse of a real or interval matrix. In case of an interval matrix the inverse of all real matrices contained in the interval matrix are included. Additionally the non-singularity of the real matrix resp. all real matrices contained in the interval matrix is verified by the computer as well as the correctness of the computed bounds.

```
Matrix inversion, real or interval data.
Please enter one of the following characters :


S  to enter the matrix
R  to generate a matrix randomly
H  to generate a Hilbert matrix
L  to see intermediate results
N  to omit intermediate results
A  to see the floating-point approximation computed by Gauss-Jordan
I  to see this information panel
Q  to quit the package

* S

Enter dimension.

* 2

Enter the matrix row by row. The dimension is 2.

* 1 1
* 9 [9,9.1]

1.  Floating-point approximation :

-3.47629999999E+02 1.11111111111E+02
 4.00000000000E+02-1.25000000000E+02


2.  Result of the new algorithm :

No inclusion could be computed. Probaly the interval matrix
contains an singular matrix.


Enter H,S,R or Q. Enter I for general information.

* S

Enter dimension.

* 2

Enter the matrix row by row. The dimension is 2.

* 941664 665857
* 665857 470832
```

```
1.  Floating-point approximation :

-1.66666666667E+05 2.35702260396E+05
 2.35702260396E+05-3.33333333333E+05


2.  Result of the new algorithm :

It has been verified that the matrix is not singular and that its
inverse is included in the following bounds :

   1st column :
[ -4.70832000000E+05,  -4.70832000000E+05]
[  6.65857000000E+05,   6.65857000000E+05]

   2nd column :
[  6.65857000000E+05,   6.65857000000E+05]
[ -9.41664000000E+05,  -9.41664000000E+05]


Enter H,S,R or Q. Enter I for general information.

* Q
```

In the first example the floating-point approximation to the inverse of the midpoint matrix of the interval input has been computed. However, it is no problem to approximate "the inverse" of a singular matrix with a floating-point algorithm.

In the second example the approximate inverse is falsified by cancellation and rounding errors. Nevertheless the inclusion of the inverse is of least significant bit accuracy. In this particular case the inclusion is a point matrix.

Eigenproblems

This package computes inclusions for eigenvalue/eigenvector pairs of a matrix with real or interval data. The correctness of the computed bounds as well as the existence and the individual uniqueness of the eigenvalue and the eigenvector within the computed bounds is automatically verified by the computer.

```
Eigenproblems, real or interval data.
Please enter one of the following characters :

S       to enter the matrix
R       to generate a matrix randomly
A       to see approximations and inclusions computed up to now
(n)     to see the nth eigenvalue/eigenvector pair
I (n)   to compute an inclusion of the nth eigenvalue/eigenvector pair
E  A    to compute inclusions of all eigenvalue/eigenvector pairs
L       to see intermediate results
N       to omit intermediate results
I       to see this information panel

Here (n) is an integer between 1 and the dimension.


* S

Enter dimension.

* 5

Enter the matrix.

* 0 1 0 0 0
* 1 0 1 0 0
* 0 1 0 1 0
* 0 0 1 0 1
* 0 0 0 1 0

Computation of approximations for the eigenvalues.

The approximations are :

     1   -1.73205080758E+00
     2    1.73205080756E+00
     3    9.99999999999E-01
     4   -1.00000000000E+00
     5    6.50343964109E-13

* E 5

5th eigenvalue :
     5    6.50343964109E-13    [                -1.0E-99,            1.0E-99]
5th eigenvector :
     1    5.77350269190E-01    [  5.77350269190E-01,  5.77350269192E-01]
     2   -3.49563760223E-12    [                -1.0E-99,            1.0E-99]
     3   -5.77350269191E-01    [ -5.77350269191E-01, -5.77350269191E-01]
     4    9.56258596272E-14    [                -1.0E-99,            1.0E-99]
     5    5.77350269190E-01    [  5.77350269190E-01,  5.77350269192E-01]

* Q
```

Rounding error and cancellation

A graph of the following polynomial will be displayed:

$$2020x{*}{*}4-5741x{*}{*}3-x{*}{*}2+11482x-8118.$$

The polynomial will be evaluated using Horner's scheme in floating-point arithmetic.

```
Give the degree of the polynomial to be displayed and then the coefficients
starting with the leading coefficient.

* 4
* 2030 -5741 -1 11482 -8118

Give the range where the polynomial shall be displayed.

* 1.41 1.42
```

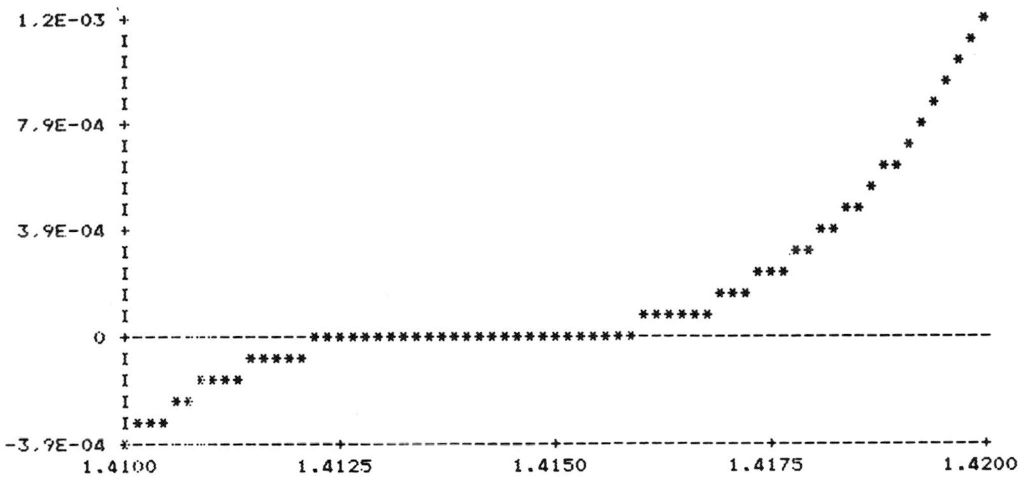

Give a new range, give P for a new polynomial or Q for quit.

*** 1.414 1.415**

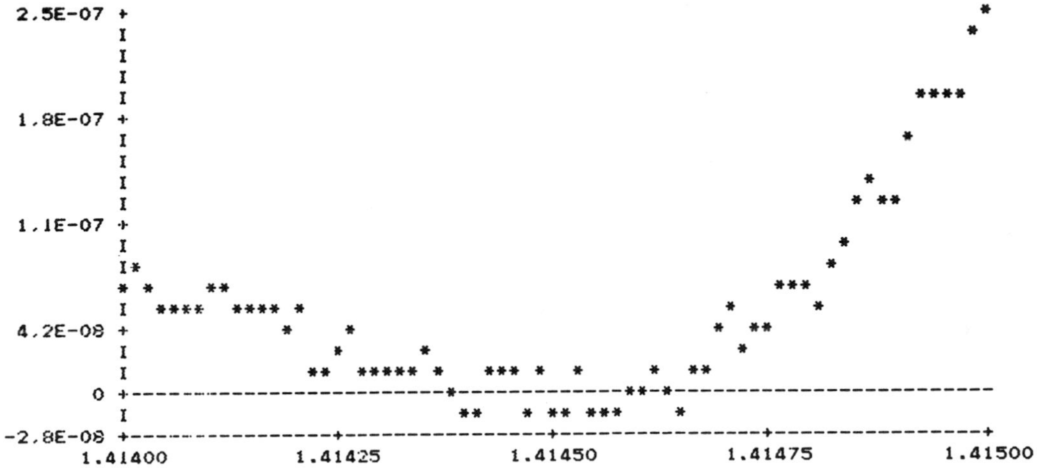

Give a new range, give P for a new polynomial or Q for quit.

* 1.4142 1.4143

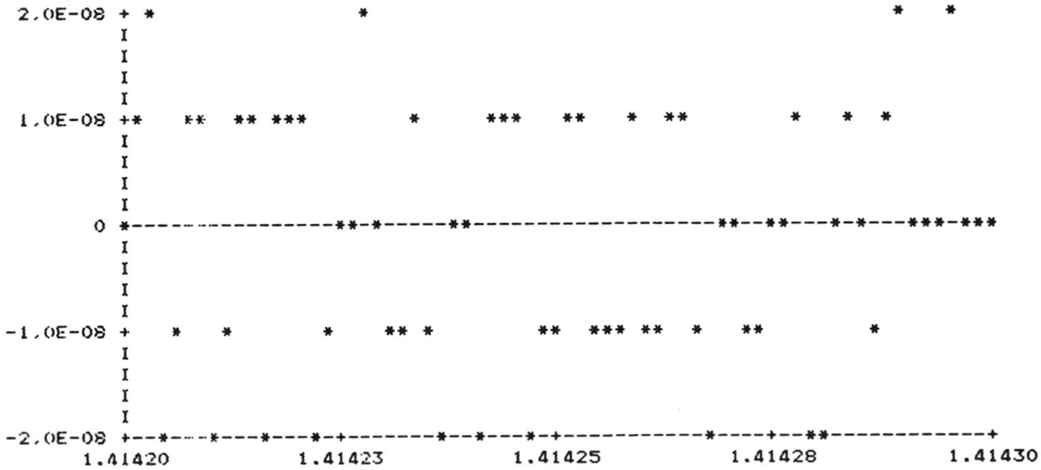

Give a new range, give P for a new polynomial or Q for quit.

* Q

A graph of the same polynomial as above will be displayed in the same ranges.

Now in contrast a new algorithm is used to evaluate the polynomial using the precise inner product. All functional values are computed with least significant bit accuracy.

```
Give the degree of the polynomial to be displayed and then the coefficients
starting with the leading coefficient.

* 4
* 2030 -5741 -1 11482 -8118

Give the range where the polynomial shall be displayed.

* 1.41 1.42


 1.2E-03 +                                                               *
         I                                                              *
         I                                                             *
         I                                                            *
         I                                                           *
 7.9E-04 +                                                          *
         I                                                         *
         I                                                       **
         I                                                      *
         I                                                    **
 3.9E-04 +                                                  **
         I                                                 **
         I                                              ***
         I                                           ***
         I                                     ******
       0 +-------------************************---------------------------
         I        *****
         I    ****
         I  **
         I***
-3.9E-04 *---------------+---------------+---------------+---------------+
      1.4100          1.4125          1.4150          1.4175          1.4200
```

Give a new range, give P for a new polynomial or Q for quit.

*** 1.414 1.415**

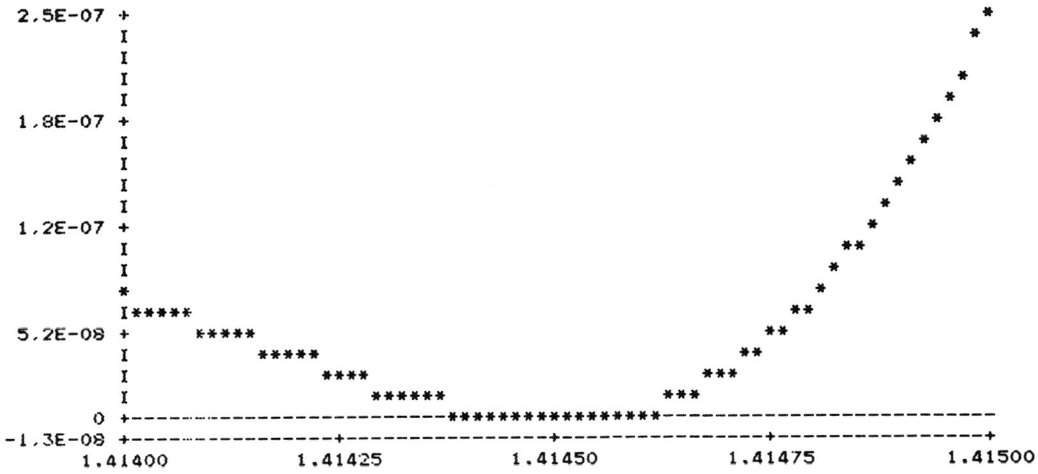

Give a new range, give P for a new polynomial or Q for quit.

* 1.4142 1.4143

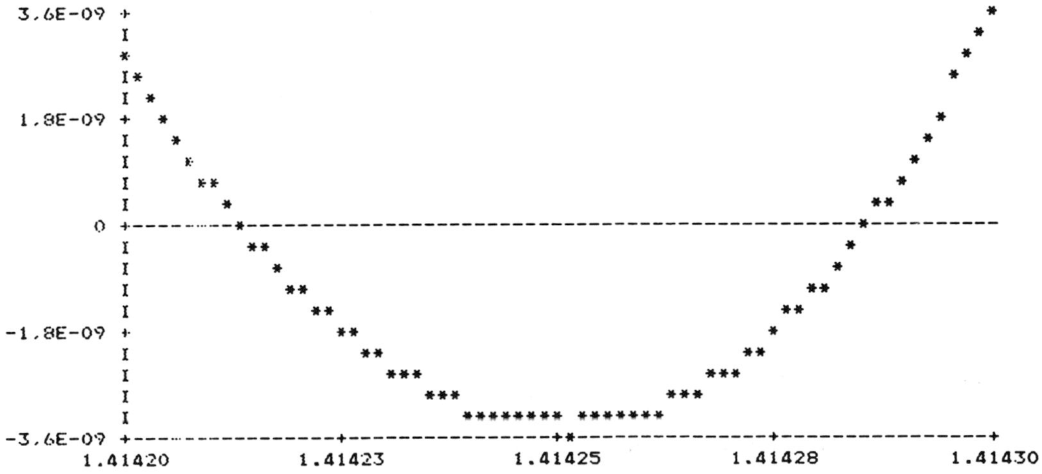

Give a new range, give P for a new polynomial or Q for quit.

* Q

Evaluation of a polynomial

The implemented arithmetic is best possible with respect to smallest possible rounding error. Nevertheless the result of an ordinary floating-point computation may be arbitrarily false due to rounding and cancellation errors:

```
Evaluation of the polynomial

p(x) = 543339720 x**3 - 768398401 x**2 - 1086679440 x + 1536796802


for  x =  1.41420000000E+00

Floating-point Horner scheme :    2.80000000000E-01
Interval Horner scheme :        [           2.7E-01,              3.0E-01]
New algorithm :                 [  2.82673919360E-01,  2.82673919361E-01]

for  x =  1.41421356238E+00

Floating-point Horner scheme :    1.00000000000E-02
Interval Horner scheme :        [          -1.0E-02,              1.0E-02]
New algorithm :                 [  7.32719247117E-14,  7.32719247118E-14]

for  x =  1.41421356100E+00

Floating-point Horner scheme :   -1.00000000000E-02
Interval Horner scheme :        [          -1.0E-02,              1.0E-02]
New algorithm :                 [  2.89746134368E-09,  2.89746134369E-09]
```

In the second and third example the evaluation with naive interval arithmetic gives no overestimation with respect to ordinary floating-point arithmetic.

In all cases the new algorithm computes the value of the polynomial with least significant accuracy.

Zero of a polynomial

The Newton algorithm is applied to the following polynomial:

$$P(x) = 67872320568\ x{*}{*}3 - 95985956257\ x{*}{*}2 - 135744641136\ x + 191971912515$$

for the starting value $x_0 = 2.0$.

The polynomial and its derivative are evaluated by Horner's scheme and floating-point arithmetic. In the left column the iteratives, in the right column the difference between two adjacent iteratives is displayed.

The procedure iterates monotonically (!) towards 1.41421353154 where there is a fixed point of the floating-point iteration. If this computation were exact (without rounding errors) this would imply a zero of the polynomial.

2.00000000000E+00	
1.73024785661E+00	2.698E-01
1.57979152125E+00	1.505E-01
1.49923019011E+00	8.056E-02
1.45733317058E+00	4.190E-02
1.43593403289E+00	2.140E-02
1.42511502231E+00	1.082E-02
1.41967473598E+00	5.440E-03
1.41694677731E+00	2.728E-03
1.41558082832E+00	1.366E-03
1.41489735833E+00	6.835E-04
1.41455549913E+00	3.419E-04
1.41438453509E+00	1.710E-04
1.41429903606E+00	8.550E-05
1.41425628589E+00	4.275E-05
1.41423488841E+00	2.140E-05
1.41422414110E+00	1.075E-05
1.41421847839E+00	5.663E-06
1.41421582935E+00	2.649E-06
1.41421353154E+00	2.298E-06
1.41421353154E+00	0.000E+00
1.41421353154E+00	0.000E+00
1.41421353154E+00	0.000E+00
1.41421353154E+00	0.000E+00
1.41421353154E+00	0.000E+00
1.41421353154E+00	0.000E+00

The truth is that there is no zero of the polynomial for positive values of x and, computed by the new algorithm, the value of the polynomial for $x = 1.41421353154$ is greater than 1:

Value of the polynomial at 1.41421353154:

$$1.00018250381 < P(1.41421353154) < 1.00018250382.$$

Polynomial package

```
Give the degree of the polynomial and then the coefficients starting with
the leading coefficient.
When entering R after the degree a random polynomial will be generated.


* 3
* 408 -577 -816 1154


Enter one of the following characters.

R       to generate a random polynomial
G       for graphical output
P       to enter a new polynomial
E       to compute an inclusion of a root of the polynomial
L       to see intermediate results
N       to omit intermediate results
V (x)   to evaluate the polynomial at x
B       to compute a root bound
C       to see the coefficients of the polynomial
H       evaluation for graphical output with Horner's scheme
S       evaluation for graphical output with new algorithm
I       to see this information panel
Q       to quit the package


* H

* G

Give the range where the polynomial shall be displayed.

* 1.41421 1.41422

 3.0E-08 +                                                                *****
         I
         I
         I
         I
 2.3E-08 +
         I
         *****                                                    *******
         I
         I
 1.5E-08 +
         I
         I
         I    ************                              **********
         I
 7.5E-09 +
         I
         I
         I
         I
       0 +----------------******************************-----+----------------+
  1.4142100        1.4142125        1.4142150        1.4142175        1.4142200
```

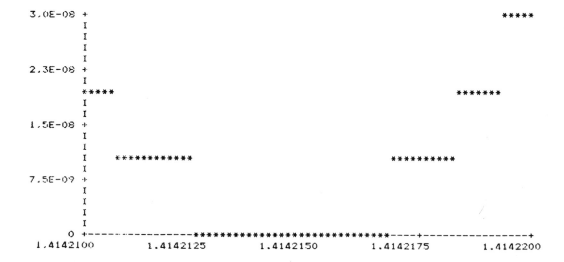

```
* S

* O

Give the range where the polynomial shall be displayed.

* 1.41421 1.41422
```

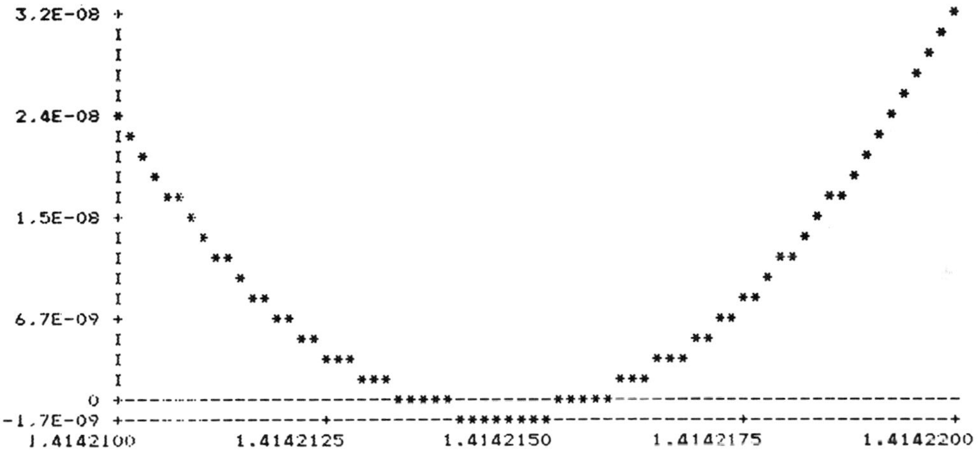


```
* E

Enter an approximation for a zero.

* 1.4

It has been verified that in the interval [ 1.41421356237E+00,  1.41421356238E+00]
there is one and only one zero of the polynomial in use.

* 1.5

It has been verified that in the interval [ 1.41421568627E+00,  1.41421568628E+00]
there is one and only one zero of the polynomial in use.

* Q
```

Evaluating the polynomial using Horner's scheme in floating-point arithmetic gives a discontinuous graph of the function. Using the new algorithm even shows two zeros of the polynomial.

Despite the rough approximations the two zeros of the polynomial are included with least significant bit accuracy.

Arithmetic expressions

```
This package computes the value of an arbitrary arithmetic expression with
maximum accuracy. An arithmetic expression consists of +,-,*,/,(,),↑, real
or integer constants and single-letter variables. In a real constant the
mantissa is separated from the exponent by &. Exponentiation (indicated by
↑ or **) is allowed only for positive integer exponents. The multiplication
sign * may be omitted. Real constants are to be separated by an operator.

Example :   -3.7&-3X + 4Y**3 - (X+Y)**5 / (3Y-1.5)


Please enter one of the following characters :

N    to enter a new arithmetic expression
F    to compute a floating-point approximation of the value of the
          arithmetic expression
E    to compute a proved inclusion of the value of the arithmetic expression
D    to display last problem
I    to see this information panel
Q    to quit the program


* N

Enter an arithmetic expression.

* X**4 + 2Y**2 - 4Y**4

Enter N,E,D or Q. Enter I for general information.


* E


Enter the values for the variables.

Value for X :
* 665857

Value for Y :
* 470832


Floating-point approximation :   -3.00000000000E+12
Naive interval arithmetic :    [              -3.0E+12,              5.0E+12]

New algorithm :                [  1.00000000000E+00,  1.00000000000E+00]


Enter N,E,D or Q. Enter I for general information.

* Q
```

Due to cancellation and rounding errors the floating-point approximation to the value of a simple arithmetic expression may be completely false. The new algorithm computes the value with least significant bit accuracy. In fact, the result is a point.

Systems of non-linear equations

Package for systems of non-linear equations. The equations may be arbitrary arithmetic expressions consisting of + ,−,∗,/,(,),↑, real or integer constants and single letter variables.

The program computes rounds for a solution of the non-linear system with automatic verification of correctness. Moreover the existence and uniqueness of a solution is verified by the computer without any effort on the part of the user.

```
Enter the dimension of the non-linear system.

* 4

Enter the 1st equation of the non-linear system.

* A+B+D-1

Enter the 2nd equation of the non-linear system.

* B-0.6D/C

Enter the 3rd equation of the non-linear system.

* C+D-1

Enter the 4th equation of the non-linear system.

* D-0.3AC


Enter an initial approximation. The dimension is 4.

Value for A :
* 1

Value for B :
* 1

Value for C :
* 1

Value for D :
* 1


The dimension of the non-linear system is 4. The system is :

  A+B+D-1 = 0
 B-0.6D/C = 0
    C+D-1 = 0
  D-0.3AC = 0


Approximation for A :  1.00000000000E+00
Approximation for B :  1.00000000000E+00
Approximation for C :  1.00000000000E+00
Approximation for D :  1.00000000000E+00
```

```
Enter one of the following characters :

G  to start floating-point Newton iteration
L  to see intermediate results
N  to omit intermediate results
P  to enter a new floating-point approximation
E  to compute an inclusion of a solution of the non-linear system
D  to see the non-linear system in use
C  to see further comments
S  to enter a new non-linear system
I  to see this information panel
Q  to quit the package


Enter E,G,S or Q. Enter I for general information.

* E


It has been verified that the system of non-linear equations is solvable.

Moreover it has been verified that there is a solution within the following bounds :

[  7.00322250679E-01,  7.00322250680E-01]
[  1.26058005122E-01,  1.26058005123E-01]
[  8.26380255801E-01,  8.26380255802E-01]
[  1.73619744198E-01,  1.73619744199E-01]

The correctness of these bounds and moreover the existence and uniqueness of a solution
of the non-linear system within these bounds has been computationally verified.


Enter E,G,S or Q. Enter I for general information.

* Q
```

Differential equations

The solution of the following ODE is $y = \cos x$.

```
Y" + Y = 0

 Y(0) = 1
Y'(0) = 0


It has been verified by the algorithm that there is one and only one solution Y(X) to the ODE
and this is included within the following interval function :

 V(X) =  + V[ 0]*X↑ 0 + V[ 1]*X↑ 1 + V[ 2]*X↑ 2 + V[ 3]*X↑ 3 + V[ 4]*X↑ 4 +
           V[ 5]*X↑ 5 + V[ 6]*X↑ 6 + V[ 7]*X↑ 7 + V[ 8]*X↑ 8 + V[ 9]*X↑ 9 +
           V[10]*X↑10 + V[11]*X↑11 + V[12]*X↑12

with the interval coefficients :

V[ 0] = [  9.99999999999E-01,  1.00000000001E+00]
V[ 1] = [            -1.0E-99,             1.0E-99]
V[ 2] = [ -5.00000000001E-01, -4.99999999999E-01]
V[ 3] = [            -2.0E-99,             2.0E-99]
V[ 4] = [  4.16666666606E-02,  4.16666666607E-02]
V[ 5] = [            -2.0E-99,             2.0E-99]
V[ 6] = [ -1.38888886079E-03, -1.38888886078E-03]
V[ 7] = [            -2.0E-99,             2.0E-99]
V[ 8] = [  2.48015230275E-05,  2.48015230276E-05]
V[ 9] = [            -2.0E-99,             2.0E-99]
V[10] = [ -2.75495951432E-07, -2.75495951430E-07]
V[11] = [            -2.0E-99,             2.0E-99]
V[12] = [  2.04071075133E-09,  2.04071075135E-09]


Sample inclusions for X in -1..1 :

V(-1.00) = [    5.4030230586E-01,    5.4030230588E-01]
V(-0.90) = [    6.2160996826E-01,    6.2160996829E-01]
V(-0.80) = [    6.9670670934E-01,    6.9670670936E-01]
V(-0.70) = [    7.6484218728E-01,    7.6484218730E-01]
V(-0.60) = [    8.2533561490E-01,    8.2533561493E-01]
V(-0.50) = [    8.7758256188E-01,    8.7758256191E-01]
V(-0.40) = [    9.2106099400E-01,    9.2106099402E-01]
V(-0.30) = [    9.5533648912E-01,    9.5533648914E-01]
V(-0.20) = [    9.8006657784E-01,    9.8006657786E-01]
V(-0.10) = [    9.9500416527E-01,    9.9500416529E-01]
V( 0.00) = [  9.99999999999E-01,  1.00000000001E+00]
V( 0.10) = [    9.9500416527E-01,    9.9500416529E-01]
V( 0.20) = [    9.8006657784E-01,    9.8006657786E-01]
V( 0.30) = [    9.5533648912E-01,    9.5533648914E-01]
V( 0.40) = [    9.2106099400E-01,    9.2106099402E-01]
V( 0.50) = [    8.7758256188E-01,    8.7758256191E-01]
V( 0.60) = [    8.2533561490E-01,    8.2533561493E-01]
V( 0.70) = [    7.6484218728E-01,    7.6484218730E-01]
V( 0.80) = [    6.9670670934E-01,    6.9670670936E-01]
V( 0.90) = [    6.2160996826E-01,    6.2160996829E-01]
V( 1.00) = [    5.4030230586E-01,    5.4030230588E-01]
```

SOLVING ALGEBRAIC PROBLEMS WITH HIGH ACCURACY

Siegfried M. Rump*

Institute for Applied Mathematics
University of Karlsruhe
Karlsruhe, West Germany

The paper gives a synopsis of new methods for solving algebraic problems with high accuracy. Examples of such problems are the solving of linear systems, eigenvalue/eigenvector determination, computing zeros of polynomials, sparse matrix problems, computation of the value of an arbitrary arithmetic expression (in particular the value of a polynomial at a point), non-linear systems, linear, quadratic and convex programming, etc. over the field of real or complex numbers as well as over the corresponding interval spaces.

We begin by demonstrating the effect of roundoff errors in numerical computation. We use several examples to show that in fact the error of a numerical computation may be arbitrarily large. Some of the examples can be performed on a pocket calculator. As a first step for avoidance of these errors we develop the fundamentals of a computer arithmetic including the precise dot product (Kulisch/Miranker theory).

Next we develop the fundamentals of our new methods. Every result given by an algorithm based on one of the new methods is automatically verified to be correct by the algorithm itself. A result includes an error bound. We say that the result is of high accuracy if the maximum relative error of each component is small. For this purpose we need a precise definition of the arithmetic of the computer in use.

All the algorithms based on our new methods have some key properties in common:

- every result is automatically verified to be correct by the algorithm
- the results are of high accuracy; the error of every component of the result is of the magnitude of the relative rounding error unit
- moreover the solution of the given problem is automatically shown to exist and to be unique within the given error bounds
- the computing time is of the same order as comparable (purely) floating-point algorithm (the latter, of course, offers none of the new features).

The key property of the algorithms is that error control is performed automatically by the computer without any requirement on the part of the user (such as estimating spectral radii. The efficiency of the algorithms will be shown, for instance, by inverting a Hilbert 15×15 matrix in a 12 decimal digit floating-point system. This (after multiplying with a proper factor) the Hilbert matrix of largest dimension which can be stored without rounding error in this floating-point system. The error bounds for all components of the inverse of the Hilbert 15×15 matrix are as small as possible, i.e., left and right bounds differ only by one in the 12th place of the mantissa of each component. We call this least significant bit accuracy (lsba). Our experience shows that the results of the algorithms using our new methods very often have the lsba-property for every component of the solution.

**Present address: IBM Deutschland, Entwicklung und Forschung, 7030 Boeblingen, West Germany*

ISBN 0-12-428660-7

CONTENTS

INTRODUCTION

In this paper we deal with errors in numerical computations and discuss possibilities for their elimination. The problems we have in mind may consist of exactly representable data on a given computer ("point problems") or may be subjected to a certain error margin. Our aim is to give a solution with an error bound such that existence and uniqueness of the solution within these bounds is automatically verified. If this verification process fails a signaling message shall be given. Further the aim is to achieve least significant bit accuracy for point problems and smallest possible bounds for problems with uncertainties in the coefficients. The algorithms presented demonstrate that even for extremely ill-conditioned problems, such as inverting the Hilbert 21×21 matrix on a 14 hexadecimal digit computer, bounds with the least significant bit accuracy property can be found. The condition number of the 21×21 Hilbert matrix is approximately 10^{30}, and is the Hilbert matrix of largest dimension exactly storable on that computer.

To achieve this accuracy and, especially, to give only true results a precisely defined arithmetic is necessary. On a UNIVAC 1108 we have for instance

$$134217728.0 - 134217727.0 = 2.0$$

with exactly representable operands. Therefore, we define a computer arithmetic in the first chapter. The theoretical background and the required algorithms for certain classes of problems are given in the succeeding chapters. The algorithms have been implemented on a minicomputer based on Z80 with a 64 k Byte memory. There, a decimal arithmetic with 12 digits in the mantissa and a PASCAL-SC compiler is implemented. The minicomputer has been developed at the Institute for Applied Mathematics at the University of Karlsruhe and the Fachbereich Informatik of the University at Kaiserslautern (Professors Kulisch and Wippermann). Further, the algorithms are implemented on a UNIVAC 1108 and IBM 370/168. However, we cannot mention every detail of the implementation in the succeeding description of the algorithms. We give explicit algorithms for solving systems of linear equations and systems of nonlinear equations and the verification of the nonsingularity of a (real or complex) matrix. Algorithms for the other problems discussed in the succeeding chapters can be derived easily from the stated theorems and corollaries as well as from the implementation hints for the explicitly given algorithms.

1. COMPUTER ARITHMETIC

Let T be one of the sets $\mathbb{R}$ (real Numbers), $V\mathbb{R}$ (real vectors with n components), $M\mathbb{R}$ (real $n \times n$ matrices), $\mathbb{C}$ (complex numbers), $V\mathbb{C}$ (complex vectors with n components) or $M\mathbb{C}$ (complex $n \times n$ matrices). Here and in the following the letter n is reserved to specify the length of a vector or the number of rows of a quadratic matrix. If the length of a vector is other than n this will be displayed in the corresponding set by an index (e.g. $V_{n+1}\mathbb{C}$). If the number of rows of a quadratic matrix is other than n we write, for instance, $M_{n+1}\mathbb{R}$, if the matrix is not quadratic this is made visible by two indices separated by a comma (e.g. $M_{\ell,m}\mathbb{R}$).

In the power set $\mathbb{P}T$, operations are defined by (with well-known restrictions for /)

$$A*B := \{a*b \mid a\in A \wedge b\in B\} \quad \text{for } A,B\in\mathbb{P}T, \quad *\in\{+,-,\bullet,/\}.$$

Other operations such as $\bullet : \mathbb{P}M\mathbb{C} \times \mathbb{P}V\mathbb{R} \to \mathbb{P}V\mathbb{C}$ are defined similarly. The order relation $\leq$ in $\mathbb{R}$ is extended to $V\mathbb{R}$ and $M\mathbb{R}$ by

$$\forall A,B\in V\mathbb{R}: \quad A\leq B :\Leftrightarrow A_i\leq B_i \quad \text{for } 1\leq i\leq n \text{ and}$$

$$\forall A,B\in M\mathbb{R}: \quad A\leq B :\Leftrightarrow A_{ij}\leq B_{ij} \quad \text{for } 1\leq i,j\leq n.$$

The order relation $\leq$ in $\mathbb{C}$ is defined by

$$\forall\, a+bi,\, c+di\in\mathbb{C}: \quad a+bi\leq c+di :\Leftrightarrow a\leq c \wedge b\leq d$$

and similarly extended to $V\mathbb{C}$ and $M\mathbb{C}$ (componentwise).

The sets $\mathbb{I}T$ of intervals over $\mathbb{R}, V\mathbb{R}, M\mathbb{R}, \mathbb{C}, V\mathbb{C}$ and $M\mathbb{C}$ are defined by

$$[A,B]\in\mathbb{I}T :\Leftrightarrow [A,B] = \{x\in T \mid A\leq x\leq B\} \quad \text{for } A,B\in T.$$

Therefore $I\!I T \subseteq I\!P T$. We consider (see [23], [24]) a rounding $o: I\!P T \rightarrow I\!I T$ with the properties

(R) $\forall A \in I\!P T$: $oA = \cap\{B \in I\!I T \mid A \subseteq B\}$

(R1) $\forall A \in I\!I T$: $oA = A$

(R2) $\forall A, B \in I\!P T$: $A \subseteq B \Rightarrow oA \subseteq oB$

(R3) $\forall A \in I\!P T$: $A \subseteq oA$

(R4) $\forall \emptyset \neq A \in I\!P T$: $o(-A) = -(oA)$.

(R1), (R2) and (R3) are (together) equivalent to (R). Operations $\circledast$: $I\!I T \times I\!I T \rightarrow I\!I T$ for $* \in \{+, -, \bullet, /\}$ are defined by (see [21], [24])

(RG) $\forall A, B \in I\!I T$: $A \circledast B := o(A * B)$ $(= \cap\{C \in I\!I T \mid A * B \subseteq C\})$.

By means of semimorphisms it can be shown (cf. [21], [24]) that the operations in $I\!I T$ are well-defined (with well-known restrictions for /).

The operations are to be executed from left to right respecting the priorities and considering the canonical embeddings $T \subseteq I\!I T \subseteq I\!P T$ and $\mathbb{R} \subseteq \mathbb{C}$, $V\mathbb{R} \subseteq V\mathbb{C}$, $M\mathbb{R} \subseteq M\mathbb{C}$. To be perfectly clear we give the following example. Let $c \in \mathbb{C}$, $V \in I\!I V\mathbb{R}$, $A \in M\mathbb{R}$, $W \in I\!I V\mathbb{C}$. Then $c \odot V + A \odot W$ is well-defined. Following the rules of priorities first $X := c \bullet V$ is computed with $\bullet: \mathbb{C} \times I\!I V\mathbb{R} \rightarrow I\!P V\mathbb{C}$ and then rounded with o: $I\!P V\mathbb{C} \rightarrow I\!I V\mathbb{C}$. Then $Y := A \bullet W$ is computed with $\bullet: M\mathbb{R} \times I\!I V\mathbb{C} \rightarrow I\!P V\mathbb{C}$ and then rounded with o: $I\!P V\mathbb{C} \rightarrow I\!I V\mathbb{C}$. Finally $Z := oX + oY$ is computed with $+ : I\!I V\mathbb{C} \times I\!I V\mathbb{C} \rightarrow I\!P V\mathbb{C}$. It is well-known, that in fact $Z \in I\!I V\mathbb{C}$ (cf. [2], [24]). Moreover, in this specific case

$$c \bullet V + A \bullet W = c \odot V + A \bullet W \subseteq c \odot V + A \odot V = c \odot V \oplus A \odot W$$

For further details (cf. [21], [24]).

With S denoting a subset of $\mathbb{R}$ (e.g. the set of single-precision floating-point numbers on a computer) we consider the set $\mathbb{C}S$ of pairs over S, VS of n-tuples over S, MS of n^2-tuples over S, the set $V\mathbb{C}S$ of n-tuples over $\mathbb{C}S$ and $M\mathbb{C}S$ of n^2-tuples over $\mathbb{C}S$. Again, if the length of a vector or the number of rows of a quadratic matrix is other than n, resp. the matrix is not quadratic, the corresponding set has one, resp. two indices (e.g.

$V_{n+1}S$, $M_{\ell,m}\mathbb{C}S$). Let U denote one of the sets $S, VS, MS, \mathbb{C}S, V\mathbb{C}S$ or $M\mathbb{C}S$. Then intervals over one of these sets U are defined by

$$[A,B] \in \mathbb{I}U :\Longleftrightarrow [A,B] = \{x \in T \mid A \leq x \leq B\} \quad \text{for } A,B \in U,$$

where T is the set corresponding to U. The order relation $\leq$ is defined canonically by regarding U as a subset of T. We consider a rounding $\Diamond$: $\mathbb{I}T \to \mathbb{I}U$ with the properties (R), (R1), (R2) and (R3), (cf.,[22]). If U is symmetric ($U = -U$), then (R4) is also satisfied. The operations $\circledast$: $\mathbb{I}U \times \mathbb{I}U \to \mathbb{I}U$ for $* \in \{+,-,\bullet,/\}$ are defined by (cf. [21], [24])

$$A \circledast B := \Diamond\,(A \circledast B) \text{ for } A,B \in \mathbb{I}U. \tag{1.1}$$

It can be shown that

- $\Diamond$ is well-defined
- $\circledast$ is well-defined for $* \in \{+,-,\bullet,/\}$
- $\circledast$ is effectively implementable on a computer and
- $A \circledast B = \cap\{C \in \mathbb{I}U \mid A \circledast B \subseteq C\}$ for $A,B \in \mathbb{I}U$.

These important properties are shown by means of algebraic and order isomorphism $\mathbb{I}V\mathbb{R} \longleftrightarrow V\mathbb{I}\mathbb{R}$, $\mathbb{I}M\mathbb{C}S \longleftrightarrow M\mathbb{I}\mathbb{C}S$ etc. (the operations in $V\mathbb{I}\mathbb{R}, M\mathbb{I}\mathbb{C}S$ etc. are defined componentwise) and by explicitly giving algorithms for the operations $\circledast$ in all sets S, VS, MS, $\mathbb{I}S$, $\mathbb{I}VS$, $\mathbb{I}MS$, $\mathbb{C}S$, $V\mathbb{C}S$, $M\mathbb{C}S$, $\mathbb{I}\mathbb{C}S$, $\mathbb{I}V\mathbb{C}S$ and $\mathbb{I}M\mathbb{C}S$ (cf. [21], [24]). For the latter purpose a precise arithmetic and Bohlender's algorithm (cf. [3]) are required. If T is the set corresponding to U then $\mathbb{I}U \subseteq \mathbb{I}T$ holds and therefore for $A,B \in \mathbb{I}U$ we have

$$\cap\{C \in \mathbb{I}U \mid A \circledast B \subseteq C\} = \cap\{C \in \mathbb{I}U \mid A * B \subseteq C\}.$$

Thus we extend the rounding $\Diamond$ to $\Diamond$: $\mathbb{P}T \to \mathbb{I}U$ by defining

$$A \in \mathbb{P}T:\ \Diamond\,(A) := \Diamond\,(oA).$$

If T is the set corresponding to U, then the monotone downwardly and monotone upwardly directed roundings $\triangledown: T \rightarrow U$ and $\triangle: T \rightarrow U$ are defined by

$$A \in T: \; \Diamond\,([A,A]) = [\triangledown A, \triangle A] \in \mathbb{I}U.$$

Similarly $\boxed{\triangledown}$ and $\boxed{\triangle}$ are defined by (1.1). Finally we consider a rounding $\square: T \rightarrow U$ with the properties (R1), (R2) and (R3). For the rounding $\square$ and the corresponding operations $\boxed{*}: U \times U \rightarrow U$ defined by a $\boxed{*}$ $b := \square(a * b)$ it can be shown (cf. [21], [24]) that

- $\square$ is well-defined
- $\boxed{*}$ is well-defined for $* \in \{+, -, \cdot, /\}$
- $\boxed{*}$ is effectively implementable on a computer and

for any $v \in U$ and $a,b \in U$ $\; a \,\boxed{*}\, b \leq v \leq a*b$ or $a*b \leq v \leq a \,\boxed{*}\, b$ implies $v = a \,\boxed{*}\, b$.

The final property holds for every set U and is called maximum accuracy. Let A,B be elements of $\mathbb{P}T$, $\mathbb{I}T$ or $\mathbb{I}U$. Then

$$A \underset{\neq}{\subset} B: \Longleftrightarrow \quad A \subseteq B \text{ and } A \neq B,$$

where the $\neq$– sign is to be understood componentwise. $\mathring{A}$ denotes the topological interior of A and ∂A denotes the topological boundary of A. The absolute value of a vector, resp. matrix, over $\mathbb{R}$ or $\mathbb{C}$ is defined to be the vector, resp. matrix, of the absolute values of the components. For $A = [a,b] \in \mathbb{I}T$ with $a,b \in T$ the diameter $d(A) \in T$ and the absolute value $|A| \in T$ are defined by

$$d(A) := |b - a| \quad \text{and} \quad |A| := \max(|a|, |b|),$$

where the maximum is to be understood componentwise.

For $X \in \mathbb{I}U$ with $X = [A,B]$; $A,B \in U$ we have

$$\inf(X) = A \quad \text{and} \quad \sup(X) = B.$$

We define the "midpoint" of X by

$$m(X) := \inf(X) \boxplus (\sup(X) \boxminus \inf(X)) \boxslash 2 \in U. \tag{1.2}$$

It can be shown (cf. [33]) that with definition (1.2) $\inf(X) \leq m(X) \leq \sup(X)$ (as far as no over- or underflow occurs). In the following I denotes the $n \times n$ identity matrix. If the number of rows is other than n we write e.g. I_{n+1}. e_k denotes the kth unit vector and $y_k := e_k' \cdot y$ the kth component of a vector y. We consider a floating-point screen $S = S(b, \ell, e1, e2)$ where b is the base, ℓ is the length of the mantissa and $e1, e2$ are the smallest and largest possible exponent. In succeeding examples we refer e.g. to $S(10, 12, -99, 99)$ (this is the screen of our Z80-based minicomputer).

2. LINEAR SYSTEMS

Essential parts of this chapter have been introduced in [34]. Compared with the original work some theorems have been added and the proofs have been altered and/or simplified.

Let a system of linear equations $Ax = b$ with $A \in M\mathbb{R}$, $x, b \in V\mathbb{R}$ be given. For an approximate inverse $R \in M\mathbb{R}$ of A we have the residue iteration

$$x^{k+1} := x^k + R(b - Ax^k).$$

This iteration converges iff $\rho(I - RA) < 1$. If $X \in \mathbb{P}\mathbb{R}$ is some non-empty, convex, compact subset of $\mathbb{R}$ then by Brouwer's fixed point Theorem

$$X + R(b - AX) \subseteq X \text{ implies } \exists x \in X: R(b - Ax) = 0. \tag{2.1}$$

As a special non-empty, convex, compact subset of $\mathbb{R}$ we can choose $\emptyset \neq X \in \mathbb{I}V\mathbb{R}$. However, in general $d(X + R(b - AX)) > d(X)$ and $X + R(b - AX) \not\subseteq X$. Moreover, only if R is non-singular do we have an $x \in X$ with $Ax = b$. The two problems are solved by the next theorem (cf. [34]).

Theorem 2.1: Let $A, R \in M\mathbb{R}$ and $b \in V\mathbb{R}$. If then for some $X \in \mathbb{I}V\mathbb{R}$

$$Rb + \{I - RA\} \cdot X \overset{\circ}{\subseteq} X, \tag{2.2}$$

then R and A are non-singular and there is one and only one $\hat{x} \in \overset{\circ}{X}$ with $A\hat{x} = b$.

Proof: Define the function f: $V\mathbb{R} \to V\mathbb{R}$ by

$$f(x) := x + R(b - Ax) \tag{2.3}$$

and let $f(X) := \{f(x) \mid x \in X\}$. Then by (2.2) we have $f(X) \subseteq \overset{\circ}{X}$ and Brouwer's fixed point Theorem implies the existence of an $\hat{x} \in X$ with $f(\hat{x}) = \hat{x}$. Obviously $\hat{x} \in \overset{\circ}{X}$. By (2.3) we have $R(b - A\hat{x}) = 0$. For $y \in V\mathbb{R}$ with $Ay = 0$ and $\lambda \in \mathbb{R}$ brief computation shows

$$f(\hat{x} + \lambda y) = \hat{x} + \lambda y. \tag{2.4}$$

If $y \neq 0$, then a $\hat{\lambda}$ exists with $\hat{x} + \hat{\lambda} y \in \partial X$. This contradicts (2.4) and (2.2). Thus A is non-singular.

For $y \in V\mathbb{R}$ with $Ry = 0$ and $\lambda \in \mathbb{R}$ brief computation shows

$$f(\hat{x} + \lambda A^{-1} y) = \hat{x} + \lambda A^{-1} y. \tag{2.5}$$

If $y \neq 0$ then $A^{-1} y \neq 0$ and there exits a $\hat{\lambda} \in \mathbb{R}$ with $\hat{x} + \hat{\lambda} A^{-1} y \in \partial X$. This contradicts (2.5) and (2.2) hence R is non-singular. Therefore, $b - A\hat{x} \in Ker\ R = \{0\}$ and by the non-singularity of A the theorem is proved. □

In practical computation better error bounds are obtained when computing an inclusion of the difference of the exact solution and an approximate solution $\tilde{x}$ instead of computing an inclusion of the solution itself. This was observed in [34] and can be done using the following corollary.

Corollary 2.2: Let $A, R \in M\mathbb{R}$ and $\tilde{x}, b \in V\mathbb{R}$. If then for some $X \in \mathbb{I}\, V\mathbb{R}$

$$R(b - A\tilde{x}) + \{I - RA\} \cdot X \subseteq \overset{\circ}{X}, \tag{2.6}$$

then R and A are non-singular and there is one and only one $\hat{x} \in \tilde{x} + \overset{\circ}{X}$ with $A\hat{x} = b$.

Proof: Obvious. □

A direct consequence of theorem 2.1 and its proof is the extension to complex linear systems. We give a version for computing an inclusion of the difference between the exact solution and an approximate solution.

Theorem 2.3: Let $A,R \in M\mathbb{C}$ and $\tilde{x},b \in V\mathbb{C}$. If then for some $X \in \mathbb{I}\,V\mathbb{C}$

$$R(b - A\tilde{x}) + \{I - RA\} \cdot X \subseteq \overset{\circ}{X}, \tag{2.7}$$

then R and A are non-singular and there is one and only one $\hat{x} \in \tilde{x} + \overset{\circ}{X}$ with $A\hat{x} = b$.

There are no assumptions on $A,R,\tilde{x}$ or b required. The only provisions are (2.2), (2.6), (2.7).

Consider corollary 2.2 which gives a sufficient condition for existence and uniqueness of a solution of $Ax = b$ in $\tilde{x} + X$. If (2.6) does not hold one may initiate the following iteration process. Define $f\colon \mathbb{I}\,V\mathbb{R} \to \mathbb{P}\,V\mathbb{R}$ and $F\colon \mathbb{I}\,V\mathbb{R} \to \mathbb{I}\,V\mathbb{R}$ by

$$X \in \mathbb{I}\,V\mathbb{R}\colon\ f(X) := R(b - A\tilde{x}) + \{I - RA\} \cdot X,\quad F(X) := o(f(X)). \tag{2.8}$$

(In this particular case we have $F(X) = f(X)$). Let $F^0(X) := X$ and $F^{k+1}(X) := F(F^k(X))$ for $k \geq 0$. If then for some $X \in \mathbb{I}\,V\mathbb{R}$ and $k \in \mathbb{N}$ with $Y := F^k(X)$

$$f(Y) \subseteq \overset{\circ}{Y}$$

holds, then the assertions of Corollary 2.2 are valid. The question is, for which X, for which k and under which conditions this iteration will terminate. The answer is given by the following lemma.

Lemma 2.4: Let $|I - RA|$ be a primitive matrix and denote $\lambda := \rho(|I - RA|)$ and $z := R(b - A\tilde{x})$. Consider the mapping f defined by (2.8). Then the following are equivalent:

(A) For all $X \in I\!\!I V\mathbb{R}$ with $|z + A \bullet m(X) - m(X)| < \frac{1-\lambda}{2} \bullet d(X)$ there exists a $k \in \mathbb{N}$ with $f(Y) \subseteq \overset{\circ}{Y}$, where $Y := F^k(X)$.

(B) $\rho(|I - RA|) < 1$.

Proof: cf. [36]. □

The final inclusion for the solution $\hat{x}$ of $Ax = b$ is $\tilde{x} + \overset{\circ}{X}$. So X may be chosen symmetric, i.e. $X = -X$. $|I - RA|$ is primitive if, for instance, $|I - RA|$ is positive. Moreover $m(X) = 0$ in this case, so the condition in A) of the preceding lemma reduces to

$$d(X) > \frac{2}{1-\lambda} \bullet |R(b - A\tilde{x})| . \tag{2.9}$$

Therefore, $\rho(|I - RA|) < 1$ if and only if the iteration terminates for <u>every</u> X satisfying (2.9). If $\tilde{x}$ is a good approximative solution of $Ax = b$ the absolute value of the components of $R(b - A\tilde{x})$ are small.

The essential advantage of corollary 2.2 is that it is applicable on computers.

Corollary 2.5: Let $A, R \in MS$ and $\tilde{x}, b \in VS$. If then for some $X \in I\!\!I VS$

$$R \;\Diamond_{\cdot}\; (b \;\Diamond_{-}\; A \bullet \tilde{x}) \;\Diamond_{+}\; (I \;\Diamond_{-}\; R \bullet A) \;\Diamond_{\cdot}\; X \subseteq \overset{\circ}{X}, \tag{2.10}$$

then R and A are non-singular and there is one and only one $\hat{x} \in \tilde{x} \;\Diamond_{+}\; \overset{\circ}{X}$ with $A\hat{x} = b$.

Proof: Obvious by the definitions of $\Diamond: I\!\!I V\mathbb{R} \to I\!\!I VS$ and $\Diamond_{*}: I\!\!I VS \times I\!\!I VS \to I\!\!I VS$ for $* \in \{+, -, \bullet, /\}$. □

According to [22], [24] and [3] (2.10) is executable on computers using the (effectively implementable) precise scalar product e.g. $I \;\Diamond_{-}\; R \bullet A = \Diamond(o(I - RA)) = \Diamond(I - R \bullet A)$. The corollary remains true when replacing S by $\mathbb{C}S$.

Now we are ready to give an algorithm for computing an inclusion of the solution of a system of linear equations which automatically verifies the correctness of the computed bounds.

1. Compute an approximate inverse R of A using your favorite algorithm;

2. $B := I \diamondminus R \cdot A;\ \tilde{x} := R \diamonddot b;\ z := b \diamondminus A \cdot \tilde{x};,\ z := R \diamonddot z;\ X := z;$

   $k := 0;$

3. <u>repeat</u> $Y := X \diamonddot [1-\varepsilon, 1+\varepsilon];\ k := k+1; X := z \diamondplus B \diamonddot Y$

   <u>until</u> $(X \overset{\circ}{\subseteq} Y)$ <u>or</u> $(k = 10)$

4. <u>if</u> $X \overset{\circ}{\subseteq} Y$ <u>then</u> {It has been verified, that the solution $\hat{x}$ of $Ax = b$

   exists and is uniquely determined and $\hat{x} \in \tilde{x} \diamondplus X$ holds}

   <u>else</u> {It could not be verified whether A is singular or not}.

Algorithm 2.1 Linear Systems

If the floating-point screen being employed is $S(b,\ell,e1,e2)$, then $\varepsilon := b^{-\ell+1}$ such that 1 and $1 + \varepsilon$ are consecutive in S. This ε-inflation is introduced in [34], where its importance is demonstrated, too. The evaluations of B and z are executable on computers (cf. [4]). The assertion $\hat{x} \in \tilde{x} \diamondplus X$ instead of $\hat{x} \in \tilde{x} \diamondplus Y$ follows directly from (2.3) and (2.10). The including region $\tilde{x} \diamondplus X$ can be refined by

<u>repeat</u> $Y := X;\ X := (z \diamondplus B \diamonddot Y) \cap Y$ <u>until</u> $X = Y$.

This iteration terminates because $d(X) \leq d(Y)$ in each step. Obviously every $\tilde{x} \diamondplus X$ includes $\hat{x}$.

In corollary 2.5 and algorithm 2.1, A and b are supposed to be elements of MS resp. VS. If this is not the case (for instance if A and b are the output of some measurement) consider the following theorem.

Theorem 2.6: Let $\mathcal{A} \in I\!IM\mathbb{R}$, $R \in M\mathbb{R}$, $\tilde{x} \in V\mathbb{R}$ and $\mathscr{b} \in I\!I\,V\mathbb{R}$. If then for some $X \in I\!I\,V\mathbb{R}$

$$R(\mathscr{b} - \mathcal{A}\tilde{x}) + \{I - R\mathcal{A}\} \cdot X \overset{\circ}{\subseteq} X, \tag{2.11}$$

then for every $A \in \mathcal{A}$ and every $b \in \mathscr{b}$ the following is true: A and R are non-singular and there is one and only one $\hat{x} \in \tilde{x} + \overset{\circ}{X}$ with $A\hat{x} = b$.

Proof: This follows because $R(b - A\tilde{x}) + \{I - RA\} \bullet X \subseteq R(\mathscr{b} - \mathcal{A}\tilde{x}) + \{I - R\mathcal{A}\} \bullet X$ for every $A \in \mathcal{A}$ and $b \in \mathscr{b}$ permitting the application of theorem 2.1. □

Again $\mathbb{I}M\mathbb{R}$ and $\mathbb{I}V\mathbb{R}$ may be replaced by $\mathbb{P}M\mathbb{R}$, and $\mathbb{P}V\mathbb{R}$ resp. if X is assumed to be non-empty, convex and compact. This remark is important when using a circular arithmetic in $\mathbb{C}$ (cf.[29]):

Corollary 2.7: Let $\mathcal{A} \in \mathbb{P}M\mathbb{C}$, $R \in M\mathbb{C}$, $x \in V\mathbb{C}$ and $\mathscr{b} \in \mathbb{P}V\mathbb{C}$. If then for some non-empty, convex and compact $X \in \mathbb{P}V\mathbb{C}$

$$R(\mathscr{b} - \mathcal{A}x) + \{I - R\mathcal{A}\} \bullet X \subseteq \overset{\circ}{X}, \tag{2.12}$$

then for every $A \in \mathcal{A}$ and every $b \in \mathscr{b}$ the following is true: A and R are non-singular and there is one and only one $\hat{x} \in \tilde{x} + \overset{\circ}{X}$ with $A\hat{x} = b$.

With theorem 2.6 and corollary 2.7 an algorithm can be derived for computing an inclusion of the solution of every linear system $Ax = b$ with $A \in \mathcal{A}$ and $b \in \mathscr{b}$ which automatically verifies the correctness of the computed bounds. This algorithm can be obtained by replacing A by $m(\mathcal{A})$ in step 1) and A by $\mathcal{A}$ resp. b by $\mathscr{b}$ in the computation of B and z in step 2) of algorithm 2.1. However, if some matrix A in $\mathcal{A}$ is ill-conditioned then $I \ominus R\mathcal{A}$ may contain matrices of spectral radius 1.

For several improvements of the algorithms see [34].

Algorithm 2.1 can be used to verify the non-singularity of a matrix automatically on a computer. With $\tilde{x} := b := 0$ in corollary 2.2 we obtain

Corollary 2.8: Let $A, R \in M\mathbb{R}$. If then for some $X \in \mathbb{I}V\mathbb{R}$

$$(I - RA) \bullet X \subseteq \overset{\circ}{X}, \tag{2.13}$$

then A and R are not singular.

To verify whether or not a given matrix is singular is not a trivial problem. In [34] a 3×3 linear system is given which is exactly storable in the single-precision floating-point screen of the UNIVAC 1108 of the University of Karlsruhe. The system has been solved by Gaussian elimination with partial pivoting in single-precision accuracy (~8.5 decimal digits in the mantissa). Then a residue iteration was applied with double-precision evaluation of the residue (~19 decimal digits in the mantissa). The first and all iterates coincides with the initial approximation. Nevertheless, the matrix of the linear system is singular. If (2.13) does not hold for the initial X, one may use an iteration:

$$k := 0;\ \underline{\text{repeat}}\ k := k + 1;\ Y := X;\ X := (I - RA) \bullet Y\ \underline{\text{until}}\ X \subseteq \mathring{Y}; \tag{2.14}$$

The question for which X (2.14) terminates is answered by the following lemma:

Lemma 2.9: Let $A, R \in M\mathbb{R}$. If no entry of $I - RA$ is zero, then the following are equivalent:

a) For every initial $X \in \mathbb{I}V\mathbb{R}$ with $X = -X$ and $|X| > 0$ the iteration (2.14) terminates.

b) $\rho(|I - RA|) < 1$.

Proof: cf. [36]. □

The assumption $|I - RA| > 0$ may be fulfilled by replacing a zero entry by b^{e1}, if $S(b,\ell,e1,e2)$ is the screen of the computer in use. The initial X may consist of $[-1,1]$ in every component according to the preceding lemma. The assertions of corollary 2.8 and lemma 2.9 remain valid when replacing $\mathbb{R}$ by ₵. Moreover $(I - RA) \bullet X \subseteq (I \mathbin{\diamond\!\!\!-} R \bullet A) \mathbin{\diamond\!\!\!\cdot} X$, so replacing (2.13) by $(I \mathbin{\diamond\!\!\!-} R \bullet A) \mathbin{\diamond\!\!\!\cdot} X \subseteq \mathring{X}$ does not affect the assertion of corollary 2.8. So it is applicable on computers. Next we present an algorithm which verifies automatically the non-singularity of a matrix $A \in M\mathbb{R}$.

1. Compute an approximate inverse R of A using your favorite algorithm;

2. $B := I \ominus R \bullet A;\ X := ([-1,1]);\ k := 0;$

3. <u>repeat</u> $k := k + 1;\ Y := X \odot (1 + \varepsilon);\ X := B \odot Y$ <u>until</u>

   $(X \subseteq \mathring{Y})$ <u>or</u> $(k = 10);$

4. <u>if</u> $X \subseteq \mathring{Y}$ <u>then</u> {It has been verified A is not singular}

   <u>else</u> {It could not be verified, whether A is singular or not}.

Algorithm 2.2 Non-singularity of a matrix

It should be mentioned, that it is not possible to verify the singularity of a matrix without computing exactly in the field of reals (for instance using an exact integer or rational number package). This is because in any ε-neighborhood of a singular matrix there are regular matrices. Algorithm 2.2 is directly applicable to complex matrices $A \in M\mathbb{C}$ when one replaces the initial X by $([-1-i, 1+i])$. After replacing A by $\mathcal{A} \in \mathbb{I}M\mathbb{R}$ or $\mathcal{A} \in \mathbb{I}M\mathbb{C}$ it can be determined, whether every matrix $A \in \mathcal{A}$ is non-singular.

If the matrix of the linear system is symmetric, the presented algorithms can easily be improved resulting in a computing time of $\sim n^3$, i.e., one half the computing time of systems with general matrix.

Algorithm 2.2 may be used to verify, that a symmetric matrix $A \in M\mathbb{R}$ is positive definite or, that a matrix $A \in M\mathbb{R}$ or $\mathcal{A} \in \mathbb{I}M\mathbb{C}$ has only eigenvalues with positive real part. If $D := I - \|A\|^{-1} \bullet A$ for some norm $\|\bullet\|$, then $\rho(D) < 1$ implies

a) $Ax = (\lambda + i\mu)x \;\Rightarrow\; \lambda > 0$ and

b) A symmetric $\;\Rightarrow\;$ A is positive definite.

There are similar applications to $\mathcal{A} \in M\mathbb{R}$ or $M\mathbb{C}$ in case a). Bounds for the real resp. imaginary parts of the eigenvalues of $A \in M\mathbb{R}$ may be obtained by estimating the eigenvalues of the symmetric resp. antisymmetric part $\frac{1}{2}(A + A^T)$ resp. $\frac{1}{2}(A - A^T)$.

All of the preceding theorems and corollaries in this chapter remain true when replacing $\overset{\circ}{\subseteq} X$ by $\underset{\neq}{\subset} X$. For a proof of this fact cf. [36].

The computing time of algorithm 2.1 when using the Gauss-Jordan algorithm in step 1 is $2n^3 + 3n^2 + 2kn^2$. If the matrix of the linear system has special properties like symmetry or positive definiteness the computing time of the presented algorithms can be reduced significantly. In general with $k \leq 10$ the ratio of computing times between algorithm 2.1 and Gaussian elimination is $\leq 6 + 0\left(\frac{1}{n}\right)$ (cf. [34]). However, every result given by any algorithm discussed in this chapter is verified to be correct.

Next we discuss some computational results. As a small ill-conditioned example consider

$$\begin{pmatrix} 37639840 & -46099201 \\ 29180474 & -35738642 \end{pmatrix} \cdot \begin{pmatrix} x \\ y \end{pmatrix} = \begin{pmatrix} 0 \\ -1 \end{pmatrix}$$

For this system the standard built-in algorithm of a very common computer with mantissa length 16 decimal places computes the approximation

$$\begin{pmatrix} 28869851.52297299 \\ 23572135.06039856 \end{pmatrix},$$

whereas on our computer with mantissa length 12 decimal digits the following inclusion were computed:

$$\begin{pmatrix} [46099201.0, \;\; 46099201.0] \\ [37639840.0, \;\; 37639840.0] \end{pmatrix}.$$

The inclusion is a point and therefore exact. The correctness is verified by the computer automatically.

We define the

Hilbert matrix H^n by $H^n_{ij} := 1cm(1,2,\ldots,2n-1)/(i+j-1)$,

Pascal matrix P^n by $P^n_{ij} := \binom{i+j}{i}$

Pascal* matrix P^{*n} by $P^{*n}_{ij} := \binom{i+j-2}{i-1}$

the matrix Q^n by $Q^n_{ij} := \dfrac{\binom{n+j-1}{i-1} \bullet n \bullet \binom{n-1}{j-1}}{i+j-1}$.

The matrices with maximum number of rows exactly storable in the screen $S(10,12,-99,99)$ of our minicomputer are H^{15}, P^{21}, P^{*22}, Q^{16}. Here are the computational results for these matrices with right hand side $(1,...,1)$:

> Every linear system with the matrices H^n, P^n, P^{*n} and Q^n up to the maximum number of rows for which the matrices are exactly storable in $S(10,12,-99,99)$ has been solved with automatic verification of the non-singularity of the matrix and with least significant bit accuracy in every component of the solution. The condition number of the Hilbert 15×15 matrix is approximately 10^{22}, the condition number of Q^{16} is $2 \bullet 10^{24}$.

The approximations for the components of the solution computed by a (purely) floating-point algorithm may be arbitrarily false. Using Gaussian elimination, for instance, yields for one component of the solution of the linear system with Hilbert 15×15 matrix and right hand side $(1,1,...,1)$ an approximation

3471.76599106

whereas the inclusion computed by the new methods is

0.00099900099900^1_0.

Least significant bit accuracy for an inclusion of a component means that the left and right bound of the inclusion are consecutive numbers in the floating-point screen. Linear systems with dense matrix and up to 210 rows have been treated on the UNIVAC 1108 of the University of Karlsruhe. Here the size is only limited by the memory of the machine. In every case the least significant bit accuracy property holds for every component of the solution.

As a final example take a linear system with Hilbert 21×21 matrix with right hand side (1,0,0,...,0). As in the previous example the components of the matrix are multiplied by a proper factor to obtain integer entries. The computer in use is a IBM 370/168, double precision (i.e. 14 hexadecimal digits in the mantissa). The Hilbert 21×21 matrix is the matrix of largest number of rows exactly storable in that floating-point screen. Here are the results of ordinary Gaussian elimination compared with the new algorithm:

Gaussian elimination	new algorithm
$0.7176601221737417D-15$	$0.20131453392980^{30}_{29}D-14$
$-0.5463879586639182D-13$	$-0.442891974645566^{4}_{5}D-12$
$0.1300043029451921D-11$	$0.322572988200187^{6}_{5}D-10$
$-0.1384647901664727D-10$	$-0.116126275752067^{5}_{6}D-08$
$0.7307664917097330D-10$	$0.246768335973143^{5}_{4}D-07$
$-0.1742770450134503D-09$	$-0.34218542588275^{89}_{90}D-06$
$0.4670356473353298D-10$	$0.329964517815517^{6}_{5}D-05$
$0.2755923072731661D-09$	$-0.230975162470862^{2}_{3}D-04$
$0.1934981248757405D-08$	$0.120941161460437^{7}_{6}D-03$
$-0.1055453079740496D-07$	$-0.483764645841750^{4}_{5}D-03$
$0.2015964106800396D-07$	$0.149967040210942^{7}_{6}D-02$
$-0.1947461412799036D-07$	$-0.363556461117436^{7}_{8}D-02$
$0.1797737164979233D-07$	$0.692155570204350^{8}_{7}D-02$
$-0.3066524812704846D-07$	$-0.103443030272298^{5}_{6}D-01$
$0.2853218866898916D-07$	$0.120683535317681^{7}_{6}D-01$
$0.1245274389638609D-07$	$-0.108615181785913^{4}_{5}D-01$
$-0.3920468862403904D-07$	$0.738742964352720^{5}_{4}D-02$
$0.1162363936934045D-07$	$-0.366957289482397^{0}_{1}D-02$
$0.2062283997953500D-07$	$0.125538020086083^{3}_{2}D-02$
$-0.1794039988466323D-07$	$-0.264290568602280^{4}_{5}D-03$
$0.4327580718388825D-08$	$0.257997936016511^{9}_{8}D-04$

3. OVER- AND UNDERDETERMINED LINEAR SYSTEMS

Let $A \in M_{\ell,m}\mathbb{R}$ and $b \in V_\ell\mathbb{R}$. For $\ell > m$, the linear system is overdetermined with, in gneral, no solution and the vector $\hat{x} \in V_m\mathbb{R}$ is to compute such that $\| b - A\hat{x} \|$ is minimal.[†]

If $\ell < m$ we have an underdetermined system. In general, there are infinitely many solutions and the vector $\hat{y} \in V_m\mathbb{R}$ is to compute for which $A\hat{y} = b$ and $\|\hat{y}\|$ is minimal. If the rank of A is maximal, the solution for both problems is uniquely determined. It is well-known (cf. [39]), that if

$$\ell > m \text{ and } rg(A) = m \text{ then } \hat{x} \text{ is the solution of } A^T A x = A^T b \tag{3.1}$$

$$\ell < m \text{ and } rg(A) = \ell \text{ then } \hat{y} = A^T x, \text{ where } AA^T x = b. \tag{3.2}$$

The linear systems occurring in (3.1) and (3.2) may be solved with algorithm 2.1 of the preceding chapter. However, in general $A^T A$ and AA^T are real matrices and not elements of MS. Thus the interval version using some $\mathcal{A} \supseteq A^T A$ resp. $A \supseteq AA^T$ with $\mathcal{A} \in \mathbb{I}MS$ would have to be used. This is out of the question because $A^T A$ and AA^T are in general ill-conditioned. Instead we use the following linear systems:

$$\begin{pmatrix} A & -E \\ 0 & A^T \end{pmatrix} \cdot \begin{pmatrix} x \\ y \end{pmatrix} = \begin{pmatrix} b \\ 0 \end{pmatrix} \quad \text{for } \ell > m, \tag{3.3}$$

$$\begin{pmatrix} A^T & -E \\ 0 & A \end{pmatrix} \cdot \begin{pmatrix} x \\ y \end{pmatrix} = \begin{pmatrix} 0 \\ b \end{pmatrix} \quad \text{for } \ell < m. \tag{3.4}$$

Then a short computation demonstrates the following theorems.

[†] In this chapter $\| \bullet \|$ denotes the Euclidean norm.

Theorem 3.1: Let $A \in M_{\ell,m}\mathbb{R}$, $b \in V_\ell\mathbb{R}$, $\ell > m$ and define $C \in M_{\ell+m,\ell+m}\mathbb{R}$ to be the square matrix in (3.3). Define $b^* \in V_{\ell+m}\mathbb{R}$ to be the vector $(b, 0)$[††] and let $\tilde{z} \in V_{\ell+m}\mathbb{R}$, $R \in M_{\ell+m,\ell+m}\mathbb{R}$. If then for some $Z \in \mathbb{IV}_{\ell+m}\mathbb{R}$

$$R(b^* - C\tilde{z}) + \{I - RC\} \cdot Z \subseteq \overset{\circ}{Z}, \tag{3.5}$$

then there is an $\hat{x} \in \tilde{x} + X$ with the following property:

For any $x \in V_m\mathbb{R}$ with $x \neq \hat{x}$ holds $\| b - A\hat{x} \| < \| b - Ax \|$,

where x resp. X are the first m components of $\tilde{z}$ resp. Z and I is the $(\ell + m) \times (\ell + m)$ unit matrix. Further the matrix A has maximum rank m.

Theorem 3.2: Let $A \in M_{\ell,m}\mathbb{R}$, $b \in V_\ell\mathbb{R}$, $\ell < m$ and define $C \in M_{\ell+m,\ell+m}\mathbb{R}$ to be the square matrix in (3.4). Define $b^* \in V_{\ell+m}\mathbb{R}$ to be the vector $(0,b)$ and let $\tilde{z} \in V_{\ell+m}\mathbb{R}$, $R \in M_{\ell+m,\ell+m}\mathbb{R}$. If then for some $Z \in \mathbb{IV}_{\ell+m}\mathbb{R}$

$$R(b^* - C\tilde{z}) + \{I - RC\} \cdot Z \subseteq \overset{\circ}{Z}, \tag{3.6}$$

then there is an $\hat{y} \in \tilde{y} + Y$ with the following properties:

a) $A\hat{y} = b$

b) if $Ay = b$ for some $y \in V_m\mathbb{R}$ with $y \neq \hat{y}$ then $\| \hat{y} \| < \| y \|$,

where $\tilde{y}$ resp. Y are the last m components of $\tilde{z}$ resp. Z and I is the $(\ell + m) \times (\ell + m)$ unit matrix. Further the matrix A has maximum rank ℓ.

Both theorems are applicable on computers. Here one replaces $\mathbb{R}$ by S, the conditions (3.5) and (3.6) by $R \mathbin{\Diamond\!\!\!\!\cdot\;} (b^* \mathbin{\Diamond\!\!\!\!-} C \cdot \tilde{z}) \mathbin{\Diamond\!\!\!\!+} \{I \mathbin{\Diamond\!\!\!\!-} R \cdot C\} \mathbin{\Diamond\!\!\!\!\cdot\;} Z \subseteq \overset{\circ}{Z}$, resp. and $\tilde{x} + X$ resp. $\tilde{y} + Y$ by $\tilde{x} \mathbin{\Diamond\!\!\!\!+} X$ resp. $\tilde{y} \mathbin{\Diamond\!\!\!\!+} Y$. Here, for instance, $b^* \mathbin{\Diamond\!\!\!\!-} C \cdot \tilde{z} = \Diamond(o(b^* - C\tilde{z}))$ is effectively computable (cf. [21], [24], [3]). However, the computing time for these algorithms is $2(\ell + m)^3$ compared with pq^2 for the orthogonalization method (cf. [39]) where

---

[††] $(b, 0)$ is the vector in $V_{\ell+m}\mathbb{R}$ such that the first ℓ components are those of b and the remaining m components are zero. We use similar notations frequently.

$p = \max(\ell,m)$, $q = \min(\ell,m)$. Utilizing symmetries and the fact that not every component of R and $I - RC$ has to be computed explicitly (this is an extended escalator method) results in a

$$\text{computing time } 3pq^2 + 2p^2q + q^3/3$$

(cf. [7]). Therefore the time for computing guaranteed bounds for the solution with automatic verification of both correctness and maximum rank of A is for $p = 2q$ approximately seven times the computing time for a usual floating-point computation. As is well-known, least square approximation problems are in general ill-conditioned. Therefore guaranteed bounds gain in significance. As well as for the least square approximating polynomial bounds for the coefficients of the interpolation polynomial can be included. In the latter case the computing time can be reduced from $16n^3$ to $5n^3$ by utilizing the special structure of the matrix.

As in the case of linear systems with square matrices the algorithm can be extended to matrices $\mathcal{A} \in I\!\!M_{\ell,m}S$ and vectors $\mathscr{b} \in I\!\!V_\ell S$. In this case with corresponding conditions (3.5) and (3.6) every least square problem $Ax = b$ with $A \in \mathcal{A}$ and $b \in \mathscr{b}$ is solvable, every matrix $A \in \mathcal{A}$ has maximum rank and the uniquely determined solution is included in the corresponding interval $\tilde{x} \diamond X$ resp. $\tilde{y} \diamond Y$. Once again automatic verification of correctness is accomplished by the algorithm without any additional effort on the part of the user.

As in the case of linear systems with equal number of rows and unknowns there are extensions in the field of complex numbers in the cases of underdetermined and overdetermined linear systems. The corresponding theorems can be extended to $\mathbb{C}S$ in the same manner as in chapter 2.

The methods and algorithms described in this chapter can be used to include the pseudoinverse of a matrix with automatical verification of correctness.

Numerical examples for least square approximation are taken from [39], Beispiel 3.6. The problem is to find a polynomial P of degree n through $N+1$ given points (x_i,y_i), $i = 1(1)N+1$ which minimizes $\Sigma(P(x_i) - y_i)^2$. We choose the abzissas $x_i = i$. The elements A_{ij} of the matrix A are given by $A_{ij} = i^{j-1}$, $i = 1(1)N + 1$, $j = 1(1)n + 1$. According to (3.19) in [39] the ratio κ between the largest and smallest eigenvalue of $A^T A$ can be estimated by

$$\kappa \geq \max_k A_{kk} / \min_k A_{kk}.$$

Therefore, in our particular case

$$\kappa \geq \left(\sum_{i=1}^{N+1} i^{2n} \right) / (N + 1).$$

In the following we give a table for the minimal value of κ for different n,N.

$n \backslash N$	10	11	12	13	14	15
10	$7.1_{10}19$	$3.8_{10}20$	$1.8_{10}21$	$7.7_{10}21$	$2.9_{10}22$	*
11		$5.4_{10}22$	*	*	*	*

The maximal component of A is $(N + 1)^n$. An asterisk in the table above indicates that some components of the specific matrix A are not exactly storable in our computer with the floating-point screen $S(10,12, -99,99)$. Computational results:

> Every least-square problem for the values of n,N without entry * in the table above has been solved by the new algorithm to least significant bit accuracy. This means that the left and right bound of every component are consecutive numbers in the floating-point screen of the computer. The correctness of every result is verified by the computer automatically.

As seen from the table the examples are ill-conditioned, but nevertheless solved with automatic error control on a computer with a 12 decimal digit floating-point screen.

As a small ill-conditioned example consider

$$\begin{pmatrix} 665857 & -941664 \\ 470832 & -665857 \\ 470833 & -665857 \end{pmatrix} \cdot \begin{pmatrix} x \\ y \end{pmatrix} = \begin{pmatrix} 1 \\ 0 \\ 665858 \end{pmatrix}$$

The computed inclusion of the vector with smallest sum of the norms of the residues is

$$\begin{pmatrix} [665858.000000, & 665858.000001] \\ [470832.707107, & 470832.707108] \end{pmatrix},$$

whereas the floating-point approximation is

$$\begin{pmatrix} 665700.0 \\ 470900.0 \end{pmatrix}.$$

4. LINEAR SYSTEMS WITH BAND MATRICES

The inverse of a band matrix is, in general, dense. Thus the algorithms presented in chapter 2 are too time consuming in this case . There is another possibility for computing an approximate inverse $R \in M\mathbb{R}$ of a matrix $A \in M\mathbb{R}$. Instead of R itself a LU-decomposition of A is computed with lower and upper triangular matrices $L, U \in M\mathbb{R}$. Then $R = (LU)^{-1} = U^{-1}L^{-1} = A^{-1}$. Of course, neither L, U nor R is determined exactly by the computer; they are merely approximated.

All theorems and corollaries of chapter 2 remain valid when one replaces R by $U^{-1} \cdot L^{-1}$. Let us consider the typical condition

$$R(b - A\tilde{x}) + \{I - RA\} \cdot X \subseteq \mathring{X}. \tag{4.1}$$

$L^{-1} \cdot c$ and $U^{-1} \cdot c$ for $c \in V\mathbb{R}$ can always be computed by backward substitution provided that the diagonal elements of L and U are not zero. We denote this process by $L^{-1} * c$ resp. $U^{-1} * c$. Thus (4.1) may be replaced by

$$U^{-1} * (L^{-1} * (b - A\tilde{x})) + \{I - U^{-1} * (L^{-1} * A)\} \cdot X \subseteq \mathring{X}, \tag{4.2}$$

where in $L^{-1}*A$ and $U^{-1}*(L^{-1}*A)$ backward substitution is applied columnwise.

Next we describe backward substitution over S with the rounding $\Diamond$. Suppose $L = (L_{ij})$ with $L_{ij} = 0$ for $i<j$, $L_{ii}\neq 0$. Then for $c\in V\mathbb{R}$

$$L^{-1}*c = v \text{ with } v = (v_i)\in V\mathbb{R} \text{ and } v_i = (c_i - \sum_{j=1}^{i-1} L_{ij}v_j)/L_{ii}. \tag{4.3}$$

v_i is computed for $i = 1(1)n$. If $L\in MS$ and $c\in VS$, then

$$L^{-1} \circledast c := v \text{ with } v = (v_i)\in \mathbb{I}VS \text{ and } v_i := (c_i \ominus \sum_{j=1}^{i-1} L_{ij}v_j) \oslash L_{ii}. \tag{4.4}$$

The definition for $U^{-1} \circledast c$ is similar. The operations in (4.4) are executable on computers (cf. [21], [24], [3]). It is easy to see that

$$L^{-1}*c \subseteq L^{-1} \circledast c. \tag{4.5}$$

For $A\in MS$ we define $L^{-1} \circledast A$ to be the matrix consisting of columns $L^{-1} \circledast A_i$, where A_i are the columns of A.

Therefore the theorems and corollaries of chapter 2 can be reformulated using a LU-decomposition instead of R. As an example we give such a version for corollary 2.5.

Theorem 4.1: Let $A,L,U\in MS$ and $\tilde{x},b\in VS$ where L resp. U are lower resp. upper triangular with non-zero diagonal elements. If then for some $X\in \mathbb{I}VS$

$$U^{-1} \circledast (L^{-1} \circledast (b \ominus A\cdot\tilde{x})) \oplus \{I - U^{-1} \circledast (L^{-1} \circledast A)\} \odot X \subseteq \mathring{X}, \tag{4.6}$$

then L,U and A are non-singular and there is one and only one $\hat{x}\in\tilde{x} + X$ with $A\hat{x} = b$.

Remark: The subtraction of $U^{-1} \circledast (L^{-1} \circledast A)$ from the unit matrix is integrated in $U^{-1} \circledast B$, where $B := L^{-1} \circledast A$. Therefore the minus-sign in the braces is written without a rounding.

For applications to band matrices theorem 4.1 has the advantage, that the band structure is not destroyed. Therefore the computing time is significantly reduced. If for the band matrix A having $A_{ij} = 0$ for $|i - j| > m$ then computation of a LU-decomposition of A

costs about nm^2. Therefore the total computing time for (4.6) is $\sim 4nm^2$. If the matrix has special properties like symmetry or positive definiteness the time can again be reduced significantly. In general there is a factor 4 in cost compared with a usual floating-point algorithm. In return one gains the automatic verification of the non-singularity of the given matrix. Therefore the solvability of the system is demonstrated by the algorithm without any effort on the part of the user.

Similar to chapter 2 the presented methods are applicable in the field of complex numbers as well as for $\mathscr{A} \in \mathbb{I}MS$, $b \in \mathbb{I}VS$ resp. $\mathscr{A} \in \mathbb{I}M\mathbb{C}S$, $b \in \mathbb{I}V\mathbb{C}S$. In the two latter cases the non-singularity of every $A \in \mathscr{A}$ is automatically verified, and in this case for every $A \in \mathscr{A}$, $b \in \mathscr{b}$ there is an $\hat{x} \in \tilde{x} \diamond\!\!\!+ X$ with $A\hat{x} = b$.

A disadvantage of (4.6) is that $U^{-1} \diamond\!\!\!* L^{-1} \diamond\!\!\!* c$ and $I - U^{-1} \diamond\!\!\!* (L^{-1} \diamond\!\!\!* A)$ with $c := b \diamond\!\!\!- A\tilde{x}$ is computed with more than one rounding. Therefore extremely ill-conditioned systems are not solvable. The author's experience showed that in a floating-point screen $S(10,12,-99,99)$ linear systems with matrices up to condition numbers $5 \cdot 10^8$ are solvable with least significant bit accuracy for each component of the solution.

5. SPARSE LINEAR SYSTEMS

Let $A \in M\mathbb{R}$, $b \in V\mathbb{R}$. We consider an iteration scheme (cf. [41])

$$x^{i+1} := x^i + B^{-1} \cdot (b - Ax^i) \tag{5.1}$$

for some initial $x^0 \in V\mathbb{R}$ and an iteration matrix $B \in M\mathbb{R}$. (5.1) converges iff $\rho(I - B^{-1}A) < 1$. Let $A := L + D + U$, where $L, U, D \in M\mathbb{R}$ are lower, upper and diagonal matrices. Then we get for $B := D$ the

Jacobi method: $$x^{i+1} := x^i + D^{-1} \cdot (b - Ax^i), \tag{5.2}$$

for $B := D + L$ the

Gauss – Seidel method: $$x^{i+1} := x^i + (D + L)^{-1} \cdot (b - Ax^i) \tag{5.3}$$

and for $B := \omega^{-1}(D + \omega L)^{-1}$, $\omega \in \mathbb{R}$ the

$$\text{relaxation method:} \qquad x^{i+1} := x^i + \omega \cdot (D + \omega L)^{-1} \cdot (b - Ax^i). \tag{5.4}$$

The methods are well-defined if the diagonal elements of A do not vanish. We distinguish the three methods by

$$B_1 := D;\ B_2 := D + L;\ B_3 := \omega^{-1}(D + \omega L) \text{ for fixed } 0 \neq \omega \in \mathbb{R}. \tag{5.5}$$

Each of the three methods can be used to compute inclusions of the solution of a linear system as one sees by replacing R by B_1, B_2 or B_3 in Theorem 2.1. However, the computing time would be approximately $n^3/3$ which is out of the question. Next we present methods based on (5.2), (5.3), (5.4), resp. for computing bounds for the solution of a linear system in computing time $n^2/2$ per step.

Theorem 5.1: Let $\mathscr{A} \in M\mathbb{R}$, $b \in V\mathbb{R}$ with $A = L + D + U$ for lower, upper and diagonal matrices $L, U, D \in M\mathbb{R}$. Suppose the diagonal elements of A do not vanish. If then for some $X \in \mathbb{IV}\mathbb{R}$ one of the following conditions is satisfied:

1) $D^{-1} \cdot \{b - A\tilde{x} - (L + U)X\} \subseteq \overset{\circ}{X}$
2) $(D + L)^{-1} \cdot \{b - A\tilde{x} - UX\} \subseteq \overset{\circ}{X}$
3) $(\omega^{-1}D + L)^{-1} \cdot \{b - A\tilde{x} - (U + D - \omega^{-1}D)X\} \subseteq \overset{\circ}{X}$ for some $0 \neq \omega \in \mathbb{R}$,

then the matrix A is not singular and there exists an $\hat{x} \in \tilde{x} + X$ such that $A\hat{x} = b$.

Proof: Define the three functions f_i: $V\mathbb{R} \to V\mathbb{R}$, $i = 1(1)3$ by

$$x \in V\mathbb{R}:\ f_i(x) := B_i^{-1} \cdot b + \{I - B_i^{-1} \cdot A\} \cdot x \tag{5.6}$$

for fixed $0 \neq \omega \in \mathbb{R}$ in case $i = 3$.

($B_i, i = 1(1)3$ is given in (5.5)). Then a brief computation shows that if assumption i) is satisfied, $i = 1(1)3$ then with $Y := \tilde{x} + X$, $Y \in \mathbb{IV}\mathbb{R}$ we have

$$f_i(Y) \subseteq \overset{\circ}{Y}.$$

Whence applying theorem 2.1 we have the non-singularity of A and the assertion $\hat{x} \in Y = \tilde{x} + X$ with $A\hat{x} = b$ if 1), 2) or 3) is satisfied. □

The preceding theorem is applicable on computers as is shown in the following corollary.

Corollary 5.2: Let $A \in MS$, $b \in MS$ with $A = L + D + U$ for lower, upper and diagonal matrices $L,U,D \in MS$. Suppose the elements of the diagonal of A do not vanish. If then for some $X \in \mathbb{I}VS$ one of the following conditions is satisfied:

A) $D^{-1} \circledast \Diamond \{b - A \cdot \tilde{x} - (L + U) \cdot X\} \subseteq \overset{\circ}{X}$

B) $(D + L)^{-1} \circledast \Diamond \{b - A \cdot \tilde{x} - U \cdot X\} \subseteq \overset{\circ}{X}$

C) $(\omega^{-1} D + L)^{-1} \circledast \Diamond \{b - A\tilde{x} - (U + D - \omega^{-1} D) \cdot X\} \subseteq \overset{\circ}{X}$

for fixed $0 \neq \omega \in \mathbb{R}$ in case $i = 3$,

then the matrix A is non-singular and there exists an $\hat{x} \in \tilde{x} \oplus X$ with $A\hat{x} = b$.

Here $\Diamond \{b - A\tilde{x} - (L + U) \cdot X\} \in \mathbb{I}VS$, $\Diamond \{b - A\tilde{x} - U \cdot X\} \in \mathbb{I}VS$ and $\Diamond \{b - A\tilde{x} - (U + D - \omega^{-1} D) \cdot X\} \in \mathbb{I}VS$ are effectively computable using one of the algorithms in [3] (cf. [4], too). The symbol $\circledast$ is defined in the previous chapter.

As demonstrated in chapter 2 the above methods are applicable in the field of complex numbers as well as for $\mathcal{A} \in \mathbb{I}MS$, $\boldsymbol{b} \in \mathbb{I}VS$ resp. $\mathcal{A} \in \mathbb{I}M\mathbb{C}S$, $\boldsymbol{b} \in \mathbb{I}V\mathbb{C}S$. In the two latter cases the non-singularity of every $A \in \mathcal{A}$ is verified automatically and for every $A \in \mathcal{A}$, $b \in \boldsymbol{b}$ there is an $\hat{x} \in \tilde{x} \oplus X$ with $A\hat{x} = b$.

Research on sparse linear systems is in progress. Up to now we have little experience on the range applicability of the algorithms. As a numerical example consider linear system (8.4.5) on page 246, [41]. The matrix derives from a discretization of the Dirichlet boundary value problem $-u_{xx} - u_{yy} = f(x,y)$ for $0 < x,y < 1$ and $u(x,y) = 0$ for $(x,y) \in \partial\Omega$ with $\Omega := \{(x,y) \mid 0 < x,y < 1\}$. Let $N = 32$ which corresponds to a linear system with 1024 unknowns approximately 11 steps were needed using (5.4) with optimal relaxation factor to reduce the relative error of an approximate solution to 1/10 (this corresponds to (8.4.9) in [41]). Using the extended relaxation method C) in corollary 5.2 with optimal relaxation factor we achieved least significant bit accuracy for every component of the solution with automatic verification of correctness. Of course, the iteration could be terminated earlier if this high accuracy is not required.

6. MATRIX INVERSION

For $A \in M\mathbb{R}$ we consider the problem of finding a matrix $\hat{C}$ with $A \cdot \hat{C} = I$. Given an initial approximation $R^0 \in M\mathbb{R}$, a Newton-like iteration can be carried out (Schulz-procedure, cf. [2]):

$$R^{i+1} := R^i + R^i(I - AR^i), \quad i \geq 0. \tag{6.1}$$

By methods analogous to those derived in chapter 2 an inclusion of the inverse $\hat{C}$ can be computed.

Theorem 6.1: Let $A, R \in M\mathbb{R}$. If then for some $X \in \mathbb{I}M\mathbb{R}$

$$R(I - A \cdot R) + \{I - R \cdot A\} \cdot X \subseteq \mathring{X}, \tag{6.2}$$

then the matrices A and R are non-singular and there is a $\hat{C} \in R + \mathring{X}$ with $A \cdot \hat{C} = I$.

Proof: The non-singularity of A and R follows as in the proof of theorem 2.1. Further for $f\colon M\mathbb{R} \rightarrow M\mathbb{R}$ with $f(D) := D + R(I - AD)$, $D \in M\mathbb{R}$ we conclude after a brief computation (6.2) that

$$f(Y) \subseteq \mathring{Y} \quad \text{where} \quad Y := R + X.$$

Therefore by Brouwer's fixed point theorem there is a $\hat{C} \in Y = R + \mathring{X}$ with $f(\hat{C}) = \hat{C}$. This implies $R(I - A\hat{C}) = 0$ and by the non-singularity of R we have $A\hat{C} = I$. □

The preceding theorem can be extended to all of the cases A in MS, $\mathbb{I}MS$, $M\mathbb{C}$, $M\mathbb{C}S$ and $\mathbb{I}M\mathbb{C}S$ in a manner similar to that demonstrated in chapter 2.

An inclusion of $\hat{C}$ can be obtained columnwise by applying the theorems and corollaries of chapter 2 to the linear systems $AX = e_i$, where e_i are the columns of I. The computing time for both processes is the same namely $\sim 4n^3$. The iteration (6.1) has the advantage, that in every iteration step a new improved inverse is used. This is not the case when using (2.1).

If, on the other hand, only a few elements of the inverse are to be computed, the second method is preferable, because it is faster and needs less memory.

As an example consider

$$A := \begin{pmatrix} 941664 & 665857 \\ 665857 & 470832 \end{pmatrix}.$$

Inverting A using the Gauss-Jordan procedure in $S(10,12,-99,99)$ yields an approximate inverse

$$R = \begin{pmatrix} -166666.666667 & 235702.260396 \\ 235702.260396 & -333333.333333 \end{pmatrix}.$$

The new algorithm computes an inclusion of A^{-1} to least significant bit accuracy with automatic verification of both correctness and non-singularity of A:

$$A^{-1} \in \begin{pmatrix} [-470832.0, -470832.0] & [665857.0, 665857.0] \\ [665857.0, 665857.0] & [-941664.0, -941664.0] \end{pmatrix}.$$

In this special case the resulting left and right bounds coincide, i.e. the inclusion of A^{-1} is a point matrix. Consider

$$\mathcal{A} := \begin{pmatrix} [(941664, 941664.000001] & 665857 \\ 665857 & 470832 \end{pmatrix} \in \mathbb{IM}S.$$

In this case only the first component of A has been replaced by to an interval of smallest possible diameter in the screen $S(10,12,-99,99)$. The computed inclusion is

$$A \in \mathcal{A} \Rightarrow A^{-1} \in \begin{pmatrix} [-8.9_{10}5, -4.7_{10}5] & [1.2_{10}5, 6.6_{10}5] \\ [1.2_{10}5, 6.6_{10}5] & [-9.5_{10}5, -1.7_{10}5] \end{pmatrix}.$$

These bounds are as small as possible and display the fact that A is ill-conditioned.

For every matrix H^n, P^n,P^{*n} and Q^n defined in chapter 2 up to the highest number of rows for which the matrix is exactly storable in $S(10,12,-99,99)$ the inverse is included by the new algorithm. All bounds for every component are of least significant bit accuracy. The condition number of H^{15} is approximately 10^{22}. Matrices with up to 210 rows were inverted on the UNIVAC 1108. In every case the correctness and non-singularity of the given matrix is verified automatically by the computer.

7. NON-LINEAR SYSTEMS

Consider a function f: $V\mathbb{R} \rightarrow V\mathbb{R}$ with continuous first derivative. We desire to find small bounds for regions containing a zero of f. The existence and uniqueness of a zero within the bounds should be verified automatically by the computer. For this purpose consider the following theorem.

Theorem 7.1: Let f: $V\mathbb{R} \rightarrow V\mathbb{R}$ be a function with continuous first derivative and let $R \in M\mathbb{R}, \tilde{x} \in V\mathbb{R}$. Denote the Jacobian matrix of f by $f' \in M\mathbb{R}$ and for $X \in \mathbb{I}V\mathbb{R}$ define $f'(X) := \cap \{Y \in \mathbb{I}M\mathbb{R} \mid f'(x) \in Y \text{ for all } x \in X\}$. If then for some $X \in \mathbb{I}V\mathbb{R}$

$$\tilde{x} - R \cdot f(\tilde{x}) + \{I - R \cdot f'(\tilde{x} \underline{\cup} X)\} \cdot (X - \tilde{x}) \subseteq \mathring{X}, \tag{7.1}$$

then there exists an $\hat{x} \in \mathring{X}$ with $f(\hat{x}) = 0$.

Proof: In every ε-neighborhood of a matrix $C \in M\mathbb{R}$ there is a non-singular matrix $\overline{C} \in M\mathbb{R}$ (this can be proved by regarding the determinant of C as a polynomial in n^2 variables which is continuous and not identically vanishing, since all coefficients are ± 1). Therefore according to (7.1) a non-singular matrix $\overline{R} \in M\mathbb{R}$ exists with

$$\tilde{x} - \overline{R} \cdot f(\tilde{x}) + \{I - \overline{R} \cdot f'(\tilde{x} \underline{\cup} X)\} \cdot (X - \tilde{x}) \subseteq \mathring{X}. \tag{7.2}$$

Define the function g: $V\mathbb{R} \rightarrow V\mathbb{R}$ by

$$x \in V\mathbb{R}: \; g(x) := x - \overline{R} \cdot f(x). \tag{7.3}$$

g is a function with continuous first derivative. A brief computation using the n-dimensional mean-value theorem yields:

$$\forall x \in X:\ g(x) \in \tilde{x} - \overline{R} \cdot f(\tilde{x}) + \{I - \overline{R} \cdot f'(\tilde{x} \underline{\cup} X)\} \cdot (X - \tilde{x})$$

(however, for $\tilde{x} \in X$ there is, in general, no $\xi \in \tilde{x} + X$ with

$g(x) = \tilde{x} - \overline{R} \cdot f(\tilde{x}) + \{I - \overline{R} \cdot f'(\xi)\} \cdot (X - \tilde{x})$). Therefore by (7.2) and $g(X) := \{g(x) \mid x \in X\}$

$$g(X) \subseteq \overset{\circ}{X}. \qquad (7.4)$$

Now by Brouwer's fixed point theorem there is an $\hat{x} \in \overset{\circ}{X}$ with $g(\hat{x}) = \hat{x}$. By the definition (7.3) of g and the non-singularity of $\overline{R}$ this implies $f(\hat{x}) = 0$. □

According to the preceding proof one cannot replace $f'(\tilde{x} \underline{\cup} X)$ by $\{f'(x) \mid x \in \tilde{x} \underline{\cup} X\}$ in theorem 7.1. It is not possible to replace $f'(\tilde{x} \underline{\cup} X)$ by $f'(X)$ as can be demonstrated by counterexamples. Again it is preferable not to include a zero $\hat{x}$ of f itself but the difference between an approximate zero $\tilde{x}$ and $\hat{x}$. Calculating an inclusion of $\hat{x}$ or $\hat{x} - \tilde{x}$ requires the same computing time.

Corollary 7.2: Let f: $V\mathbb{R} \to V\mathbb{R}$ be a function with continuous first derivative and let $R \in M\mathbb{R}$, $\tilde{x} \in V\mathbb{R}$. Define f' and $f'(X)$ for $X \in \mathbb{I}V\mathbb{R}$ as in theorem 7.1. If then for some $X \in \mathbb{I}V\mathbb{R}$ with $0 \in X$

$$- R \cdot f(\tilde{x}) + \{I - R \cdot f'(\tilde{x} + X)\} \cdot X \subseteq \overset{\circ}{X}, \qquad (7.5)$$

then there exists an $\hat{x} \in \tilde{x} + \overset{\circ}{X}$ with $f(\hat{x}) = 0$.

The assertions of this corollary can be sharpened under slightly weaker assumptions. To prove the stronger result we need the following lemma.

Lemma 7.3: Let $Z \in \mathbb{I}V\mathbb{R}$, $\mathscr{A} \in LM\mathbb{R}$ and $X \in \mathbb{I}V\mathbb{R}$. If then

$$Z + \mathscr{A} \cdot X \subsetneq X, \qquad (7.6)$$

then for every matrix $A \in \mathscr{A}$ holds $\rho(A) \le \rho(|A|) < 1$.

Proof: Using (7.6) and (18), p. 153 in [2] we obtain

$$|\mathscr{A}| \cdot d(X) \le d(\mathscr{A} \cdot X) < d(X).$$

Therefore by corollary 3, p.18 in [42] we have $\rho(|\mathcal{A}|)<1$ and by Perron-Frobenius Theory for every $A\in\mathcal{A}$: $\rho(A)\le\rho(|A|)\le\rho(|\mathcal{A}|)<1$. □

A proof for this lemma is given in [34]; the presented proof is due to Alefeld. Next we give a theorem improving Corollary 7.2.

Theorem 7.4: Let f: $V\mathbb{R}\rightarrow V\mathbb{R}$ be a function with continuous first derivative and let $R\in M\mathbb{R}$, $\tilde{x}\in V\mathbb{R}$. Define f' and $f'(X)$ for $X\in \mathbb{I}V\mathbb{R}$ as in theorem 7.1. If then for some $X\in \mathbb{I}V\mathbb{R}$ with $0\in X$

$$-R\cdot f(\tilde{x})+\{I-R\cdot f'(\tilde{x}+X)\}\cdot X \underset{\neq}{\subset} X, \tag{7.7}$$

then the matrix R and each $B\in M\mathbb{R}$ with $B\in f'(\tilde{x}+X)$ is non-singular and there is one and only one $\hat{x}\in\tilde{x}+\mathring{X}$ with $f(\hat{x})=0$.

Proof: Applying lemma 7.3 to (7.7) yields the non-singularity of R and the matrices $B\in M\mathbb{R}$ with $B\in f'(\tilde{x}+X)$. Define the function g: $V\mathbb{R}\rightarrow V\mathbb{R}$ by

$$x\in V\mathbb{R}:\ g(x):=x-R\cdot f(\tilde{x}+x). \tag{7.8}$$

Then g is continuous and differentiable. As in the proof of theorem 7.1 brief computation using the n-dimensional mean-value theorem yields:

$$\forall x\in X:\ g(x)\in -R\cdot f(\tilde{x})+\{I-R\cdot f'(\tilde{x}+X)\}\cdot X.$$

Then with $g(X):=\{g(x)\,|\,x\in X\}$ we have $g(X)\underset{\neq}{\subset}X$ and by Brouwer's fixed point theorem there is an $\hat{x}\in\mathring{X}$ with $g(\hat{x})=\hat{x}$. This implies by (7.8) and the non-singularity of R that $f(\tilde{x}+\hat{x})=0$. Suppose there is a $\hat{y}\in X$ with $f(\tilde{x}+\hat{y})=0$. Then applying the n-dimensional mean-value theorem there is a matrix $B\in f'(\tilde{x}+X)$ with

$$f(\tilde{x}+\hat{y})=f(\tilde{x}+\hat{x})+B\cdot(\tilde{x}+\hat{y}-\tilde{x}-\hat{x}).$$

This implies $B(\hat{y}-\hat{x})=0$ and by the non-singularity of B the theorem is proved. □

The preceding theorem gives under fewer assumptions than in corollary 7.2, in addition the non-singularity of R and every $B \in f'(\tilde{x} + X)$. For the generalization to complex function we need a complex version of the mean-value theorem. This is given in the next lemma, which is due to Böhm (see [6]).

Lemma 7.5: Let $G \in \mathbb{P}V\mathbb{C}$ be convex and non-empty. Then for a holomorphic function $f: G \rightarrow \mathbb{C}$ and $z, z_0 \in G$ there are $t_1, t_2 \in \mathbb{R}$ with $0 \leq t_1, t_2 \leq 1$ such that with $\zeta_i := z_0 + t_i(z - z_0)$, $i = 1,2$

$$f(z) = f(z_0) + \text{Re}\ \{f'(\zeta_1) \cdot (z - z_0)\} + j \cdot Im\{f'(\zeta_2) \cdot (z - z_0)\}. \tag{7.9}$$

Remark: f' denotes the gradient of f, Re resp. Im of a vector denotes the vector of real resp. imaginary parts of the components; j is the imaginary unit.

Proof of Lemma 7.5: For $z := x + jy$, $x_0 = z_0 + jy_0$ with $x, y, x_0, y_0 \in V\mathbb{R}$ we have $f(z) = u(x,y) + j \cdot v(x,y)$ with $u,v: V\mathbb{R} \rightarrow \mathbb{R}$. By the (real) mean-value theorem there are $t_1, t_2 \in \mathbb{R}$ with $0 \leq t_1, t_2 \leq 1$ such that with $\zeta_i := x_0 + t_i(x - x_0)$ and $\mu_i := y_0 + t_i(y - y_0)$, $i = 1,2$ we have

$$u(x,y) = u(x_0,y_0) + u_x(\xi_1,\mu_1)(x - x_0) + u_y(\xi_1,\mu_1)(y - y_0)$$
$$v(x,y) = v(x_0,y_0) + v_x(\xi_2,\mu_2)(x - x_0) + v_y(\xi_2,\mu_2)(y - y_0).$$

Now short computation using the Cauchy-Riemann differential equations proves the lemma. □

In the following for a function $f: V\mathbb{C} \rightarrow V\mathbb{C}$ we denote the Jacobian matrix of f by f'.

Theorem 7.6 (Böhm): Let $z_1, z_2 \in V\mathbb{C}$ and define $Z := \{z \in V\mathbb{C} \mid z_1 \leq z \leq z_2\}$. Let $G \in \mathbb{P}V\mathbb{C}$ with $\tilde{z} \,\underline{\cup}\, Z \subseteq G$ and let $f: G \rightarrow V\mathbb{C}$ be a holomorphic function. Define $f'(Z) := \cap \{Y \in \mathbb{I}M\mathbb{C} \mid \inf(Y) \leq f(z) \leq \sup(Y) \text{ for } z \in Z\}$. Then for all $z \in Z$

$$f(z) \in f(\tilde{z}) + f'(\tilde{z} \,\underline{\cup}\, Z) \cdot (Z - \tilde{z}). \tag{7.10}$$

Proof: This is a consequence of the preceding lemma. □

According to [36] the assertions of lemma 7.3 remain valid after replacing $\subsetneq X$ by $\subseteq \mathring{X}$ or replacing $\mathbb{R}$ by $\mathbb{C}$. Combining this with the preceding corollary yields

Theorem 7.7: Let $G \in \mathbb{P}V\mathbb{C}$ and let $f\colon G \to V\mathbb{C}$ be a holomorphic function. Let $R \in M\mathbb{C}$ and $\tilde{z} \in V\mathbb{C}$. Define f' to be the Jacobian matrix of f and define $f'(Z) := \cap\{Y \in \mathbb{I}V\mathbb{C} \mid \inf(Y) \le f'(z) \le \sup(Y) \text{ for all } z \in Z\}$ for $Z \in \mathbb{I}V\mathbb{C}$. If then for some $Z \in \mathbb{I}V\mathbb{C}$ with $\tilde{z} + Z \subseteq G$ and $0 \in Z$

$$- R \cdot f(\tilde{z}) + \{I - R \cdot f'(\tilde{z} + Z)\} \cdot Z \subseteq \mathring{Z}, \tag{7.11}$$

then the matrix R and each matrix $B \in M\mathbb{C}$ with $B \in f'(\tilde{z} + Z)$ is non-singular and there is one and only one $\hat{z} \in \tilde{z} + Z$ with $f(\hat{z}) = 0$.

Proof: Similar to the proof of theorem 7.4.

Both theorem 7.4 and the preceding theorem are applicable on computers as stated in th$ following two corollaries.

Corollary 7.8: Let $f\colon V\mathbb{R} \to V\mathbb{R}$ be a function with continuous first derivative and let $R \in MS$, $\tilde{x} \in VS$. Let $\langle f\rangle\colon V\mathbb{R} \to \mathbb{I}VS$ be a function satisfying $x \in V\mathbb{R} \Rightarrow f(x) \in \langle f\rangle(x)$. Define f' to be the Jacobian matrix of f and for $X \in \mathbb{I}VS$ define $f'(X) := \cap\{Y \in \mathbb{I}VS \mid f'(x) \in Y \text{ for all } x \in X\}$. If then for some $X \in \mathbb{I}VS$ with $0 \in X$

$$- R \;\langle\cdot\rangle\; \langle f\rangle(\tilde{x}) \;\langle+\rangle\; \diamond\,\{I - R \cdot f'(\tilde{x} \langle+\rangle X)\} \;\langle\cdot\rangle\; X \subsetneq X, \tag{7.12}$$

then the matrix R and each matrix $B \in M\mathbb{R}$ with $B \in f'(\tilde{x} \langle+\rangle X)$ is non-singular and there is one and only one $\hat{x} \in \tilde{x} \langle+\rangle X$ with $f(\hat{x}) = 0$.

Corollary 7.9: Let $G \in \mathbb{P}V\mathbb{C}$ and let $f\colon G \to V\mathbb{C}$ be a holomorphic function. Let $R \in M\mathbb{C}S$ and $\tilde{z} \in V\mathbb{C}S$. Let $\langle f\rangle\colon V\mathbb{C} \to \mathbb{I}V\mathbb{C}S$ be a function satisfying $z \in V\mathbb{C} \Rightarrow f(z) \in \langle f\rangle(z)$. Define f' to be the Jacobian matrix of f and define $f'(Z) := \cap\{Y \in \mathbb{I}V\mathbb{C}S \mid \inf(Y) \le f'(z) \le \sup(Y) \text{ for all } z \in Z\}$ for $Z \in \mathbb{I}V\mathbb{C}S$. If then for some $Z \in \mathbb{I}VS$ with $\tilde{z} + Z \subseteq G$ and $0 \in Z$

$$- R \;\langle\cdot\rangle\; \langle f\rangle(\tilde{z}) \;\langle+\rangle\; \diamond\,\{I - R \cdot f'(\tilde{z} \langle+\rangle Z)\}) \;\langle\cdot\rangle\; Z \subseteq \mathring{Z}, \tag{7.13}$$

then R and each matrix $B \in M\mathbb{C}$ with $B \in f'(\tilde{z} \oplus Z)$ is non-singular and there is one and only one $\hat{z} \in \tilde{z} \oplus Z$ with $f(\hat{z}) = 0$.

Remark: A close reading of the proof of theorem 7.4 yields the non-singularity of each matrix $B \in M\mathbb{R}$ with $B \in f'(\tilde{x} \oplus X)$ resp. each matrix $B \in M\mathbb{C}$ with $B \in f'(\tilde{z} \oplus Z)$. Moreover, the same proof for the uniqueness of $\hat{x}$ resp. $\hat{z}$ in $\tilde{x} \oplus X$ resp. $\hat{z} \oplus Z$ instead in $\tilde{x} + X$ resp. $\tilde{z} + Z$ can be applied for the preceding two lemmata. □

(7.12) and (7.13) are (effectively) executable on computers according to [21], [24] and [3]. This is true especially for $\Diamond \{I - R \bullet f'(\tilde{x} \oplus X)\}$ resp. $\Diamond \{I - R \bullet f'(\tilde{z} \oplus Z)\}$. Next we give an algorithm to compute an inclusion of a zero of a system of non-linear equations with automatic verification of existence and uniqueness.

1. Use your favorite floating-point algorithm to compute an approximate zero $\tilde{x}$ of f.

2. Use your favorite floating-point algorithm to compute an approximate inverse of $f'(\tilde{x})$.

3. $Y := ([0])$; $k := 0$; $Z := \Diamond f(\tilde{x})$; $Z := \ominus R \odot Z$;

   <u>repeat</u> $k := k + 1$; $X := Y \underline{\cup} 0$; $D := \Diamond f'(\tilde{x} \oplus X)$;

   $Y := Z \oplus \Diamond \{I - R \bullet D\} \odot X$;

   <u>until</u> $Y \subseteq \mathring{X}$ <u>or</u> $k > 10$;

4. <u>if</u> $Y \subseteq \mathring{X}$ <u>then</u> {It has been verified, that there exists one and only one $\hat{x} \in \tilde{x} \oplus \mathring{X}$ with $f(\hat{x}) = 0$}

   <u>else</u> {No verification}.

Algorithm 7.1. Non-linear Systems of Equations

Here $\Diamond f: VS \rightarrow \mathbb{I}VS$ is any function satisfying $x \in VS \Rightarrow f(x) \in \Diamond f(x)$ and $\Diamond f': \mathbb{I}VS \rightarrow \mathbb{I}VS$ is a function satisfying $X \in \mathbb{I}VS \Rightarrow \{\forall x \in X$ holds $\Diamond f'(x) \in \Diamond f'(X)\}$. There is a similar algorithm for complex systems of non-linear equations and there are similar extensions as in chapter 2, to functions $\boldsymbol{f}$: $I\!PV\mathbb{R} \rightarrow I\!PV\mathbb{R}$ and $\boldsymbol{f}$: $I\!PV\mathbb{C} \rightarrow I\!PV\mathbb{C}$. In the two latter cases every function f with $f(x) \in \boldsymbol{f}(x)$ for every $x \in \tilde{x} \mathbin{\Diamond\!\!\!\!+} X$ resp. $\tilde{z} \mathbin{\Diamond\!\!\!\!+} Z$ has exactly one zero in $\tilde{x} \mathbin{\Diamond\!\!\!\!+} \mathring{X}$ resp. $\tilde{z} \mathbin{\Diamond\!\!\!\!+} \mathring{Z}$ (in the terminology of corollaries 7.8 and 7.9).

The above algorithm can be used after a floating-point algorithm to determine the accuracy of the computed approximation. The computing time is $(k + 1)n^3$ plus one evaluation of $\Diamond\!\!\!\!f$ and k evaluations of $\Diamond\!\!\!\!f\,'$, where the computing time for evaluating $\Diamond\!\!\!\!f$ resp. $\Diamond\!\!\!\!f\,'$ is roughly the same as for f resp. f'. As shown in the following examples the algorithm terminates almost always with $k = 1$. This automatic verification process could replace efforts on the part of the user to make an approximation plausible. The question, for which initial X algorithms 7.1 terminates is answered by lemma 2.9.

Theorem 7.4 resp. theorem 7.7 can be regarded as an extension of the Kantorovish Lemma.

Experience showed, that if the approximation $\tilde{x}$ is of the magnitude of a solution of the non-linear system, then algorithm 7.1 terminates for $k \leq 1$ with results of least significant bit accuracy.

The following computational results are from the UNIVAC 1108 at the University of Karlsruhe. Here the floating-point screen S is (2,27,-128,127). So the mantissa length is approximately $8\frac{1}{2}$ decimal digits. We treated Example 7 in [1], Problem 1 in [31] and Problem 2 in [31]. For more examples see [36]. In the following table we display from left to right

- the number of the problem
- n: number of functions and variables
- Newton-steps: number of Newton iterations starting with the initial guess prescribed in the cited literature
- k: defined in step 3 of algorithm 7.1
- succeeded: yes indicates $Y \subseteq \mathring{X}$ in step 3 of algorithm 7.1

- digits guaranteed: least number of digits for which the left and right bound coincide; here an additional l.s.b.a. means least significant bit accuracy, i.e. that the left and right bound of the inclusion of <u>each</u> component are consecutive numbers in the floating-point screen.

Example 7 [1]:

Discretization of $3\ddot{y}y + \dot{y}^2 = 0$, $y(0) = 0$, $y(1) = 20$.

$$f_1 = 3x_1(x_2 - 2x_1) + x_2^2/4$$

$$f_i = 3x_i(x_{i+1} - 2x_i + x_{i-1}) + (x_{i+1} - x_{i-1})^2/4 \qquad 2 \leq i \leq n-1$$

$$f_n = 3x_n(20 - 2x_n + x_{n-1}) + (20 - x_{n-1})^2/4$$

Solution $10t^{3/4}$; initial guess $x_i = 10$ for $i \leq i \leq n$.

Example 1 in [31]:

Discretization of $u''(t) = \frac{1}{2}(\bar{u}(t) + t + 1)^{3,}$ $0<t<1$, $\bar{u}(0) = \bar{u}(1) = 0$

$$x_k = \bar{u}(t_k),\ f_k(x) = 2x_k - x_{k-1} + \frac{1}{2}h^2(x_k + t_k + 1)^3 \qquad 1 \leq k \leq n$$

$$x_0 = x_{k+1} = 0,\ t_k = k \bullet h;\ h = (n+1)^{-1}$$

initial guess $x \equiv (\xi_i)$, $\xi_i = t_i(t_i - 1) \qquad 1 \leq i \leq n.$

Example 2 in [31]:

$$\bar{u}(t) + \int_0^1 H(s,t)(\bar{u}(s) + s + 1)^3 ds = 0$$

$$H(s,t) = \begin{cases} s(1-t) & s \leq t \\ t(1-s) & s > t \end{cases}$$

$$x_k = \bar{u}(t_k), f_n(x) \equiv x_n + \frac{1}{2}\left\{(1 - t_k)\sum_{j=i}^{k} t_j(x_j + t_j + 1)^3 + \right.$$

$$\left. + t_k \sum_{j=k+1}^{n} 1 - t_j)(x_j + t_j + 1)^2 \right\}$$

$x_0 = x_{n+1} = 0$, $t_j = jh$; $h = (n+1)^{-1}$; initial guess $x_i = t_i(t_i - 1)$.

Problem	n	Newton – steps	k	succeeded	digits guaranteed
1)	10	8	2	yes	$8\frac{1}{2}$ (l.s.b.a.)
	20	6	2	yes	$8\frac{1}{2}$ (l.s.b.a.)
	50	6	3	yes	8
	100	7	3	yes	8
2)	10	4	1	yes	$8\frac{1}{2}$ (l.s.b.a.)
	20	6	1	yes	$8\frac{1}{2}$ (l.s.b.a.)
	50	6	1	yes	8
3)	10	3	1	yes	$8\frac{1}{2}$ (l.s.b.a.)
	20	4	1	yes	$8\frac{1}{2}$ (l.s.b.a.)
	50	4	1	yes	8

The additional computing time required to obtain results with verification of correctness is about k times the computing time for one usual floating-point Newton iteration. However, any recomputing with slightly altered entries to gain in security is unnecessary.

For further improvements of the above algorithm with a reduction of k to 1 in every of the mentioned examples see chapter 11 of this article.

8. THE ALGEBRAIC EIGENVALUE PROBLEM

The eigenproblem (cf. [34]) can be formulated as a non-linear system. For $A \in M\mathbb{R}$ let

$$\begin{aligned} &Ax - \lambda x = 0 \\ &e'_k x - \zeta = 0. \end{aligned} \tag{8.1}$$

Here e'_k is the kth unitvector, $1 \leq k \leq n$. If $\zeta \neq 0$ then any pair (x, λ) with $x \in V\mathbb{R}$, $\lambda \in \mathbb{R}$ is an eigenvector/eigenvalue pair of A. In the following the proofs first given in [34] are shortened and completed by lemma 8.3. Finally theorem 8.8 is added. From the preceding chapter a theorem can be derived to compute an inclusion for an eigenvector/eigenvalue pair of A satisfying (8.1):

Theorem 8.1: Let $A \in M_n\mathbb{R}$, $R \in M_{n+1}\mathbb{R}$, $\tilde{x} \in V_n\mathbb{R}$ and $\tilde{\lambda}, \zeta \in \mathbb{R}$ with $\zeta \neq 0$. Define for $Y \in \mathbb{IV}_n\mathbb{R}$, $M \in \mathbb{IR}$ the function G: $\mathbb{IV}_{n+1}\mathbb{R} \rightarrow \mathbb{PV}_{n+1}\mathbb{R}$ by

$$G\begin{pmatrix} Y \\ M \end{pmatrix} := \begin{pmatrix} \tilde{x} \\ \tilde{\lambda} \end{pmatrix} - R \cdot \begin{pmatrix} A\tilde{x} - \tilde{\lambda}\tilde{x} \\ e'_k\tilde{x} - \zeta \end{pmatrix}$$
$$+ \left\{ I_{n+1} - R \cdot \begin{pmatrix} A - (\tilde{\lambda} \,\underline{\cup}\, M) I_n - (\tilde{x} \,\underline{\cup}\, Y) \\ e'_k \quad\quad 0 \end{pmatrix} \right\} \cdot \begin{pmatrix} Y - \tilde{x} \\ M - \tilde{\lambda} \end{pmatrix}. \tag{8.2}$$

If then for some $X \in \mathbb{IV}_n\mathbb{R}$, $\Lambda \in \mathbb{IR}$

$$G(T) \subseteq \mathring{T} \quad \text{for} \quad T = \begin{pmatrix} X \\ \Lambda \end{pmatrix} \tag{8.3}$$

holds, then there exists one and only one eigenvector/eigenvalue pair $(\hat{x}, \hat{\lambda})$ with $\hat{x} \in X$ and $\hat{\lambda} \in \Lambda$.

Here I_n is the $n \times n$ unit matrix and $\tilde{\lambda} \,\underline{\cup}\, M := \square(\tilde{\lambda} \cup M)$, $\tilde{x} \,\underline{\cup}\, Y := \square(\tilde{x} \cup Y)$ as defined in chapter 1. There are similar extensions to complex eigenvector/eigenvalue problems and problems with uncertain data. Next we will improve the assertions of theorem 8.1 under weaker assumptions. Instead of the Jacobian matrix used in (8.2), consider

$$S(X) := \begin{pmatrix} A - \tilde{\lambda} I_n & -X \\ e'_k & 0 \end{pmatrix} \in \mathbb{PM}_{n+1}\mathbb{R} \quad \text{with } X \in \mathbb{IV}_n\mathbb{R}. \tag{8.4}$$

Define

$$Z := \begin{pmatrix} \tilde{x} \\ \tilde{\lambda} \end{pmatrix} - R \cdot \begin{pmatrix} A\tilde{x} - \tilde{\lambda}\tilde{x} \\ e'_k\tilde{x} - \zeta \end{pmatrix} \in \mathbb{PV}_{n+1}\mathbb{R}. \tag{8.5}$$

We will show that using the function G^*: $\mathbb{IV}_{n+1}\mathbb{R} \rightarrow \mathbb{PV}_{n+1}\mathbb{R}$ defined by

$$G^*\begin{pmatrix} Y \\ M \end{pmatrix} := Z + (I_{n+1} - R \cdot S(Y)) \cdot \begin{pmatrix} Y - \tilde{x} \\ M - \tilde{\lambda} \end{pmatrix} \tag{8.6}$$

instead of the G in theorem 8.1, it follows from (8.3) that there exists exactly one eigenvector $\hat{x}$ of A in X, there exists exactly one eigenvalue $\hat{\lambda}$ of A in Λ, $A\hat{x} = \hat{\lambda}\hat{x}$ holds and that $\hat{\lambda}$ is of multiplicity one. Obviously

$$G^*\begin{pmatrix} Y \\ M \end{pmatrix} \subseteq G\begin{pmatrix} Y \\ M \end{pmatrix},$$

and thus the assumptions are weaker.

Lemma 8.2: Define G^*: $\mathbb{IV}_{n+1}\mathbb{R} \to \mathbb{PV}_{n+1}\mathbb{R}$ by (8.6) for $R \in M_{n+1}\mathbb{R}$; $\tilde{x} \in V_n\mathbb{R}$; $\tilde{\lambda}, \zeta \in \mathbb{R}$; $\zeta \neq 0$.

If

$$G^*(T) \subseteq \mathring{T} \quad \text{with} \quad T = \begin{pmatrix} X \\ \Lambda \end{pmatrix} \quad \text{for some} \quad X \in \mathbb{IV}_n\mathbb{R} \quad \text{and} \quad \Lambda \in \mathbb{IR}, \tag{8.7}$$

then there exists an eigenvector $\hat{x}$ of A with $\hat{x} \in X$ and an eigenvalue $\hat{\lambda}$ of A with $\hat{\lambda} \in \Lambda$ and

$$A\hat{x} = \hat{\lambda}\hat{x}. \tag{8.8}$$

Proof: Define f: $V_{n+1}\mathbb{R} \to V_{n+1}\mathbb{R}$ by

$$\begin{aligned} f\begin{pmatrix} x \\ \lambda \end{pmatrix} &:= \begin{pmatrix} x \\ \lambda \end{pmatrix} - R \cdot \begin{pmatrix} Ax - \lambda x \\ e'_k x - \zeta \end{pmatrix} \\ &= Z + \left\{ I_{n+1} - R \cdot \begin{pmatrix} A - \tilde{\lambda} I_n & -x \\ e'_k & 0 \end{pmatrix} \right\} \cdot \begin{pmatrix} x - \tilde{x} \\ \lambda - \tilde{\lambda} \end{pmatrix}. \end{aligned} \tag{8.9}$$

Then

$$y \in Y,\ \mu \in M \Rightarrow f\begin{pmatrix} y \\ \mu \end{pmatrix} \in G^*\begin{pmatrix} Y \\ M \end{pmatrix}. \tag{8.10}$$

By (8.9), (8.7) and the fixed point theorem of Brouwer there is a $(\hat{x},\hat{\lambda}) \in (X,\Lambda)$ with $f(\hat{x},\hat{\lambda}) = (\hat{x},\hat{\lambda})$, and by (8.9)

$$\begin{pmatrix} A\hat{x} - \hat{\lambda}\hat{x} \\ e'_k x - \zeta \end{pmatrix} \in \ker R.$$

By lemma 7.3, R is non-singular and using $e'_k\hat{x} = \zeta \neq 0$ the proof is completed. □

Our aim is to prove the uniqueness of $\hat{x}$ in X and $\hat{\lambda}$ in Λ separately. To this end we first derive the uniqueness of the pair $(\hat{x},\hat{\lambda})$ in (X,Λ).

Lemma 8.3: With the assumptions of lemma 8.2 the eigenvector/eigenvalue pair $(\hat{x},\hat{\lambda})$ is uniquely determined in (X,Λ).

Proof: Define f_z: $V_{n+1}\mathbb{R} \to V_{n+1}\mathbb{R}$ by

$$f_z\begin{pmatrix} w \\ \sigma \end{pmatrix} := \begin{pmatrix} w \\ \sigma \end{pmatrix} - R \cdot \begin{pmatrix} Aw - \tilde{\lambda}w - \sigma z + \tilde{\lambda}z \\ e'_k w - \zeta \end{pmatrix}. \tag{8.11}$$

Then short computation yields for arbitrary $x \in V_n\mathbb{R}$

$$f_z\begin{pmatrix} w \\ \sigma \end{pmatrix} = \begin{pmatrix} \tilde{x} \\ \tilde{\lambda} \end{pmatrix} - R \cdot \begin{pmatrix} A\tilde{x} - \tilde{\lambda}\tilde{x} \\ e'_k\tilde{x} - \zeta \end{pmatrix} + \left\{ I_{n+1} - R \cdot \begin{pmatrix} A - \tilde{\lambda}I_n & -z \\ e'_k & 0 \end{pmatrix} \right\} \cdot \begin{pmatrix} w - \tilde{x} \\ \sigma - \tilde{\lambda} \end{pmatrix}.$$

From (8.7) and (8.12) follows for every $z \in X$

$$f_z(T) \subseteq \mathring{T} \text{ with } T = \begin{pmatrix} X \\ \Lambda \end{pmatrix}. \tag{8.13}$$

Therefore for every $z \in X$ there is a fixed point of f_z in $\overset{\circ}{T}$. Obviously (x,λ) is a fixed point of f_x.

Lemma 7.3 imples the non-singularity of every matrix

$$M_z := \begin{pmatrix} A - \tilde{\lambda} I_n & -z \\ e'_k & 0 \end{pmatrix} \in M_{n+1}\mathbb{R} \text{ for every } z \in X. \tag{8.14}$$

For the purpose of establishing a contradiction we assume $Ax = \lambda x$ and $Ay = \mu y$ with $x,y \in X$ and $\lambda,\mu \in \Lambda$ with $\lambda \neq \mu$. Suppose further $\tilde{\lambda}$ is an eigenvalue of A with eigenvector v. Then in case $e'_k \bullet v = 0$ we have $(v,0)' \in \ker M_x$, in case $e'_k \bullet v \neq 0$ w.1.o.g. $e'_k \bullet v = \zeta$ and $(v - x, \tilde{\lambda} - \lambda)' \in \ker M_x$ contradicting (8.14). Therefore especially $\tilde{\lambda} \neq \lambda$ and $\tilde{\lambda} \neq \mu$. Next we will compute a fixed point of $f_{x+\delta(y-x)}$. Short computation yields for $\delta \neq (\mu - \tilde{\lambda})/(\mu - \lambda)$

$$f_{x+}\delta(y-x)\begin{pmatrix} w \\ \sigma \end{pmatrix} = \begin{pmatrix} w \\ \sigma \end{pmatrix} \text{ for } w(\delta) = x + \frac{\delta(\lambda - \tilde{\lambda})}{N}(y-x) \text{ and}$$
$$\sigma(\delta) = \tilde{\lambda} + (\lambda - \tilde{\lambda})(\mu - \tilde{\lambda})/N \text{ with} \tag{8.15}$$
$$N = (1-\varepsilon)(\mu - \tilde{\lambda}) + \varepsilon(\lambda - \tilde{\lambda}).$$

This fixed point $(w(\delta), \sigma(\delta))'$ lies therefore on the straight line connecting x and y. By (8.7) and $x,y \in X \backslash \partial X$ follows the existence of δ_1, $\delta_2 \in \mathbb{R}$ with

$$\delta_1 < 0 < 1 < \sigma_2 \text{ and } \delta_1 \leq \delta \leq \delta_2 \Rightarrow x + \delta(y-x) \in X$$
$$\text{and } x + \delta_1(y-x) \in \partial X,\ x - \delta_2(y-x) \in \partial X. \tag{8.16}$$

For every $\delta \in [\delta_1, \delta_2]$ with $\delta \neq (\mu - \tilde{\lambda})/(\mu - \lambda)$ we have $N \neq 0$ and by (8.13)

$$w(\delta) \in \overset{\circ}{X}. \tag{8.17}$$

If $(\mu - \tilde{\lambda})/(\mu - \lambda) \in [\delta_1, \delta_2]$ then there exists $w(\delta) \ni \overset{\circ}{X}$ contradicting (8.17). Therefore (8.17) and (8.18) implies

$$\delta \in [\delta_1, \delta_2] \Rightarrow \delta_1 < \frac{\delta(\lambda - \tilde{\lambda})}{(1-\delta)(\mu - \tilde{\lambda}) + \delta(\lambda - \tilde{\lambda})} < \delta_2. \tag{8.18}$$

Suppose $N>0$. Then the left inequality with $\delta = \delta_1$ implies $\mu>\lambda$, the right inequality with $\delta = \delta_2$ implies $\lambda>\mu$. There is the same contradiction for $N<0$. For $\lambda = \mu$ we have $N = \lambda - \tilde{\lambda} \neq 0$, $w(\delta) = x + \delta(y - x)$ and $\sigma(\delta) = \lambda$ demonstrating lemma 8.3. □

Lemma 8.4: Under the assumptions of lemma 8.3 given an eigenvalue μ of A with $\mu \in \Lambda$ every eigenvector y of A corresponding to μ must in X.

Proof: Suppose $Ay = \mu y$ with $\mu \in \Lambda$. Given f from (8.9) we define g_μ: $V_n\mathbb{R} \to V_n\mathbb{R}$ to be the first n components of $f(t,\mu)$:

$$t \in V_n\mathbb{R} \Rightarrow f\begin{pmatrix} t \\ \mu \end{pmatrix} = \begin{pmatrix} g_\mu(t) \\ \nu \end{pmatrix} \quad \text{with some } \nu \in \mathbb{R}.$$

According to (8.10 and (8.7), g_μ is a continuous self-mapping of X:

$$g_\mu\colon X \to X \quad \text{and} \quad g_\mu(X) \subseteq X \setminus \partial X. \tag{8.19}$$

Brouwer's fixed point theorem states that

$$\exists z \in X\colon g_\mu(z) = z, \quad \text{i.e.} \quad f\begin{pmatrix} z \\ \mu \end{pmatrix} = \begin{pmatrix} z \\ \mu^* \end{pmatrix} \tag{8.20}$$

for some $\mu^* \in \mathbb{R}$. Define g: $V_{n+1}\mathbb{R} \to V_{n+1}\mathbb{R}$ by

$$w \in V_n\mathbb{R},\ \sigma \in \mathbb{R} \Rightarrow g\begin{pmatrix} w \\ \sigma \end{pmatrix} := Z + \left\{ I_{n+1} - R \cdot \begin{pmatrix} A - \sigma I_n & -\tilde{x} \\ e'_k & 0 \end{pmatrix} \right\} \cdot \begin{pmatrix} w - \tilde{x} \\ \sigma - \tilde{\lambda} \end{pmatrix},$$

then $g(w,\sigma) = f(w,\sigma)$ and (8.10) holds with g substituted for f. Applying lemma 7.3 yields that

$$T(\sigma) := \begin{pmatrix} A - \sigma I_n & -\tilde{x} \\ e'_k & 0 \end{pmatrix} \quad \text{is non-singular for every } \sigma \in \Lambda.$$

Since y is an eigenvector of A, $y \neq 0$. So $T(\mu) \cdot (y,0) \neq 0$ yields $y_k := e'_k \cdot y \neq 0$. Define

$$M(\nu) := \left\{ t \in V_n\mathbb{R} \mid f\begin{pmatrix} t \\ \mu \end{pmatrix} = \begin{pmatrix} t \\ \nu \end{pmatrix} \right\} \quad \text{for } \nu \in \mathbb{R}. \tag{8.21}$$

For the purpose of establishing a contradiction we assume $\mu \neq \mu^*$ and define h: $\mathbb{R} \rightarrow V_n\mathbb{R}$ by

$$\nu \in \mathbb{R} \Rightarrow h(\nu) := \zeta z + \eta y \quad \text{with} \quad \xi := (\mu - \nu)(\mu - \mu^*)^{-1},\ \eta = \zeta(1 - \xi) y_k^{-1}. \tag{8.22}$$

Then brief computation using (8.20) and (8.9) yields

$$h(\nu) \in M(\nu) \quad \text{for every} \quad \nu \in \mathbb{R}. \tag{8.23}$$

h is continuous and

$$h(\nu + \varepsilon) - h(\nu) = \varepsilon \bullet \{\zeta(\mu - \mu^*)^{-1} y_k^{-1} \bullet y - (\mu - \mu^*)^{-1} \bullet z\}. \tag{8.24}$$

If $z = \zeta y_k^{-1} \bullet y$ then $(A - \mu I)z = 0$ and $e'_k z = \zeta$ which contradicts $\mu \neq \mu^*$ according to (8.20) and (8.9). Thus $z \neq \zeta y_k^{-1} \bullet y$ and from (8.24) follows

$$|h(\nu)| \rightarrow \infty \quad \text{for} \quad \nu \rightarrow \infty. \tag{8.25}$$

For every $\nu \in \mathbb{R}$, any $t \in M(\nu)$ is a fixed point of g_μ. (8.20) yields $z = h(\mu^*) \in M(\mu^*)$ and $z \in X$, so by (8.25) there exists a fixed point of g_μ on ∂X. This contradicts (8.19).

Thus, $\mu = \mu^*$ and the lemma is proved. □

Lemma 8.5: Under the assumptions of lemma 8.2 given an eigenvector y of A with $y \in X$ the eigenvalue μ of A corresponding to y must lie in Λ.

Proof: Suppose $Ay = \mu y$ with $y \in X$. Using f from (8.9) we define g_y: $\mathbb{R} \rightarrow \mathbb{R}$ be the $(n + 1)^{\text{-st}}$ component of $f(y,\nu)$:

$$\nu \in \mathbb{R} \Rightarrow f\begin{pmatrix} y \\ \nu \end{pmatrix} = \begin{pmatrix} z \\ g_y(\nu) \end{pmatrix} \quad \text{with some } z \in V_n\mathbb{R}. \tag{8.26}$$

According to (8.10) and (8.7) g_y is a continuous self-mapping of Λ:

$$g_y\colon\ \Lambda\rightarrow\Lambda \quad\text{and}\quad g_y(\Lambda)\subseteq\Lambda\setminus\partial\Lambda. \tag{8.27}$$

Brouwer's fixed point theorem states

$$\exists\sigma\in\Gamma\colon\ g_y(\sigma) = \sigma,\ \text{i.e. } f\begin{pmatrix} y \\ \sigma \end{pmatrix} = \begin{pmatrix} z \\ \sigma \end{pmatrix} \tag{8.28}$$

for some $z\in V_n\mathbb{R}$. Define

$$M(t) := \left\{ \nu\in\mathbb{R} \mid f\begin{pmatrix} y \\ \nu \end{pmatrix} = \begin{pmatrix} t \\ \nu \end{pmatrix} \right\} \quad\text{for } t\in V_n\mathbb{R}. \tag{8.29}$$

Assuming $\mu\neq\sigma$ we define

$$\mathrm{v} := z + \eta(y-z) \quad\text{with}\quad \eta := (\nu-\sigma)/(\mu-\sigma).$$

A brief computation using (8.28) and (8.29) yields $\sigma\in M(z)$ and $\nu\in M(\mathrm{v})$ for every $\nu\in\mathbb{R}$, whence for every $\nu\in\mathbb{R}$ there is a $\mathrm{v}\in V_n\mathbb{R}$ with $\nu\in M(\mathrm{v})$. By (8.29) and (8.26) every ν in some $M(\mathrm{v})$ is a fixed point of g_y. This contradicts (8.27). Thus $\mu = \sigma$ and by (8.28), $\mu\in\Lambda$. □

After these preparatory lemmata, we are ready to state the following theorem.

Theorem 8.6: Define $G^*\colon \mathbb{I}V_{n+1}\mathbb{R}\rightarrow\mathbb{P}V_{n+1}\mathbb{R}$ by (8.6) for $R\in M_{n+1}\mathbb{R}$; $\tilde{x}\in V_n\mathbb{R}$; $\tilde{\lambda},\zeta\in\mathbb{R}$; $\zeta\neq 0$.

If

$$G^*(T)\subseteq\mathring{T} \quad\text{with}\quad T=\begin{pmatrix} X \\ \Lambda \end{pmatrix} \quad\text{for some } X\in\mathbb{I}V_n\mathbb{R},\ \Lambda\in\mathbb{I}\mathbb{R}$$

then all of the following are true:

1) there is one and only one eigenvector $\hat{x}$ of A with $\hat{x}\in X$,

2) there is one and only one eigenvalue $\hat{\lambda}$ of A with $\hat{\lambda}\in\Lambda$,

3) these are corresponding, i.e. $A\hat{x} = \hat{\lambda}\hat{x}$ and

4) the multiplicity of $\hat{\lambda}$ is one.

Proof: The existence of an eigenvector/eigenvalue pair $(\hat{x},\hat{\lambda})$ in (X,Λ) follows by lemma 8.2. Hence we have 3) and 1) and 2) follow by lemmata 8.4 and 8.3 resp. lemmata 8.5 and 8.3. Assertion 4) follows by lemma 8.3. □

In contrast to theorem 8.1, we use $\tilde{\lambda}$ instead of $\tilde{\lambda}\,\underline{\cup}\,\Lambda$ and $-X$ instead of $-(\tilde{x}\,\underline{\cup}\,X)$, which reduces the diameter of $G^*(T)$ compared with $G(T)$ significantly. Theorem 8.6 is applicable on computers. This is shown by the following theorem. We formulate it at once for finding an inclusion of the difference between the solution $(\hat{x},\hat{\lambda})$ and an approximate solution $(\tilde{x},\tilde{\lambda})$.

Theorem 8.7: Let $A \in M_nS$, $R \in M_{n+1}S$, $\tilde{x} \in V_nS$ and $\tilde{\lambda},\zeta \in S$ with $\zeta \neq 0$. For $X \in I\!IV_nS$ define

$$Q(X) := \begin{pmatrix} A - \tilde{\lambda} I_k & -\tilde{x} - X \\ e'_k & 0 \end{pmatrix} \in I\!IM_{n+1}\mathbb{R} \text{ and}$$

$$Z := \diamondminus\; R \;\diamonddot\; \diamond \begin{pmatrix} A\tilde{x} - \tilde{\lambda}\tilde{x} \\ e'_k\tilde{x} - \zeta \end{pmatrix} \in I\!IV_{n+1}S. \tag{8.30}$$

If then for some $X \in I\!IV_nS$ and $\Lambda \in I\!I\mathbb{R}$

$$Z \;\diamondplus\; \diamond\; \{I_{n+1} - R \cdot Q(X)\} \;\diamonddot\; T \subseteq \mathring{T} \text{ with } T := \begin{pmatrix} X \\ \Lambda \end{pmatrix}, \tag{8.31}$$

then the matrix R and each matrix $B \in M_{n+1}\mathbb{R}$ and $B \in Q(X)$ is non-singular and the following are true:

1) there is one and only one eigenvector $\hat{x}$ of A with $\hat{x} \in \tilde{x} + X$,

2) there is one and only one eigenvalue $\hat{\lambda}$ of A with $\hat{\lambda} \in \tilde{\lambda} + \Lambda$,

3) they are corresponding, i.e. $A\hat{x} = \hat{\lambda}\hat{x}$ and

4) the multiplicity of $\hat{\lambda}$ is one.

Proof: This is a consequence of theorem 8.6. □

The uniqueness of $\hat{x}$ resp. $\hat{\lambda}$ cannot be guaranteed in $\tilde{x} \;\diamondplus\; X$ resp. $\tilde{\lambda} \;\diamondplus\; \Lambda$. (8.31) and especially $\diamond(\square\{I - R \cdot Q(X)\})$ is (effectively) executable on computers using the precise scalar product (cf. [3]).

Theorem 8.7 yields an algorithm to include an eigenvector/eigenvalue pair of a given matrix A with automatic verification of correctness. For the algorithm and further improvements see [34]. Of course in the actual implementation x_k need not to be stored as an additional variable. Instead (8.30) and (8.31) are rewritten in n variables. It is possible to insert some $X^* \in \mathbb{I}V_n S$, $\Lambda^* \in \mathbb{I}S$ with $X^* \supseteq X$, $\Lambda^* \supseteq \Lambda$ in (8.30), (8.31). If for both for (X,Λ) and (X^*,Λ^*) the condition (8.31) is satisfied, then it has been verified that there is no eigenvalue of A in $\Lambda^* \backslash \Lambda$. The computing time for the algorithm is approximately $2n^3$. Each additional evaluation of (8.31) with another T costs $\sim 3n^2$.

Finally we mention another version of theorem 8.7.

Theorem 8.8: Let $A \in M_n S$, $R \in M_{n+1} S$, $\tilde{x} \in V_n S$ and $\tilde{\lambda}$, $\zeta \in S$ with $\zeta \neq 0$. If the linear system $Cx = b$, $b \in \mathbb{I}B_n S$ with

$$C := \begin{pmatrix} A - \tilde{\lambda} I_n & -\tilde{x} \\ e'_k & 0 \end{pmatrix} \quad \text{and} \quad b := \Diamond \begin{pmatrix} -A\tilde{x} + \tilde{\lambda}\tilde{x} + \Lambda X \\ 0 \end{pmatrix}$$

is solved using algorithm 2.1 yielding an inclusion of the solution $(Y,M)'$, $Y \in \mathbb{I}V_n S$, $M \in \mathbb{I}S$ and, moreover,

$$Y \subseteq X \quad \text{and} \quad M \subseteq \Lambda$$

is satisfied, then all assertions of theorem 8.7 remain valid.

Remark: Algorithm 2.1 has to be used in its version for interval right hand side b as described in chapter 2. In general, $A - \tilde{\lambda} I_n$ is not an element of $M_{n+1} S$. However, algorithm 2.1 can be applied for point matrices, for instance, by splitting a product $(A - \tilde{\lambda} I_n)x$, $x \in V_n S$ in a scalar product of length $n + 1$. This assures that all assertions respecting algorithm 2.1 remain true.

Proof of theorem 8.8: A brief computation using (2.2) yields exactly that provision (8.30) in theorem 8.7 is satisfied. Therefore all assumptions of theorem 8.7 are valid. □

Theorem 8.8 extends Satz 3.7 in [47]. In this specific case we do not assume $0 \in X$, $0 \in \Lambda$

but conclude (extending the cited Satz 3.7) the non-singularity of C,the uniqueness of the eigenvector/eigenvalue pair, the uniqueness of the eigenvalue and that the multiplicity of the eigenvalue is 1. Most important is the fact that the non-singularity of C is verified by the computer and not assumed to be checked by the user (which is, in fact, hardly solvable). This makes the corresponding algorithm widely applicable especially for non-mathematicians.

There are similar extensions, as in chapter 7, to complex matrices and problems with uncertain data. In the latter case (with a matrix $\mathcal{A} \in IIM_n S$) the assertions 1), 2), 3) and 4) remain valid for any $A \in M_n \mathbb{R}$ with $A \in \mathcal{A}$.

The following numerical examples were computed on the UNIVAC 1108 at the University of Karlsruhe. We denote by

H	the Hilbert-matrix
P	the Pascal-matrix
S_1, S_2	matrices with uniformly distributed eigenvalues in $[-1,1]$, $[1,10]$, resp.
C	a matrix with clustered eigenvalues $1 + i \cdot 10^{-5}$, $i = 1(1)n$
R	a randomly generated matrix with $\lvert R_{ij} \rvert \leq 1$.

We first applied a built-in procedure to compute approximations $\tilde{x}, \tilde{\lambda}$ for the eigenvectors and eigenvalues of A, resp. Then the new algorithm was applied. The following table displays the matrix, the number of rows n, the maximum relative error $\tilde{\Delta}$ of the components of all approximations $\tilde{x}, \tilde{\lambda}$ and the number of digits guaranteed in the final inclusions for all eigenvectors and all eigenvalues of the matrix. Here an additional l.s.b.a. indicates that all components of the inclusion of eigenvectors and the inclusions of eigenvalues are of least significant bit accuracy, i.e. left and right bounds are consecutive numbers in the floating-point screen $S = (2,27,-128,127)$. The $*$ for $\tilde{\Delta}$ in H^7 indicates, that the approximation $\tilde{\lambda} = -2.17_{10^{-3}}$ for one eigenvalue was of wrong sign (the correct value is $+1.259..._{10^{-3}}$).

matrix	n	$\tilde{\Delta}$	digits guaranteed
H	6	$1.7_{10^{-2}}$	$8\frac{1}{2}$ (l.s.b.a.)
	7	*	$8\frac{1}{2}$ (l.s.b.a.)
	8	43	$8\frac{1}{2}$ (l.s.b.a.)
P	8	$1.5_{10^{-3}}$	$8\frac{1}{2}$ (l.s.b.a.)
	9	$3.0_{10^{-3}}$	$8\frac{1}{2}$ (l.s.b.a.)
S_1	20	$5.6_{10^{-2}}$	$8\frac{1}{2}$ (l.s.b.a.)
S_2	20	$3.1_{10^{-2}}$	$8\frac{1}{2}$ (l.s.b.a.)
C	20	$1.9_{10^{-1}}$	8
R	50	$6.9_{10^{-6}}$	$8\frac{1}{2}$ (l.s.b.a.)

9. REAL AND COMPLEX ZEROS OF POLYNOMIALS

Consider a polynomial p of degree n. p can be regarded as a (continuously differentiable) mapping, so the theorems and corollaries derived in chapter 7 are applicable. Here we mention two theorems for real zeros of real polynomials and complex zeros of complex polynomials. They both are formulated directly for application on the computer, for an inclusion between the difference of a zero and an approximation.

Theorem 9.1: Let $p(x) = \sum_{i=0}^{n} a_i \cdot x^i$ with $a_i \in S$ for $0 \le i \le n$ and let $\tilde{x} \in S$, $r \in S$ be given. Let $\Diamond_{p} : S \to \mathbb{I}S$ resp. $\Diamond_{p}' : \mathbb{I}S \to \mathbb{I}S$ be functions satisfying $x \in S \Rightarrow p(x) \in \Diamond_{p}(x)$ resp. $p'(x) \in \Diamond_{p}'(x)$. If then for some $X \in \mathbb{I}S$ with $0 \in X$

$$\Diamond_{-}\; r \;\Diamond_{\cdot}\; \Diamond_{p}(\tilde{x}) \;\Diamond_{+}\; \Diamond\{1 - r\cdot \Diamond_{p}'(\tilde{x} \;\Diamond_{+}\, X)\} \;\Diamond_{\cdot}\; X \subsetneqq X, \tag{9.2}$$

then there exists one and only one $\hat{x} \in \mathbb{R}$ with $\hat{x} \in \tilde{x} \;\Diamond_{+}\; X$ and $p(\hat{x}) = 0$. $\hat{x}$ is a simple zero of p.

Theorem 9.2: Let $p(z) = \sum_{i=0}^{n} c_i\, z^i$ with $c_i \in \mathbb{C}S$ for $0 \le i \le n$ and let $\tilde{z} \in \mathbb{C}S$, $r \in \mathbb{C}S$. Let $\Diamond_{p} : \mathbb{C}S \to \mathbb{I}\mathbb{C}S$ resp. $\Diamond_{p}' : \mathbb{I}\mathbb{C}S \to \mathbb{I}\mathbb{C}S$ be functions satisfying $x \in \mathbb{C}S \Rightarrow p(x) \in \Diamond_{p}(x)$ resp. $\Diamond_{p}'(x) \in \Diamond_{p}'(x)$. If then for some $Z \in \mathbb{I}\mathbb{C}S$ with $0 \in Z$

$$\Diamond_{-}\; r \;\Diamond_{\cdot}\; \Diamond_{p}(\tilde{z})\, \mathrm{i} \;\Diamond_{+}\; \Diamond\{1 - r\cdot \Diamond_{p}'(\tilde{z} \;\Diamond_{+}\, Z)\} \;\Diamond_{\cdot}\; Z \subsetneqq Z, \tag{9.2}$$

then there exists one and only one $\hat{z} \in ₵$ with $\hat{z} \in \tilde{z} \diamondplus Z$ and $p(\hat{z}) = 0$. $\hat{z}$ is a simple zero of p.

The ***proofs*** are an immediate consequence of corollaries 7.8 and 7.9. However, both theorems can be proved directly using Banach's fixed point Theorem.

The functions $\Diamond_p$ resp. $\Diamond_p'$ may be the usual interval extensions of p resp. p'. However, this is an overestimation. In chapter 11 a new method will be derived for the computation of the value of arbitrary arithmetic expressions at certain points with least significant bit accuracy. Applying this to (9.1) and (9.2) gives a significant improvement.

In [6] Böhm gave a large number of different algorithms for the inclusion of zeros of polynomials with automatic verification of correctness. Their presentation lies outside the scope of this article; we give only a few keywords. For a complete discussion cf. [6].

In [6] algorithms of higher order using higher order derivatives are given. Here the possibility is demonstrated of computing inclusions of the coefficients of quadratic factors of a polynomial. If, e.g., $p(x) = \sum_{i=0}^{n} a_i x^i$ is given, then $Ax^2 + Bx + C$, where $A,B,C \in \mathbb{I}S$ resp. $\mathbb{I}₵S$, is computed and the following is true. There exist $a,b,c \in \mathbb{R}$ resp. $₵$ with $a \in A$, $b \in B$, $c \in C$ such that $ax^2 + bx + c$ divides $p(x)$ without reminder. In this manner double zeros and, when including factors of higher degree, multiple zeros of a polynomial can be included. However, it cannot be verified that p has a zero of multiplicity greater than one.

Moreover in [6] several theorems and corresponding algorithms are derived using the Frobenius matrix of p and the transposed Frobenius matrix.

Next we briefly describe two methods for simultaneous inclusion of all complex zeros of a complex polynomial. The first one is an extension of a well-known procedure.

Theorem 9.3: (Böhm): Let $p(z) = \sum_{i=0}^{n} a_i \cdot z^i$ with $a_i \in ₵S$ for $0 \leq i \leq n$ and let $\tilde{z} = (\tilde{z}_1,\ldots,\tilde{z}_n) \in V₵S$ with $\tilde{z}_1 \neq \tilde{z}_j$ for $1 \leq i,j \leq n$ and $i \neq j$. If then for some $Z = (Z_1,\ldots,Z_n) \in \mathbb{I}V₵S$

$$\tilde{z}_1 \diamondminus \Diamond_p(\tilde{z}_i) \diamondslash \{a_n \diamonddot \prod_{\substack{j=1 \\ j \neq i}}^{n} (\tilde{z}_i \diamondminus Z_j) \subseteq Z_i \quad \text{for } 1 \leq i \leq n,$$

then for the zeros ξ_i, $1 \leq i \leq n$ of p and a suitable indexing, we have $\xi_i \in \tilde{z}_i \diamondplus Z_i$, $1 \leq i \leq n$.

Here ⓟ is defined as above. Notice, that p' is not needed.

The next theorem gives an improvement of the well-known method of Gargantini and Henrici for simultaneous inclusion of the complex zeros of complex polynomials.

Theorem 9.4: (Böhm): Let $p(z) = \sum_{i=0}^{n} a_i z^i$ with $a_i \in ¢$ for $0 \le i \le n$ and let $\tilde{z} = (\tilde{z}_1,\ldots,\tilde{z}_n) \in V¢$. Define f: $V¢ \rightarrow V¢$, $f = (f_1,\ldots,f_n)$ componentwise for $z = (z_1,\ldots,z_n) \in ¢$ by

$$f_i(z) = \tilde{z}_i - p(\tilde{z}_i)/\{p'(\tilde{z}_i) + p(\tilde{z}_i) \bullet \sum_{\substack{j=1 \\ j \neq 1}}^{n} (\tilde{z}_i - z_j)^{-1}\},\ 1 \le i \le n. \tag{9.3}$$

If then for some $Z \in I\!\!IV¢$, $Z = (Z_1,\ldots,Z_n)$ the demoninator of (9.3) does not vanish for $z \in Z$, $1 \le i \le n$ and

$$\{f(z) \mid z \in Z\} \subsetneq Z, \tag{9.4}$$

then the zeros ξ_i, $1 \le i \le n$ of p satisfy $\xi_i \in \tilde{z}_i + Z_i$ with suitable indexing. Moreover for every $k \in I\!\!N$

$$\xi = (\xi_1,\ldots,\xi_n) \in \{\tilde{z} + f_k(z) \mid z \in Z\}.$$

For the proof of the two preceding theorems cf. [6]. In contrast to the algorithm of Gargantini and Henrici an algorithm based on theorem 9.4 does not require inclusions for the zeros of p as an input. Moreover any complex arithmetic (rectangle, circular) can be used as long as the intervals are convex. The $\tilde{z}_i$ need not to be the midpoints of Z_i, in fact $\tilde{z}_i$ is not required to be an element of Z_i, $1 \le i \le n$.

Again the coefficients of the given polynomial may be intervals themselves. In this case the zeros of every point polynomial included by the interval polynomial are included.

The computing time for the new algorithms is of the same order as comparable (purely) floating-point algorithms. The latter, of course, offers none of the new features.

In the following we give some numerical examples. The algorithms are programmed on the UNIVAC 1108 of the University at Karlsruhe. There the floating-point screen is $S(2,27,-127,128)$. In fact the computational results of the algorithms derived from theorems 9.1, 9.2, 9.3 and 9.4 are almost identical, so we display the results for the last case only. The polynomials treated are:

P_1 with zeros $\pm\sqrt{2}$, 17/12, 41/29

P_2 with zeros $\pm\sqrt{2}$, 3363/2378

R_1 product of linear factors with random zeros in $[-2,2]$

R_2 coefficients randomly generated in $[-1,1]$

W $(x-1)(x-2)\cdot\ldots\cdot(x-11)-1$

L Legendre polynomial, coefficients computed in floating-point

RC_1 randomly generated coefficients in the unit square

RC_2 randomly generated zeros with $|Re\, z| \leq 1.5$, $|Im\, z| \leq 1.5$

RC_3 products of linear factors, randomly generated with zeros in $|z| \leq 2$.

As an example we give two figures displaying the zeros of RC_2 and RC_3, both for degree 49.

The zeros of R_1 are particularly ill-conditioned. A procedure from IMSL implemented on the UNIVAC 1108 generated for R_1 of degree 50 an approximation -2.1 for a real zero, whereas -1.06 is the smallest real zero. The results are displayed in the following table. For more examples see [6].

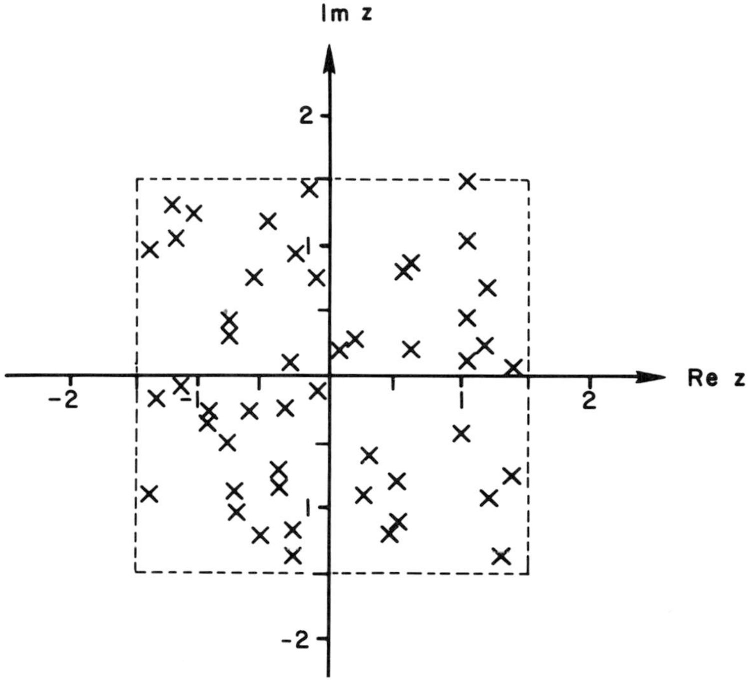
Im z
2
1
Re z
-2
-1
1
2
1
-2

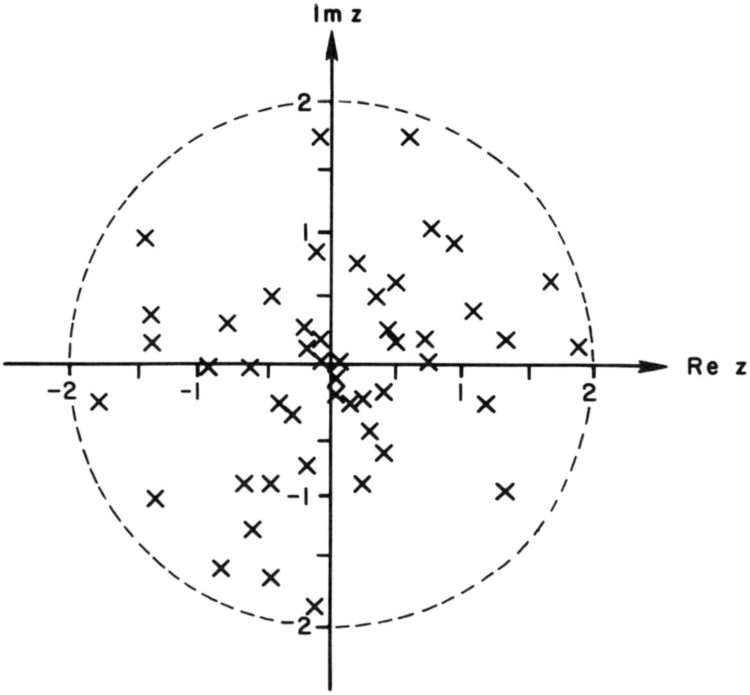
Im z
2
1
Re z
-2
-1
1
2
-1
2

polynomial	degree	k	# of digits guaranteed
P_1	4	2	$8\frac{1}{2}$ (l.s.b.a.)
P_2	3	2	$8\frac{1}{2}$ (l.s.b.a.)
R_1	25	2	$8\frac{1}{2}$ (l.s.b.a.)
	50	3	8
R_2	25	2	$8\frac{1}{2}$ (l.s.b.a.)
	50	2	$8\frac{1}{2}$ (l.s.b.a.)
W	11	2	$8\frac{1}{2}$ (l.s.b.a.)
L	15	2	$8\frac{1}{2}$ (l.s.b.a.)
RC_1	25	2	$8\frac{1}{2}$ (l.s.b.a.)
	49	2	$8\frac{1}{2}$ (l.s.b.a.)
RC_2	25	2	$8\frac{1}{2}$ (l.s.b.a.)
	49	2	$8\frac{1}{2}$ (l.s.b.a.)
RC_3	25	2	$8\frac{1}{2}$ (l.s.b.a.)
	49	2	$8\frac{1}{2}$ (l.s.b.a.)

The initial X resp. Z is an interval with left and right bound equal to a floating-point approximation. Therefore (9.4) cannot be satisfied for the initial X resp. Z and an iteration is started similar to those described in chapter 7. In the table, k is the number of iterations, $k \geq 2$. In the last column the minimum number of digits guaranteed of all inclusions of all zeros is displayed. An additional l.s.b.a. means that the left and right bounds of all inclusions were consecutive points in the floating-point screen. The zeros of P_2 are

$$\pm\ 1.414213562\ldots \quad = \pm\sqrt{2}$$

and

$$+1.414213625\ldots \quad = 3363/2378$$

Therefore between $+\sqrt{2}$ and 3363/2378 are only 4 points of the floating-point screen. Nevertheless all zeros have been included to least significant bit accuracy with automatic verification of the correctness.

10. LINEAR, QUADRATIC AND CONVEX PROGRAMMING

In this chapter our aim is to give algorithms which verify the optimality of a solution to a linear, quadratic or convex programming problem. We begin the discussion with linear programming problems. We use the same notation as in [9]. In this chapter all vectors are by definition column vectors, the transposed vector is indicated by a prime.

Let $x,p \in V_n\mathbb{R}$; $b \in V_m\mathbb{R}$, $A \in M_{m,n}\mathbb{R}$ and Q: $V_n\mathbb{R} \rightarrow \mathbb{R}$ where $Q(x) := p'x$. We suppose $m<n$. The problem is to find a vector $\hat{x} \in V_n\mathbb{R}$ with $\hat{x} \geq 0$ satisfying the condition $A\hat{x} = b$ and having the property

$$y \in V_n\mathbb{R},\ y \geq 0 \text{ and } Ay = b \Rightarrow Q(y) \geq Q(x).$$

We write this linear programming problem as follows:

$$Ax = b,\ x \geq 0 \text{ and } Q(x) = p'x = \text{Min!} \tag{10.1}$$

Let A_k, $1 \leq k \leq n$ be the column vectors of A and let Z be a set of indices in the range $1...n$. We suppose $|Z| = m$ and define $\tilde{A} \in M_{m,m}\mathbb{R}$ to be the matrix with columns A_k, $k \in Z$ and $\tilde{p} \in V_m\mathbb{R}$ to be the vector with components p_k, $k \in Z$. Let $\tilde{A}$ be non-singular and define $t' = (t_i) = \tilde{p}'\tilde{A}^{-1}A$. Define the vector $x^0 \in V_n\mathbb{R}$ in the following way: $x_j^0 := 0$ for $j \notin Z$ and the m components x_j^0, $j \in Z$ are the components of $\tilde{A}^{-1}b$ in successive order. Then the following is true (cf. [9]):

1) If $t_j \leq p_j$ for every $j \notin Z$, then x^0 is an optimal solution to (10.1).

2) If there is a $j \notin Z$ such that $t_j > p_j$ and $(\tilde{A}^{-1}A_j)_k \leq 0$ for $1 \leq k \leq m$, then (10.1) has no solution.

3) If $t_j > p_j$ for a $j \notin Z$ and a component of $\tilde{A}^{-1}A_j$ is greater than zero, then there is a feasible solution $x^1 \in V_n\mathbb{R}$ of (10.1) satisfying $Q(x^1) < Q(x^0)$.

Consider the linear systems

$$\begin{pmatrix} \tilde{A} & 0 \\ -p' & 1 \end{pmatrix} \cdot \begin{pmatrix} y_j \\ c_j \end{pmatrix} = \begin{pmatrix} A_j \\ -p_j \end{pmatrix}, \text{ where } y_j \in V_n\mathbb{R} \text{ and } c_j \in \mathbb{R}, j \notin Z. \tag{10.2}$$

Then (provided $\tilde{A}^{-1}$ exists)

$$y_j = \tilde{A}^{-1} \bullet A_j \text{ and } c_j = \tilde{p}'\tilde{A}^{-1}A_j - p_j = t_j - p_j, \quad j \notin Z.$$

So if $c_j \le 0$ for all $j \notin Z$, then $\hat{x} \in V_n\mathbb{R}$ defined such that $\hat{x}_j := 0$ for $j \notin Z$ and the m components $\hat{x}_j$, $j \in Z$ are the m components of $\tilde{A}^{-1}b$, is an optimal solution to (10.1) (cf. [9]). This leads to an algorithmic application on computers with automatic verification of correctness. For this purpose the problem is formulated as follows:

$$\begin{aligned} &A \in M_{m,n}S;\ x,p \in V_nS;\ b \in V_mS \text{ and define } Q\colon V_n\mathbb{R} \to \mathbb{R} \text{ by } Q(x) := p'x. \\ &\text{Find an } \hat{x} \in V_n\mathbb{R} \text{ with } Q(\hat{x}) = \text{Min}! \text{ under the restrictions } \hat{x} \ge 0,\ A\hat{x} = b. \end{aligned} \tag{10.3}$$

Theorem 10.1: Let the linear programming problem (10.3) be given. Suppose inclusions $Y_j \in \mathbb{I}V_mS$ for y_j, resp. $C_j \in \mathbb{I}S$ for c_j, $j \notin Z$ in (10.2) using algorithm 2.1 have been computed. Then A is non-singular and the following is true:

1) If $\sup(C_j) \le 0$ for every $j \notin Z$, then the vector $\hat{x} \in V_n\mathbb{R}$ defined such that $\hat{x}_j := 0$ for $j \notin Z$ and the m components $\hat{x}_j$, $j \in Z$ are the m components of $\tilde{A}^{-1}b$, is an optimal solution to (10.3).

2) If there is a $j \notin Z$ such that $\inf(C_j) > 0$ and $(\sup(Y_j))_k \le 0$ for $1 \le k \le m$, then (10.3) has no solution.

3) If $\inf(C_j > 0$ for a $j \notin Z$ and $(\inf(Y_j))_k > 0$ for some $1 \le k \le m$, then a base vector $A_j, j \in Z$ has to be exchanged.

Proof: This is a consequence of corollary 2 and the preceding discussion. □

The linear system (10.2) always involves the same matrix for all right hand sides, so the total computing time reduces to $m^3 + 2(n-m)m^2$. The optimal solution can be included by setting the right hand side of (10.2) equal to $(b,0)$. Moreover the last component of this solution vector includes the optimal value $Q(\hat{x})$. Of course, the restriction $m<n$ can be omitted and dual problems can be treated in a similar way.

Next we discuss convex programming problems. Again we use the same notation as in [9]. Consider $F\colon V_n\mathbb{R}\rightarrow\mathbb{R}$, $f_j\colon V_n\mathbb{R}\rightarrow\mathbb{R}$ for $1\le j\le m$. Suppose F, f_j for $1\le j\le m$ to be convex, having first partial derivate. Then for $x\in V_n\mathbb{R}$

$$F(x) = \text{Min}\,! \text{ with the restrictions } x\ge 0 \text{ and } f_j(x)\le 0,\ 1\le j\le m \tag{10.4}$$

is a convex programming problem. We define the Lagrange function $\phi(x,u)\colon V_{n+m}\mathbb{R}\rightarrow\mathbb{R}$, $x\in V_n\mathbb{R}$, $u\in V_m\mathbb{R}$ by (cf. [9])

$$\phi(x,u) := F(x) + u'\bullet f(x) \text{ with } f := (f_1,\ldots,f_m). \tag{10.5}$$

If ϕ_x resp. ϕ_u denotes the gradient of ϕ with respect to x resp. u:

$$\phi_x = \left(\frac{\partial\phi}{\partial x_1},\ldots,\frac{\partial\phi}{\partial x_n}\right)',\quad \phi_u = \left(\frac{\partial\phi}{\partial u_1},\ldots,\frac{\partial\phi}{\partial u_n}\right)', \tag{10.6}$$

then the following is true (Kuhn-Tucker, cf. [9]):

If there exists a feasible point $\bar{x}\in V_n\mathbb{R}$ with $f(\bar{x})<0$,
then $\hat{x}\in V_n\mathbb{R}$ with $\hat{x}\ge 0$ is an optimal solution of (10.4) (10.7)
if and only if there is a $u\in V_m\mathbb{R}$ with $u\ge 0$ satisfying

$$\begin{array}{ll} \phi_x(\hat{x},u)\ge 0 & \hat{x}'\bullet\phi_x(\hat{x},u)=0 \\ \phi_u(\hat{x},u)\le 0 & u'\bullet\phi_u(\hat{x},u)=0. \end{array} \tag{10.8}$$

Since $x,u,>0$ condition (10.8) is equivalent to

$\phi_x(\hat{x},u)\ge 0,\qquad \phi_u(\hat{x},u)\le 0$

for every $1\le j\le n$ $\hat{x}_j$ or $(\phi_x(\hat{x},u))_j$ equals zero (10.9)

for every $1\le j\le m$ u_j or $(\phi_u(\hat{x},u))_j$ equals zero.

Suppose the convex programming problem (10.4) is to be solved on a computer including automatic verification of correctness of the result. Assume there are functions $\diamond f$: $V_n\mathbb{R}\rightarrow I\!IV_nS$, $\diamond F$: $V_n\mathbb{R}\rightarrow I\!IS$, $\diamond\phi_x$: $V_{n+m}\mathbb{R}\rightarrow I\!IV_nS$ and $\diamond\phi_u$: $V_{n+m}\mathbb{R}\rightarrow I\!IV_mS$ satisfying

$$
\begin{aligned}
& x \in V_n\mathbb{R} \Rightarrow f(x) \in \Diamond f\,(x) \text{ and } F(x) \in \Diamond F\,(x) \\
& (x,u) \in V_{n+m}\mathbb{R} \Rightarrow \phi_x(x,u) \in \Diamond \phi_x\,(x,u) \text{ and } \phi_u(x,u) \in \Diamond \phi_u\,(x,u).
\end{aligned}
\tag{10.10}
$$

Let $\bar{x} \in V_n S$ with $\bar{x} \geq 0$ and $\sup(\Diamond f\,(\bar{x})) < 0$ be given and let floating-point approximations $\tilde{x} \in V_n S$ and $\tilde{u} \in V_m S$ for a solution of (10.8) be given. Let I resp. J be the set of indices i resp. j for which $\tilde{x}_i$ resp. $\tilde{u}_j$ is approximately zero. Consider the following system of non-linear equations.

$$
\begin{aligned}
& \phi_{x_i}(x,u) \cdot x_i = 0 \quad \text{for } 1 \leq i \leq n \text{ and } i \notin I \\
& \phi_{u_j}(x,u) \cdot u_j = 0 \quad \text{for } 1 \leq j \leq m \text{ and } j \notin J.
\end{aligned}
\tag{10.11}
$$

These are $n + m - |I| - |J|$ equations in the same number of unknowns when omitting x_i for $i \in I$, u_j for $j \in J$ in the computation of $\phi_{x_i}(x,u)$, $\phi_{u_j}(x,u)$. To this non-linear system algorithm 7.1 is applicable.

Theorem 10.2: Solve the non-linear system (10.11) using Algorithm 7.1 and let X_i, $1 \leq i \leq n$, $i \notin I$ and U_j, $1 \leq j \leq m$, $j \notin J$ be the computed inclusions for the solutions. Define $X_i := 0$ for $i \in I$ and $U_j := 0$ for $j \in J$ and let $X := (X_1, \ldots, X_n) \in I\!IV_n S$ and $U := (U_1, \ldots, U_m) \in I\!IV_m S$. If then

$$
\begin{aligned}
& \inf(X_i) \geq 0 \text{ and } \inf(U_j) \geq 0 \text{ for } 1 \leq i \leq n,\ 1 \leq j \leq m \text{ and} \\
& \inf\{\Diamond \phi_x\,(X,U)\} \geq 0 \text{ and } \sup\{\Diamond \phi_u\,(X,U)\} \leq 0,
\end{aligned}
$$

then the convex programming problem (10.4) has an optimal solution $\hat{x} \in X$.

Next we discuss quadratic programming problems. Again we use the same notation as in [9]. To solve a quadratic programming problem with automatic verification of correctness of the result, theorem 10.2 could be used. However, taking advantage of the special structure of the problem finally leads to a system of <u>linear</u> equations as will be shown now.

Let $A \in M_{m,n}\mathbb{R}$; $x, p \in V_n\mathbb{R}$, $b \in V_m\mathbb{R}$ and let $C \in M_{n,n}\mathbb{R}$ be a symmetric, positive definite matrix. Then a quadratic programming problem is given by (cf. [9]):

$$
Q: V_n\mathbb{R} \rightarrow \mathbb{R} \text{ with } Q(x) := p'x + x'Cx = \text{Min!} \text{ with } Ax \leq b,\ x \geq 0. \tag{10.12}
$$

A specialization of the Kuhn-Tucker Theorem yields (cf. [9]):

A vector $\hat{x} \in V_n\mathbb{R}$ with $\hat{x} \geq 0$ is an optimal solution to (10.12) if and only if there exist $u \in V_m\mathbb{R}$, $\mathrm{v} \in V_n\mathbb{R}$, and $y \in V_m\mathbb{R}$ such that

$$\begin{aligned} &A\hat{x} + y = b,\ \mathrm{v} - 2Cx - A'u = p \\ &u \geq 0,\ \mathrm{v} \geq 0,\ y \geq 0 \end{aligned} \qquad (10.13)$$

and

$$x\mathrm{v} + yu = 0. \qquad (10.14)$$

The assumption of the existence of an $\bar{x} \in V_n\mathbb{R}$ with $f(\bar{x}) < 0$, as in the convex case, can be omitted because the restrictions are affine-linear. Condition (10.14) means because of $x,\mathrm{v},y,u \geq 0$ that for $1 \leq i \leq n$ resp. $i \leq j \leq m$ either $x_i = 0$ or $\mathrm{v}_i = 0$ resp. either $y_j = 0$ or $u_j = 0$. Thus we can proceed in the following way.

Consider the system of non-linear equations

$$Ax + y = b,\ \mathrm{v} - 2Cx - A'u = p, x_i\mathrm{v}_i = 0 \ \text{ for } \ 1 \leq i \leq n,\ y_j u_j = 0 \ \text{ for } \ 1 \leq j \leq m \qquad (10.15)$$

in $2n + 2m$ unknowns (x,v,y,u). Let $(\tilde{x},\tilde{\mathrm{v}},\tilde{y},\tilde{u})$ be an approximate solution of (10.15). It follows from (10.15) that for every $1 \leq i \leq n$ either x_i or v_i equals zero and for every $1 \leq j \leq m$ either y_j or u_j equals zero. Delete in (10.15) every variable x_i,v_i,y_j,u_j for which $\tilde{x}_i,\tilde{\mathrm{v}}_i,\tilde{y}_j,\tilde{u}_j$ is approximately zero. Then $n + m$ equations

$$Ax^* + y^* = b \text{ and } \mathrm{v}^* - 2Cx^* - A'u^* = p \qquad (10.16)$$

remain, where in x^*,v^*,y^*,u^* have on the whole $n + m$ fewer components than x,v,y,u. The system (10.16) is linear.

Theorem 10.3: Let $A \in M_{m,n}S$, $p \in V_nS$, $b \in V_mS$ and let $C \in M_{n,n}$ be a symmetric, positive definite matrix. Define Q: $V_n\mathbb{R} \to \mathbb{R}$ by

$$x \in V_n\mathbb{R}: Q(x) := p'x + x'Cx.$$

If the linear system (10.16) has been solved using algorithm 2.1 with including intervals

$X^*, V^* \in \mathbb{IV}_n S$ and $Y^*, U^* \in \mathbb{IV}_m S$ of the solution, then the following is true:

> If $\inf(X^*) \geq 0$, $\inf(V^*) \geq 0$, $\inf(Y^*) \geq 0$ and $\inf(U^*) \geq 0$, then the quadratic programming problem (10.12) has an optimal solution $\hat{x} \in V_n \mathbb{R}$. The non-zero components of $\hat{x}$ are included in X^*, the others are zero respective to the procedure generating (10.16) described above.

the presented theorems 10.1, 10.2 and 10.3 lead to algorithms for automatic verification of the optimality of an approximate solution to a linear, convex and quadratic programming problem as in chapters 2 and 7. The presented theorems and the corresponding algorithms can easily be extended to uncertain data. They are similarly applicable to dual optimization problems.

Computational results of the corresponding algorithms are for instance those presented in chapters 2 and 7 of this article for linear and non-linear problems.

11. ARITHMETIC EXPRESSIONS

Single precision floating-point computations may yield an arbitrarily false result due to cancellation and rounding errors. This is true even for very simple, structured expressions such as Horner's scheme for polynomial evaluation. A simple procedure will be presented for fast calculation of the value of an arithmetic expression to least significant bit accuracy in single-precision computation. For this purpose in addition to the usual floating-point arithmetic, only a precise scalar product is required. If the approximation computed by usual floating-point arithmetic is good enough, the computing time for the new algorithm is approximately the same as for usual floating-point computation. If not, the essential advantage of the algorithm presented here is that the inaccuracy of the approximation is recognized and corrected. An inclusion with least significant bit accuracy for the value of the arithmetic expression is computed with automatic verification of correctness. Following we give a brief description of the procedure. For more details cf. [5], [33].

Let S be the floating-point screen of the computer in use. Then elements of S are named constants. Arithmetic expressions consist of constants and $x, -, \cdot, /, (\ ,\ )$. An arithmetic expression can be transformed to the quotient of two arithmetic expressions where in the numerator and denominator quotients may occur but only with constants in the denominator. Further an expression can be altered in such a way that in every product at most one factor is an expression itself, the others are constants. Example:

$$a^2-b+\frac{4a^2}{b(b-a)} \rightarrow \frac{(a^2-b)b-(a^2-b)a+4a^2/b}{b-a}. \tag{11.1}$$

This process can be performed automatically (cf. [8]). Therefore we consider arithmetic expressions which can be obtained by applying the following rules:

1) A constant is an expression.

2) The sum and difference of two expressions is an expression.

3) The product of an expression and a constant is an expression.

4) An expression divided by a constant is an expression.

Such expressions are called simple (arithmetic) expressions.

When evaluating a simple expression, each rule 1...4 corresponds to the evaluation of an intermediate result. Let a,b,c be constants and x,y be values of subterms. Then a new intermediate result z is obtained in one of the following ways:

1) $z=a$

2) $z=x \pm y$

3) $z=x \bullet a$

4) $z=x/a$ or $a \bullet z = x$.

Thus a simple arithmetic expression can be regarded as a system of linear equations. The variables are the intermediate results. Applying only rules 1), 2), 3) and 4) may result in many variables. The number of variables can be reduced. Example for (11.1):

$$\begin{array}{ll} x_1=a & x_4=x_3 \bullet b \\ x_2=x_1 \bullet a & b \bullet x_5 = x_2 \\ x_3=x_2-b & x_6=-a \bullet x_3+x_4+4x_5 \end{array} \tag{11.2}$$

For calculating the value of a polynomial $p(\xi) = \sum_{i=0}^{n} a_i \cdot \xi^{n-i}$ we obtain the linear system

$$x_0 = a_0;\ x_{i+1} = \xi \cdot x_i + a_{i+1} \text{ for } 0 \leq i \leq n-1.$$

Obviously the linear system corresponding to a simple arithmetic expression is lower triangular. This it can be solved by forward substitution and, which is important, this process can be iterated. Moreover not only approximations are achieved but also an inclusion of the value of the arithmetic expression with automatic verification of correctness.

The following remarks hold for arbitrary lower triangular linear systems and especially for linear systems corresponding to simple arithmetic expressions.

Let $L \in MS$, $b \in VS$ with $L_{ij} = 0$ for $i<j$, $1 \leq i, j \leq n$ and consider

$$Lx = b \quad \text{with} \quad L_{ii} \neq 0,\ 1 \leq i \leq n. \tag{11.3}$$

Then (11.3) can be solved using Bohlender's algorithm (cf. [3]) by

$$\hat{x}_k \in X_k, \quad 1 \leq k \leq n. \tag{11.4}$$

This process can be iterated using the residue iteration. The corresponding algorithm is given in [33]. Let X^k, $k \geq 1$ denote the computed inclusion vector ($X^k \in \mathbb{I}VS$) for $\hat{x}$ after the k^{-th} iteration step. Then in [33] the following is proved.

Lemma 11.1: Let ε be the unit of relative rounding error. If then no over- or underflow occurred during computation then for some norm $\| \bullet \|$

$$\| d(X^{k+1}) \| \leq \varepsilon \bullet \eta \bullet \| d(X^k) \| \quad \text{with} \quad \eta = 5n^2 + 0(\varepsilon).$$

The constant η is given explicitly in [33]. Lemma 11.1 covers *all* rounding errors due to arithmetic operations in the floating-point screen. If ℓ is the length of the mantissa of the computer in use then, by the preceding lemma, the diameter of the inclusion improves in every iteration step by a factor of at least $5n^2 \bullet B^{-\ell+1}$, where B is the base of the floating-point screen in use.

There are direct extensions to arithmetic expressions consisting of complex numbers. The result is a complex interval with least significant bit accuracy. The arithmetic may be rectangular, circular, segment or any other.

In the following we give some computational results. For more examples see [5], [33].

1) $y^2(4x^4 + y^2 - 4x^2) - 8x^6$ for $x = 470832$, $y = 665857$

2) $\sum_{i=1}^{5} x_i^2 - \frac{1}{5} \cdot (\sum_{i=1}^{5} x_i)^2$ for $x_i = 7.951_{10}7 + i - 3$, $i = 1(1)5$.

Remark: Expressions like this occur in least square approximation.

3) $f(x) = ((543339720x - 768398401)x - 1086679440)x + 1536796802$

a) $x = 1.4142$

b) $x = 1.41421356238$

c) $x = 1.414213561$

4) $(f(x-h) - 2f(x) + f(x+h))/h^2$ with $f(x) = \dfrac{2734x - 2761}{4556x^2 - 9247x + 4692}$ for $x = 1$.

This is an approximation for $f''(1)$.

5) $\sum_{i=0}^{90} (-1)^i \cdot \frac{x^i}{i!}$ for $x = 20$.

This is an approximation to e^{-20}.

The following table shows computational results computed on a minicomputer based on Z80 with floating-point screen (10,12, − 99,99). In the columns of the table are displayed from left to right:

- The number of the example
- The floating-point approximation $\tilde{x}$
- The correct value $\hat{x}$ of the expression rounded to 12 decimal digits

- The number k of iterations

- The final result X of the new algorithm.

	$\tilde{x}$	$\hat{x}$	k	X
1)	$+5.0_{10}23$	$+1.0$	2	$+1.0$
2)	-100000.0	$+10.0$	1	$+10.0$
3a)	$+0.2800$	$+0.282673919360$	1	$+0.282673919360$
3b)	$+0.01$	$+7.32719247117_{10}{-14}$	2	$+7.3271924711^{8}_{7}{}_{10}{-14}$
3c)	-0.01	$+2.89746134369_{10}{-9}$	2	$+2.8974613436^{9}_{8}{}_{10}{-9}$

Example 4) resp. 5) are approximations to $f''(1)$ resp. e^{-20}. It seems not to be meaningful to give the exact value of an approximation. Therefore in the following table we display only the leading digits of $\tilde{x}$ and X to demonstrate their discrepancy. However, all inclusions X were computed with least significant bit accuracy. The exact value of $f''(1)$ in example 4) is 54. Notice, that the final summand in Example 5) is $20^{90}/90$!

Example		$\tilde{x}$	X
4)	$h = 10^{-1}$	5645	5645
	$h = 10^{-2}$	3788500	378500
	$h = 10^{-3}$	1184	1185
	$h = 10^{-4}$	125.0	65.21
	$h = 10^{-5}$	4900	54.11
	$h = 10^{-6}$	380000	54.001
	$h = 10^{-7}$	-10000000	54.00001
	$h = 10^{-12}$	0	54.0000000002, 53.9999999999
5)		$1.188_{10}{-4}$	$2.0611536224^{4}_{3}{}_{10}{-9}$

The initial floating-point approximation is the result when evaluating the expression using usual floating-point arithmetic. Either, the final result with automatic verification of correctness is a point interval or else (in examples 3b, 3c) identical digits of the left and right bound are displayed only once.

If the initial approximation is "good enough", one iteration is executed to achieve least significant bit accuracy. In this case the total computing time is of the same order as usual floating-point evaluation.

The algorithm for computing the value of an arithmetic expression to least significant bit accuracy gives a significant improvement of the algorithm for non-linear systems presented in chapter 7 (applicable to those functions not consisting of transcendental functions) and of the algorithms for including real or complex zeros of a polynomial presented in chapter 9. It is obvious that if (in the latter case) the value of a polynomial is not correct computable, then arbitrarily false results may be computed. Consider the following example:

$$P(x) = 67872320568x^3 - 95985956257x^2 - 135744641136x + \\ + 191971912515.$$

Newton's procedure was applied to P with starting point $x = 2$, where $P(x)$ and $P'(x)$ were evaluated using Horner's scheme and usual floating-point arithmetic. The floating-point screen in use is (10,12, − 99,99) in which the coefficients of P and P' are storable without rounding error. The arithmetic in use satisfies (R), (R1), (R2), (R4) and (R6) given in chapter 1.

In the following table in the left column (cf. [37]) the

x^k	$x^{k+1}-x^k$
1.73024785661	$2.698_{10}-01$
1.57979152125	$1.505_{10}-01$
1.49923019011	$8.056_{10}-02$
1.45733317058	$4.190_{10}-02$
1.43593403289	$2.140_{10}-02$
1.42511502231	$1.082_{10}-02$
1.41967473598	$5.440_{10}-03$
1.41694677731	$2.728_{10}-03$
1.41558082832	$1.366_{10}-03$
1.41489735833	$6.835_{10}-04$
1.41455549913	$3.419_{10}-04$
1.41438453509	$1.710_{10}-04$
1.41429903606	$8.550_{10}-05$
1.41425628589	$4.275_{10}-05$
1.41423488841	$2.140_{10}-05$
1.41422414110	$1.075_{10}-05$
1.41421847839	$5.663_{10}-06$
1.41421582935	$2.649_{10}-06$
1.41421353154	$2.298_{10}-06$
1.41421353154	0
1.41421353154	0
1.41421353154	0
1.41421353154	0
1.41421353154	0

iterates x^k are displayed and in the right column the difference $x^{k+1} - x^k$ of two successive iterates. The iteration is monotone, the difference of two successive iterates decreases and

$$\bar{x} = 1.41421353154$$

is a fixed point. Computation in $\mathbb{R}$ would imply $P(\bar{x}) = 0$. However, with the new algorithm to compute the value of an arithmetic expression to least significant bit accuracy we obtain

$$P(\bar{x}) \in 1.0001825038^2_1,$$

and in fact the smallest value for $P(x)$ for $x>0$ is approximately 1.

CONCLUSIONS

In the preceding chapters the theoretical background and corresponding algorithms have been given for several problems in numerical analysis. The algorithms for solving linear systems (dense, band, overdetermined, underdetermined and sparse), inversion of matrices and evaluation of arithmetic expressions compute an inclusion of the solution with automatic verification of correctness, existence and uniqueness; all this in a self-contained manner. The other algorithms for non-linear systems, algebraic eigenvalue problems, zeros of real and complex polynomials and linear, quadratic and convex programming problems provide an approximation of the solution for use in computing an inclusion of the solution with automatic verification of correctness, existence and uniqueness. This approximation can be obtained by any floating-point algorithm. Therefore the latter procedures estimate the error of an approximate solution. In other words they verify the correctness of an error margin (in addition to verification of existence and uniqueness). These "verification-algorithms" could replace additional tests such as altering input data, recomputing in higher precision etc. These tests would have to be developed and utilized for each individual problem by the programmer. The new algorithms perform the verification automatically without any effort on the part of the user, without any knowledge about the condition of the problem and, most importantly, without either a deep mathematical background or an extensive investigation. This, of course, is also true for the algorithms which compute an inclusion of the solution directly without initial approximation. The automatic error control is a key property of all the algorithms presented here.

The efficiency of the algorithms has been demonstrated by inverting the Hilbert 21×21 matrix on a 14 hexadecimal digit computer. This is (after multiplying with a proper factor) the Hilbert matrix of largest dimension which can be stored without rounding errors in this floating-point system. The error bounds for all components of the inverse of the Hilbert 21×21 matrix are as small as possible, i.e., left and right bounds differ only by one in the last place of the mantissa of each component. We call this least significant bit accuracy (1sba). Our experience shows that the results of the algorithms using our new methods very often have the 1sba-property for every component of the solution.

REFERENCES

[1] Abbott, J. P., Brent, R. P. (1975). Fast Local Convergence with Single and Multistep Methods for Nonlinear Equations, Austr. Math. Soc. 19 (series B), 173-199.

[2] Alefeld, G. Herzberger, J. (1974). Einführung in die Intervallrechnung, Bibl. Inst. Mannheim, Wien, Zürich.

[3] Bohlender, G. (1977). Floating-point computation of functions with maximum accuracy. IEEE Trans. Comput. C-26, No. 7, 621-632.

[4] Bohlender, G., Grüner, K. Gesichtspunkte zur Implementierung einer optimalen Arithmetik. "Wissenschaftliches Rechnen und Programmiersprachen", Herausgeber U. Kulisch and Ch. Ullrich, B. G. Teubner Stuttgart.

[5] Böhm, H. Auswertung arithmetischer Ausdrücke mit maximaler Genauigkeit. "Wissenschaftliches Rechnen und Programminersprachen", Herausgeber U. Kulisch and Ch. Ullrich, B. G. Teubner Stuttgart.

[6] Böhm, H. (1980). Berechnung von Schranken für Polynomwurzeln mit dem Fixpunktsatz von Brouwer. Interner Bericht des Inst. f. Angew. Math., Universität Karlsruhe.

[7] Böhm, H. Private Communication.

[8] Böhm, H. (1981). Automatische Umwandlung eines arithmetischen Ausdrucks in eine zur exakten Auswertung geeignete Form. Interner Bericht des Inst. f. Angew. Math., Universität Karlsruhe.

[9] Collatz, L., Wetterling, W. (1966). Optimierungsaufgaben. Heidelberger Taschenbücher, Band 15, Springer-Verlag, Berlin-Heidelberg-New York.

[10] Collatz, L. (1968). *Funktionalanalysis und Numerische Mathematik,* Springer-Verlag.

[11] Forsythe, G. E., Moler, C. B. (1967). *Computer Solution of Linear Algebraic Systems,* Prentice-Hall.

[12] Forsythe, G. E. (1970). Pitfalls in computation, or why a Math book isn't enough, Technical Report No. CS147, Computer Science Department, Stanford University, 1-43.

[13] Gastinel, N. (1972). *Lineare Numerische Analysis.* F. Vieweg & Sohn, Braunschweig.

[14] Hansen, E. Interval Arithmetic in Matrix Computations, Part 1. SIAM J. Numer. Anal. 2, 308-320 (1965), Part II. SIAM J. Numer. Anal. 4, 1-9 (1967).

[15] Heuser, H. (1967). *Funktionalanalysis.* Mathematsche Leitfäden, B. G. Teubner, Stuttgart.

[16] Kaucher, E., Rump, S. M. (1980). Generalized iteration methods for bounds of the solution of fixed point operator equations, Computing 24, 131-137.

[17] Kaucher, E., Rump, S. M. (1982). E-methods for Fixed Point Equations $f(x) = x$, Computing 28, p. 31-42.

[18] Knuth, D. (1969). "The Art of Computer Programming", Vol. 2, Addison-Wesley, Reading, Massachusetts.

[19] Krawczyk, R. (1969). Newton-Algorithmen zur Bestimmung von Nullstellen mit Fehlerschranken, Computing, 4, 187-120.

[20] Köberl, D. (1980). The Solution of Non-linear Equations by the Computation of Fixed Points with a Modification of the Sandwich Method, Computing, 25, 175-178.

[21] Kulisch, U., Miranker, W. L. (1981). Computer Arithmetic in Theory and Practice. Academic Press, New York.

[22] Kulisch, U. (1969). Grundzüge der Intervallrechnung, Überblicke Mathematik 2, Herausgegeben von D. Laugwitz, Bibliographisches Institut, Mannheim, S. 51-98.

[23] Kulisch, U. An Axiomatic Approach to Rounded Computations, Mathematics Research Center, University of Wisconsin, Madison, Wisconsin, TS Report No. 1020, 1-29 (1969), and Numer. Math. 19, 1-17 (1971).

[24] Kulisch, U. (1976). Grundlagen des numerischen Rechnens (Reihe Informatik, 19). Mannheim-Wien-Zürich: Bibliographisches Institut.

[25] Kulisch, U., Wippermann, H.-W. PASCAL-SC, PASCAL für wissenschaftliches Rechnen, Gemeinschaftsentwicklung von Institut für Angewandte Mathematik, Universität Karlsruhe (Prof. Dr. U. Kulisch), Fachbereich Informatik, Universtät Kaiserslautern (Prof. Dr. H.-W. Wippermann).

[26] Martinez, J. M. (1980). Solving Non-linear Simultaneous Equations with a Generalization of Brent's Method, BIT, 20, 501-510.

[27] McShane, E. J., Botts, T. A. (1959). *Real Analysis.* Von Nostrand.

[28] Meinardus, G. (1964). *Approximation von Funktionen und ihre numerische Behandlung,* Berlin-Göttingen-Heidelberg-New York; Springer, 180 S.

[29] Moore, R. E. (1966). *Interval Analysis.* Prentice-Hall.

[30] Moore, R. E. (1977). A Test for Existence of Solutions for Non-Linear Systems, SIAM J. Numer. Anal., 4.

[31] Moré, J. J., Cosnard, M. Y. (1979). Numerical Solution of Non-Linear Equations. ACM Trans. on Math. Software, Vol. 5, No. 1, 64-85.

[32] Ortega, J. M., Reinboldt, W. C. (1970). *Iterative Solution of Non-linear Equations in several Variables.* Academic Press, New York-San Francisco-London.

[33] Rump, S. M., Böhm, H. Least Significant Bit Evaluation of Arithmetic Expressions in Single-precision, to appear in Computing.

[34] Rump, S. M. (1980). Kleine Fehlerschranken bei Matrixproblemen, Dr.-Dissertation, Inst. f. Angew. Math., Universität Karlsruhe.

[35] Rump, S. M. (1979). Polynomial Minimum Root Separation, Math. of Comp. Vol. 33, No. 145, 327-336.

[36] Rump, S. M. (1982). Solving Non-linear Systems with Least Significant Bit Accuracy, Computing 29, 183-200.

[37] Rump, S. M. Rechnervorführung, Pakete für Standardprobleme der Numerik, "Wissenschaftliches Rechnen und Programmiersprachen", Herausgeber U. Kulisch and Ch. Ullrich, B. G. Teubner Stuttgart.

[38] Rump, S. M. Lösung linearer und nichtlinearer Gleichungssysteme mit maximaler Genauigkeit. "Wissenschaftliches Rechnen und Programmiersprachen", Herausgeber U. Kulisch and Ch. Ullrich, B. G. Teubner Stuttgart.

[39] Schwarz, H. R., Rutischauser, H., Stiefel, E. (1972). *Matrizen-Numerik,* B. G. Teubner Stuttgart.

[40] Stoer, J. (1972). *Einführung in die Numerische Matematik I.* Heidelberger Taschenbücher, Band 105, Springer-Verlag, Berlin-Heidelberg-New York.

[41] Stoer, J., Bulirsch, R. (1973). *Einführung in die Numerische Mathematik II.* Heidelberger Taschenbücher, Band 114, Springer-Verlag, Berlin-Heidelberg-New York.

[42] Varga, R. S. (1962). *Matrix Iterative Analysis.* Prentice-Hall, Englewood Cliffs, New Jersey.

[43] Wilkinson, J. H. (1969). *Rundungsfehler.* Springer-Verlag.

[44] Wongwises, P. Experimentelle Untersuchungen zur numerischen Auflösung von linearen Gleichungssystemen mit Fehlererfassung, Interner Bericht 75/1, Institüt für Praktische Mathematik, Universität Karlsruhe.

[45] Yohe, J. M. (1973). Interval Bounds for Square Roots and Cube Roots, Computing 11, 51-57.

[46] Yohe, J. M. (1973). Roundings in Floating-Point Arithmetic, IEEE Trans. on Comp., Vol. C12, No. 6, 577-586.

[47] Alefeld, G. (1979). Intervallanalytische Methoden bein nicht-linearen Gleichungen, In "Jahrbuch Überblicke Mathematik 1979", B. I. Verlag, Zürich.

[48] Rump, S. M., Kaucher, E. (1980). Small Bounds for the Solution of Systems of Linear Equations, Computing Suppl. 2.

EVALUATION OF ARITHMETIC EXPRESSIONS WITH MAXIMUM ACCURACY

Harald Böhm

Institute for Applied Mathematics
University of Karlsruhe
Karlsruhe, West Germany

The usual way of evaluating an arithmetic expression approximately with floating-point arithmetic is to replace every operation in the expression by the corresponding floating-point operation. This may result in relative rounding errors of arbitrary magnitude. Here we describe a new algorithm for evaluating an arithmetic expression with maximum accuracy. This algorithm uses floating-point operations with directed roundings and a scalar product of maximum accuracy in addition to the usual floating-point operations. The computing time of the new algorithm is of the same order as for conventional floating-point calculation, assuming the latter does not fail completely. In this case additional computing time and storage are needed.

INTRODUCTION

Our aim is to use a computer with floating-point arithmetic to obtain guaranteed sharp bounds for the value of an arithmetic expression. For this purpose a floating-point arithmetic with strictly defined properties is required. In the following, we use an arithmetic as described in [3]. We also use the terminology of [3]. Therefore, we give only a short summary of the symbols to be employed.

S denotes the set of floating-point numbers; VS the set of vectors over S, where the dimension will be obvious from the context. $\square$ is a monotonic antisymmetric rounding, ∇ and $\triangle$ the downwardly and upwardly directed roundings. Floating-point operations in S and VS with these roundings are denoted by $\boxed{*}$, $\overset{\nabla}{*}$ resp. $\overset{\triangle}{*}$ for $* \in \{+,-,\cdot,/\}$.

$\mathbb{I}S$ resp. $\mathbb{I}VS$ is the set of intervals with bounds in S resp. VS. $\diamond\!\!\!\!*$ with $* \in \{+,-,\cdot,/\}$ are the interval operations in $\mathbb{I}S$ and $\mathbb{I}VS$.

The base of the floating-point system will be denoted by B and the number of digits in the mantissa by ℓ. Then with $\varepsilon := B^{1-\ell}$ the following error estimates are satisfied as long as no over- or underflow occurs (cf. [1]):

ISBN 0-12-428660-7

$$|a*b - a \boxed{*}\, b| \leq \varepsilon \cdot |a \boxed{*}\, b|\ , \qquad (1)$$

$*\in\{+,-,\cdot,/\}$ (for division, $b\neq 0$ assumed).

Since this is true for every monotonic rounding, it holds for $\triangledown$ and $\triangle$ also.

The scalar product of two vectors $a,b\in VS$ is considered as a single operation which is performed with only one rounding. Then the error estimate (1) is also valid for $a,b\in VS$ when $*$ denotes the scalar product. Effective implementation of all operations mentioned above is possible (cf. [1], [3]).

We construct our arithmetic expressions according to the syntax of PASCAL. They consist of the four operation symbols, $+,-,\cdot,/$, parentheses, variables and constants. It is not necessary to distinguish between variables and constants because we only deal with the values of the variables. The constants must be exactly representable floating-point numbers. If a constant requires more than ℓ digits in the mantissa it can be replaced by a sum of two or more constants.

Now we define what we understand by maximum accuracy.

Definition 1

Let S be a floating-point screen and $x\in\mathbb{R}$. $\tilde{x}\in S$ is called an approximation of maximum accuracy for x if

$$\forall y\in S\colon (x\leq y\leq \tilde{x} \text{ or } \tilde{x}\leq y\leq x) \Rightarrow y = \tilde{x}.$$

For the definition of an interval of maximum accuracy, we first need the definition of the successor of a floating-point number.

Definition 2

Let S be a floating-point screen and $x\in S$. The successor $z = \mathrm{succ}(x)\in S$ is defined by

$z>x$ and

$\forall y\in S\colon y>x \Rightarrow y\geq z.$

An interval $X = [x_1, x_2] \in I\!IS$ is an inclusion of $x \in \mathbb{R}$ of maximum accuracy if $x \in X$ and

$\text{succ}(\tilde{x}_1) \geq \tilde{x}_2$ or

$\text{succ}(\text{succ}(\tilde{x}_1)) \geq x_2$ and $\tilde{x}_1 < x < \tilde{x}_2$.

This definition asserts that an interval of maximum accuracy is either a point or an interval whose bounds are consecutive floating-point numbers, or else an interval containing precisely one floating-point number in its interior. In the latter case x cannot equal one of the bounds. Our algorithm will be formulated to yield an interval of maximum accuracy. No essential change is necessary if one is interested in an approximation of maximum accuracy.

1. EVALUATION OF POLYNOMIALS

We know that the linear operations in S, VS and MS can be performed with maximum accuracy (cf. [3]). The key for this is the exact scalar product (cf. [1]). The first step toward achieving maximum accuracy for the value of an arithmetic expression is to reduce the problem to a linear problem. The second step is to solve the linear problem. Both steps can easily be surveyed in a special case, the evaluation of polynomials.

Let

$$p(z) = \sum_{i=0}^{n} p_{n-i} z^i, \; p_i \in S \text{ for } i = 0,\ldots,n$$

be a polynomial of degree n with coefficients in S. The evaluation of p at point $t \in S$ using Horner's scheme yields the expression

$$(\ldots(p_0 t + p_1)t + \ldots + p_{n-1})t + p_n.$$

The intermediate results occurring during the evaluation are:

$$x_0 := p_0,$$
$$x_i := x_{i-1} \cdot t + p_i, \; i = 1,\ldots,n.$$

These are simultaneous linear equations for the unknowns x_i, $i = 0,\ldots,n$. The coefficients and components of the right-hand side are elements of S. With

$$A := \begin{pmatrix} 1 & & & & 0 \\ -t & 1 & & & \\ & \cdot & \cdot & & \\ & & \cdot & \cdot & \\ & & & \cdot & \cdot \\ 0 & & & -t & 1 \end{pmatrix}, \quad x := \begin{pmatrix} x_0 \\ \cdot \\ \cdot \\ \cdot \\ x_n \end{pmatrix}, \quad p := \begin{pmatrix} p_0 \\ \cdot \\ \cdot \\ \cdot \\ p_n \end{pmatrix}$$

the simultaneous equations can be written in the form

$$Ax = p,$$

where x_n is the value of the polynomial.

Here the first step, the reduction to a linear problem, is just a new interpretation of the intermediate results of Horner's scheme. The second step, solving the simultaneous equations with maximum accuracy, could be carried out by one of the methods given in [4]. However, these methods are designed for general matrices and do not take advantage of the very simple structure of the present problem. Furthermore, we do not need maximum accuracy for all components of the solution but only for the last one. Thus, it is worthwhile to look for a more economic way to solve the simultaneous linear equations.

It is easy to calculate an approximation $x^0 \in VS$ for the solution $\hat{x} := A^{-1} p \in V\mathbb{R}$ by forward substitution using floating-point arithmetic:

$$x_0^0 := p_0,$$

$$x_i^0 := x_{i-1}^0 \boxdot t \boxplus p_i, \; i = 1,\ldots,n.$$

This is precisely the floating-point evaluation about which we know that its result may be completely wrong. The advantage is that defect-iteration can be used to improve the accuracy and to compute bounds for the solution.

In the kth iteration we have computed k vectors, x^0 through x^{k-1}, the sum of which is an approximation for the solution $\hat{x}$ of $Ax = p$. For this approximation we include the defect in the interval vector

$$d^k := \Diamond\left(p - \sum_{j=0}^{k-1} Ax^j\right) \in IIVS.$$

This is computable because the components of d^k originate from scalar products.

Then we solve the simultaneous linear equations

$$AX^k = d^k,$$

i.e., we compute an interval $X^k \in IIVS$ which contains $A^{-1}\tilde{d}$ for all $\tilde{d} \in d$. This is done by forward substitution, where each operation is performed as an interval-operation in IIS:

$$X_0^k := d_0^k,$$

$$X_i^k := X_{i-1}^k \quad \diamond\!\!\!\!\cdot \; t \; \diamond\!\!\!\!+ \; d_i^k, \; i = 1,\ldots,n.$$

In particular, X^k contains $A^{-1}\left(p - \sum_{j=0}^{k-1} Ax^j\right)$ and therefore

$$\hat{x} = A^{-1}p \in X^k + \sum_{j=0}^{k-1} Ax^j.$$

This holds for arbitrary x_k^j. In order to achieve convergence to zero for the diameter of X^k we choose

$$x^j \in X^j, \; j \geq 1.$$

It is convenient to take the midpoint of X^j. Care must be taken if the midpoint of an interval $[a,b] \in IIS$ is computed with floating-point arithmetic with a base different from 2. The expression $a \boxplus (b \boxminus a) \boxslash 2$ should be used to ensure that the value is inside the interval. The notation mid $([a,b])$ will be used for this value. For an interval vector $X \in IIVS$, mid(X) is to be understood componentwise.

The iteration is continued until the diameter of X_n^k is small enough to guarantee maximum accuracy for the interval $\Diamond\left(X_n^k + \sum_{j=0}^{k-1} x_n^j\right)$. Before we give a proof of convergence we summarize the procedure in an algorithm.

Algorithm 1

Evaluation of a polynomial $p(z) = \sum_{i=0}^{n} p_{n-i} x^i$, $p_i \in S$ at a point $t \in S$ with maximum accuracy.

1. {Approximate solution}

$x_0^0 := p_0$,

$x_i^0 := x_{i-1}^0 \boxdot t \boxplus p_i$, $i = 1,...n$.

2. {Iteration}

$k := 0;$

<u>repeat</u> $k := k + 1;$

<u>if</u> $k > 1$ <u>then</u> $x^k := \text{mid}(X^k)$;

$d^k := \Diamond\left(p - \sum_{i=0}^{k-1} Ax^k\right)$; {defect}

$X_0^k := d_0^k;$

$X_i^k := X_{i-1}^k \; \Diamond\!\!\!\cdot \; t \; \Diamond\!\!\!+ \; d_i^k$, $i = 1,...,n$; {solve $AX^k = d^k$};

$Y := \Diamond\left(X_n^k + \sum_{j=0}^{k-1} x_n^j\right)$

<u>until</u> $(\triangle(\text{diam}(X_n^k)) < \delta \cdot |Y|)$ <u>or</u> $(Y = 0)$

<u>or</u> (underflow <u>and</u> $k \geq 10$);

3. It has been proven that $p(t) \in Y$. If no underflow occurred Y is of maximum accuracy.

For an interval $X \in I\!\!I S$, diam(X) denotes the diameter of X. δ must be chosen such that the interval $X := X_n^k + \sum_{i=0}^{k-1} x_n^i \in I\!\!I\mathbb{R}$ cannot include two different floating-point numbers if the stopping criterium is fulfilled. If a floating-point screen with base B and mantissa-length ℓ is used, the appropriate value is $\delta = B^{-\ell}$. Depending on whether or not the interior of X contains a floating-point number, the bounds of Y differ by two or only one unit in the last digit of the mantissa. In both cases, maximum accuracy is achieved. A floating-point approximation of maximum accuracy can easily be computed by

$$y := \triangle\left(\inf X_n^k + \sum_{j=0}^{k-1} x_n^j\right).$$

This expression, as well as Y in the algorithm, can be computed using the exact scalar product.

In case of underflow the error estimations (1) are no longer valid. Thus, the number of iterations has been limited in this case to ensure termination. The bounds for the value of the polynomial are also valid in this case, but the resulting interval might not be of maximum accuracy.

Now we require a theorem which proves that the diameter of X_n^k decreases in every iteration. It is assumed that no over- or underflow occurs.

Theorem 1

For the intervals X_n^k, $k \geq 1$, computed with Algorithm 1, the following estimation is valid:

$$\text{diam}(X_n^k) \leq \frac{1}{2}\lambda^{k+1} \max_{0 \leq j \leq n} |p_{n-j}t^j| \quad \text{with}$$

$$\lambda := 4(n+1)^2 \cdot (1+\varepsilon)^{2n+2} \cdot \varepsilon.$$

This estimate includes all rounding errors.

Proof:

The proof is accomplished by forward error analysis. We give only a short sketch.

First it can be shown by induction with respect to i that for $i = 0,\ldots,n$ and $k \geq 1$

$$|t|^{n-i} \cdot \text{diam}(X_i^k) \leq 2 \cdot (i+1)^2 \cdot (1+\varepsilon)^{2i+1} \cdot \varepsilon \cdot s^k \leq \frac{1}{2} \cdot (1+\varepsilon)^{-1} \cdot \lambda \cdot s^k \quad \text{with}$$
$$s^k := \max_{0 \leq i \leq n} |d_i^k| \cdot |t|^{n-i} \qquad (2)$$

The same estimation holds for $k = 0$ when we set $d_i^0 := p_i$ and replace $\text{diam}(X_i^0)$ by the maximum error in the x_i^0. Using (2) and the fact that the approximation $\sum_{j=0}^{k} x^j$ cannot differ from the exact solution $A^{-1}p$ by more than the diameter of X^k we obtain the estimation

$$|t|^{n-i} \cdot |d_i^{k+1}| \leq (1+\varepsilon)\left\{ |t|^{n-i} \cdot \text{diam}(X_i^k) + |t|^{n-i-1} \cdot \text{diam}(X_{i-1}^k) \right\}$$
$$\leq \lambda \cdot s^k \quad \text{for} \quad i = 1,\ldots,n.$$

Therefore we have

$$s^k \leq \lambda^k \cdot s^0.$$

In view of (2) we conclude that

$$|t|^{n-i} \operatorname{diam}(X_i^k) \leq \frac{1}{2}\lambda^{k+1} \bullet s^0.$$

Setting $i = n$ proves the theorem. ■

For reasonable n and ε, this theorem gives an upper bound for the number of iterations, dependent on the relation between the value of the polynomial and the size of the coefficients. This estimate, however, is rather pessimistic. Experience shows that the diameter of X_n^k decreases by about $B^{-\ell}$ in every iterative step. The following example demonstrates the effect of the iterations.

Example 1

We use a decimal arithmetic with two digits, i.e., $B = 10$ and $\ell = 2$. We seek to compute the value of

$$p(t) = 82\,t^3 - 58\,t^2 - 41\,t + 29$$

for $t = 0.71$. We obtain

$$x_3^0 = 0$$

$$Y^2 = [-0.01, 0.01]$$

$$Y^2 = [0.0009, 0.0010]$$

$$Y^3 = [0.00090.0.00091].$$

To achieve this accuracy by conventional floating-point calculation, a mantissa-length of at least seven decimal digits would be needed.

2. EVALUATION OF ARBITRARY ARITHMETIC EXPRESSIONS

Our method for polynomials is directly applicable to certain arithmetic expressions.

Example 2

For $a,b,c,d,e \in S$. the expression

$$(a + b) \bullet c - d/e$$

is equivalent to the simultaneous linear equations

$$x_1 = a$$

$$x_2 = x_1 + b$$

$$x_3 = c \cdot x_2$$

$$x_4 = d$$

$$e \cdot x_5 = x_4$$

$$x_6 = x_3 + x_5.$$

But there are expressions where the equations become nonlinear:

Example 3

The expression

$$(a + b) \cdot (c + d), \quad a,b,c,d \in S$$

leads to the simultaneous equations

$$x_1 = a$$

$$x_2 = x_1 + b$$

$$x_3 = c$$

$$x_4 = x_3 + d$$

$$x_5 = x_2 \cdot x_4$$

which are nonlinear.

The reason is that a product of expressions occurs. Division by an expression causes the same trouble. In these cases an algebraic transformation is necessary. Any arithmetic expression can be transformed into a quotient in which the numerator and the denominator yield linear simultaneous equations. These can be evaluated with maximum accuracy. The final division causes a relative error of less than ε. The last digit of the mantissa may then be incorrect. This can be prevented if, for both the numerator and the denominator, a higher precision is simulated by representing the value as the sum of a floating-point number and an interval.

The algebraic transformation can be performed automatically. The aim is an expression with the following properties.

$$
\begin{array}{l}
-\text{For any subterm of the form } a \bullet b \\
\quad b \text{ is a constant} \\
-\text{For any subterm of the form } a/b \\
\quad b \text{ is a constant or } a/b \text{ is the whole expression.}
\end{array}
\tag{3}
$$

The following rewrite-rule system performs the desired reduction. It is assumed that neither unary minuses nor pluses occur. These can be removed before applying the following rules by changing the sign of appropriate constants. The symbol T is used for any constant, a,b,c,d are variables for terms.

Rewrite-rule System

(1)	$(a \pm b) \bullet (c \pm d)$	$\rightarrow$	$(a \pm b) \bullet c \pm (a \pm b) \bullet d$	
(2)	$T \bullet (a \pm b)$	$\rightarrow$	$(a \pm b) \bullet T$	
(3)	$(a \bullet T) \bullet (b \pm c)$	$\rightarrow$	$(a \bullet (b \pm c)) \bullet T$	
(4)	$a \bullet (b \bullet c)$	$\rightarrow$	$(a \bullet b) \bullet c$	
(5)	$a/b \pm c$	$\rightarrow$	$(a \pm b \bullet c)/b$	if $b \neq T$
(6)	$a \pm b/c$	$\rightarrow$	$(c \bullet a \pm b)/c$	if $c \neq T$
(7)	$(a/b) \bullet c$	$\rightarrow$	$(a \bullet c)/b$	
(8)	$a \bullet (b/c)$	$\rightarrow$	$(a \bullet b)/c$	
(9)	$(a/b)/c$	$\rightarrow$	$a/(b \bullet c)$	if $b \neq T$
(10)	$a/(b/c)$	$\rightarrow$	$(a \bullet c)/b$	

For brevity, we have summarized the rules for addition and subtraction in one rule. Rules (5), (6) and (9) actually contin four rules each. In (5), for instance, we have one rule each for

$b = b_1 + b_2,$

$b = b_1 - b_2,$

$b = b_1 \bullet b_2,$

$b = b_1 / b_2.$

It is easy to see that every irreducible term has the properties (3). The termination is proved in [2] where special assumptions are made about the order of application of these rules.

A simpler rewrite-rule system could be given which completely reduces any product of a constant with a sum to a sum of products, but this would increase the number of operations significantly in many cases.

The determination of the corresponding simultaneous linear equations is not difficult if one has postfix or prefix notation of the expression. For each operation, one or two equations are determined (two equations, if both operands are constants).

The number of variables can be reduced by collecting several operations in one linear equation:

Example 4

The expression

$$(a \bullet b \bullet c - d \bullet e)/f, \quad a,b,c,d,e,f \in S$$

can be transformed into the simultaneous linear equations

$$x_1 = a \bullet b$$

$$f \bullet x_2 = c \bullet x_1 - d \bullet e.$$

Here the right-hand side consists of products of constants. This is permissible because the defects d^k in the algorithm can still be computed with one scalar product.

A subterm which is duplicated when applying a rule need not be evaluated twice. The corresponding intermediate result can be used several times.

Example 5

The term

$$(a + b)\cdot(c + d)\cdot(e + f), \quad a,b,c,d,e,f \in s$$

is transformed into

$$((a + b)\cdot c + (a + b)\cdot d)\cdot e + ((a + b)\cdot c + (a + b)\cdot d)\cdot f.$$

The resulting linear system is

$$x_1 = a$$

$$x_2 = x_1 + b$$

$$x_3 = c\cdot x_2$$

$$x_4 = d\cdot x_2 + x_3$$

$$x_5 = e\cdot x_4$$

$$x_6 = f\cdot x_4 + x_5.$$

x_6 is the value of the original expression.

The matrices of simultaneous linear equations arising from arithmetic expressions are lower triangular. The algorithm for solving such a linear system with maximum accuracy is similar to Algorithm 1:

Algorithm 2

Solve the system of n linear equations $Ax = b$, $A \in MS$ lower triangular, $b \in VS$, with maximum accuracy.

1) {Approximation}

$$x_i^0 := \square \left(b_i - \sum_{j=1}^{i-1} A_{ij}\cdot x_j^0\right) \boxslash A_{ii} \quad i = 1,\ldots,n.$$

2) {Iteration}

$k := 0$

<u>repeat</u> $k := k = 1$

<u>if</u> $k>1$ <u>then</u> $x^k := \text{mid}(X^k)$

$$d^k := \lozenge\left(b - \sum_{j=0}^{k-1} Ax^j\right)$$

$$X_i^k := \lozenge\left(d_i^k - \sum_{j=1}^{i-1} A_{ij} \bullet X_j^k\right) \lozenge\ A_{ii},\ i = 1,\ldots,n$$

$$Y := \lozenge\left(X_n^k + \sum_{j=0}^{k-1} x_n^j\right)$$

<u>until</u> $(\triangle(\text{diam}(X_n^k)) < \delta\,|Y|$ <u>or</u> $(Y = 0)$

<u>or</u> (underflow <u>and</u> $k \geq 10$)

3) The value of the arithmetic expression described by $Ax = b$ is contained in Y. If no underflow occurred, Y is of maximum accuracy.

Just as in the evaluation of polynomials, we need a theorem which ensures termination. Because the following theorem holds only for the case in which no overflow or underflow occurs, the number of iterations is limited after an underflow. Overflow is assumed to terminate the algorithm.

Theorem 2

Let $A \in MS$ be a non-singular lower triangular matrix with n rows and columns, $b \in VS$, $x^* := A^1 b \in V\mathbb{R}$ and let Y^k be the value of Y in the kth iteration in Algorithm 2. With

$$M_i := \max\left\{1, \sum_{j=1}^{i-1} \frac{|A_{ij}|}{|A_{ii}|}\right\},\ i = 1,\ldots,n \text{ and}$$

$$N_i := \prod_{j=1}^{i} M_i,\ i = 1,\ldots,n$$

the following estimate is valid:

$$\operatorname{diam}(Y_n^{k+1}) \leq \{5 \bullet n^2 \bullet (1 + \varepsilon)^{2n+1} \bullet \varepsilon\}^k \bullet N_n \bullet \max_{1 \leq j \leq n} \frac{\operatorname{diam}\,(Y_j^1)}{N_j}.$$

Proof: cp. [5]. ∎

3. NUMERICAL RESULTS

The computations for all examples were performed by a minicomputer using a 12-digit decimal arithmetic.

Example 6

Evaluation of the polynomial

$$p(x) = 23616x^5 - 161522x^4 + 401773x^3 - 406754x^2 + 87511x + 66576$$

for various x between 1.780 and 1.781:

x	floating-point approximation for $p(x)$	result of new algorithm
1.7800	0	$-5.12000000000 \quad E-8$
1.7801	$-1.4 \quad E-6$	$-3.4094229723^{9}_{8} \quad E-8$
1.7802	$2.6 \quad E-6$	$-2.0535703802^{9}_{8} \quad E-8$
1.7803	$-2.3 \quad E-6$	$-1.0434985373^{2}_{1} \quad E-8$
1.7804	$3.5 \quad E-6$	$-3.58579593216 \quad E-9$
1.7805	$-1.5 \quad E-6$	$3.35013000000 \quad E-10$
1.7806	$2.2 \quad E-6$	$1.76748822016 \quad E-9$
1.7807	$-1.5 \quad E-6$	$1.26860318112 \quad E-9$
1.7808	$-1.9 \quad E-6$	$-4.87713669120 \quad E-10$
1.7809	$8.0 \quad E-7$	$-2.71055054816 \quad E-9$
1.7810	$-5.0 \quad E-7$	$-4.49198400000 \quad E-9$

Horner's scheme is used for the floating-point approximation. The results do not give any information about the real values of the polynomial, and thus do not determine whether it has zeros in the interval [1.780, 1.781]. By contrast, our new algorithm proves that there exist at least two zeros in this interval.

Example 7

Evaluation of the expression

$$y = 27a^6 - 10a^3b^3 - b^6 - 3ab - 12a^2b^2$$

for $a = 12970$ and $b = 16897$.

Conventional floating-point calculation yields the approximation.

$$\tilde{y} = 5.78173780000E13.$$

The exact value of the expression is 1, which is found by the new algorithm. The successive inclusions found in the loop of the algorithm are

$$Y^1 = [-500, 600]$$

$$Y^2 = [1.0, 1.0].$$

Example 8

Computation of a difference approximation for the second derivative of the rational function

$$f(x) = \frac{4161x - 4204}{4160x^2 - 8449x + 4290}$$

for $x = 1$, i.e., evaluation of

$$y = (f(1-h) - 2f(1) + f(1+h))/h^2.$$

The following table gives the results for various h

h	floating-point approximation	result of new algorithm
$1E-1$	9.02902174902 $E3$	[9.02902174902 $E3$, 9.02902174903 $E3$]
$1E-2$	4.37007003836 $E5$	[4.37007003836 $E5$, 4.37007003837 $E5$]
$1E-3$	1.67466590654 $E3$	[1.67418350822 $E3$, 1.67418350823 $E3$]
$1E-4$	2.20018307723 $E2$	[1.86898367489 $E2$, 1.86898367490 $E2$]
$1E-5$	6.20000508400 $E3$	[1.72148971404 $E2$, 1.72148971406 $E2$]
$1E-6$	− 4.60000004600 $E5$	[1.72001489712 $E2$, 1.72001489714 $E2$]
$1E-7$	− 6.00000006000 $E5$	[1.72000014896 $E2$, 1.72000014898 $E2$]
$1E-12$	0.0	[1.71999999999 $E2$, 1.72000000001 $E2$]

The true value for $f''(1)$ is 172. This value cannot be found by usual floating-point calculation because for a sufficiently small stepsize h, the result obtained by the method of difference approximation is completely falsified by round-off errors.

Example 9

Evaluation of the partial sum of the power series for the exponential function

$$y = \sum_{k=0}^{90} \frac{x^k}{k!}$$

for $x = -20$.

It is well known that the power series is not a suitable method to evaluate the exponential function for negative arguments. With our new algorithm the evaluation is no problem. We get the inclusion

$$Y = [2.06115362243E-9, 2.06115362244E-9]$$

whereas conventional calculation yields

$$\tilde{y} = 1.188534E-4.$$

For the purposes of demonstration, extremely ill-conditioned examples were chosen. In most of these examples the algorithm required two iterations to arrive at a result with maximum accuracy. In cases where the floating-point evaluation does not fail completely, one iteration will suffice. In this case, the number of operations used in the new algorithm is about three times the number of operations for usual floating-point calculation.

REFERENCES

[1] Bohlender, G. (1977). Floating-point computation of functions with maximum accuracy. IEEE Trans. Comput. 7, 621-632.

[2] Böhm, H. (1982). Automatische Umwandlung eines arithmetischen Ausdrucks in eine zur exakten Auswertung geeignete Form, interner Bericht des Inst. f. Angew. Math., Karlsruhe.

[3] Kulisch, U., Miranker, W. L. (1980). Computer Arithmetic in Theory and Practice, Academic Press.

[4] Rump, S. M. Solving algebraic problems with high accuracy, in this volume.

[5] Rump, S. M., Böhm, H. Least significant bit evaluation of arithmetic expressions in single precision, to appear in Computing.

SOLVING FUNCTION SPACE PROBLEMS WITH GUARANTEED CLOSE BOUNDS

Edgar Kaucher

Institute for Applied Mathematics
University of Karlsruhe
Karlsruhe, West Germany

Using arithmetical routines and methods given in [2], [8], [9] and in the present volume, it becomes possible to compute very close and reliable bounds economically for the solutions of functional, differential and integral equations. The methods are suitable for software packages. They provide automatically error control for the first time. Without any additional demand on the part of the user, an incomparably high level of software security and reliability can be achieved.

All this is achievable by using Schauder's and Darbo's fixed point theorems to compute bounds for, and to verify existence (and local uniqueness) of, the solutions of a fixed point equation set in function space. Since nearly all such practical problems can be transformed into α-condensing fixed point equations, the methods at hand are very effective and constructive in applications.

These methods are called E-methods corresponding to their properties (in German). Existenz (Existence) - Einschliessung (inclusion) - Eindeutigkeit (uniqueness).

We illustrate these methods by means of a collection of examples in the case of ordinary differential equations with mixed boundary conditions. If the problem is not too poorly conditioned the bounds for the solution barely differ in the last digit.

CONTENTS

ISBN 0-12-428660-7

1. INTRODUCTION

The growing and increasing use of enormous computers and processor systems in all areas of technology requires a significantly higher level of reliability and accuracy fo the intermediate and final results of computation. The less the opportunity of engineers, physicists and mathematicians to oversee the computational process, the greater the danger of rounding and approximation error accumulating into large and inestimable errors in the result. This accumulation can occur explosively fast, since modern computers execute millions of operations in a second.

Consequently, the urgency of error controlling computational methods has been recognized in all areas of technology, the engineering sciences and mathematics. Basic results in this direction have been recently obtained (see, for example, [1] to [17] and [20]).

But until very recently, it was nearly impossible to compute for example, sharp bounds for the solution of differential equations in an effective manner (except for certain special classes). This was primarily because,

(a) there was a lack of implementation techniques for operations of interval arithmetic,

(b) effective methods for the numerical inclusion of solutions of linear and nonlinear systems were inadequately developed, and

(c) a method for computing best approximation of a polynomial at a point was not available.

Now however, methods of interval arithmetic have been refined to such an extent that they are effective on computers. See, for example, the reports of Böhm and Rump (this volume) and [9], [10], [11], [18] and [19].

Based on these methods, it is now economically possible to contain (i.e., provide a containing interval) the solutions of functional equations within guaranteed functional bounds.

By functional equations we mean a very general class of problems including all of the following:

(a) functional equations not involving infinitesimal processes, e.g. $y(x^2 + a) = e^{y(x)}$

(b) differential equations, e.g. $y'(x) + y(x^2 + 1) = f(x,y)$

(c) integral equations, e.g. $y'(x) + \int_0^1 y(t + x)dl = f(x,y)$

(d) general parameterized problems.

For the sake of simplicity, we now apply our methods to the following relatively simple system of functional equations:

$$A(t) \bullet x(t) = b(t) \text{ for } t \epsilon [0, 1],$$

where A is a given matrix function, b a given vector function and the function x the solution vector. The functional dependence of x on t is of interest. The computation of a reliable error estimate for the solution involves the determination of a set of vector functions $X(t)$ containing the solution $x(t)$ for all t.

The computation of $X(t)$ requires new computer arithmetic such as segment arithmetic and functional ultra arithmetic. All these must be executed automatically and intrinsically on a computer without intervention by the users. Consequently we call such methods *error controlling methods* where the phrase *error controlling* refers to the following fact: If $x(t) \epsilon X(t)$, then for all approximations $\tilde{x}(t) \epsilon X(t)$, the maximal error bound $|\tilde{x}(t) - x(t)| \leq d(X(t))$ is trivially given by the diameter $d(X(t))$. In most cases, an *error controlling method* necessarily proves the existence of the solution $x(t)$ in $X(t)$ automatically. Thus all these methods realize an inherent proof of the existence of the solution (fo the theoretically given problem). Furthermore, in some problems our methods prove the local uniqueness of $x(t)$ in $X(t)$ as a by-product.

In summary, we denote all such *Error controlling methods* by the term E-methods, where E serves as an abbreviation for the following properties:

(1) For all imput data an E-method is able

- to verify, whether a solution of the theoretical (characterized by the input data) problem exists (Existenz),
- to include this solution in a set of functions with small diameter (Einschliessung),
- to verify (if possible) the uniqueness of the solution within the containing set

(Eindeutigkeit).

(2) If there is no solution, or if computational time, mantissa length or the method itself do not permit the required verification of the first two statements in (1), then the method will not produce *meaningless solutions.* Instead, the computation will automatically terminate and a message to this effect will be given.

In every case the user of such E-methods receives clear information about the situation with regard to his problem. In the case of (1) he knows that the problem possesses the computed approximation with given accuracy. In the case (2), he knows that his problem is either an ill posed problem or unsolvable, or highly ill-conditioned (as a result of poor conditioning). In the latter case much more time and cost is required in order to solve the problem with satisfactory accuracy. In no case can it happen that the user is confronted with arrays of numbers representing a so-called approximation, without a precise statement of the meaning of these numbers.

We emphasize that an economical implementation of E-methods is only possible on computers with the properties described in [14] and in this volume. Naturally, some results are achievable by simulation, but this remains a makeshift means of solution.

Another and a further point of view appears more and more following the lines and ideas arising in the methods described here. The urgency to develop a unified theory of including the solution of functional equations leads automatically to a unified description of all customary numerical methods. It turns out that discretization as well as collocation or Fourier-Taylor expansion are the same spectral (or pseudo spectral) methods with certain basis elements and a certain corresponding rounding. As a consequence it now becomes possible to compute automatically and simply, error bounds for collocation approximation as well as for Ritz-Galerkin approximation. Until recently this was nearly impossible without an enormous amount of work and nevertheless typically with error bounds of poor quality. A further advantage of such an unified theory of numerics is the use of *iterative residue correction* which is a well known method in finite dimensional numerics, but until recently not in functional problems. With *iterative residue correction* it becomes possible to compute with the use of polynomials of degree N an approximation of a solution of degree $k \bullet N$, $K > 1$. A more detailed description of this idea will be given in another paper [21].

2. MATHEMATICAL PRELIMINARIES

Fixed point theorems are the basis for all E-methods similar to those in [11]. In contrast to [11], we consider infinite dimensional function spaces; usually Banach spaces. Hence, we need modifications of the fixed point theorems of Banach, Schauder-Tychonoff or Krasnoselski-Darbo. These fixed point theorems were originally adapted in [12] so that sufficient conditions for existence of the solution and its containment in an including set are verified simultaneously, along with the computation of the including set (see [21] also).

As an example we now demonstrate how the Schauder-Tychonoff fixed point theorem must be adapted to this purpose. From the proof of this theorem we use the following Lemma:

Lemma: Let $\mathcal{M}$ be a Banach space and $\ell = \mathcal{M} \rightarrow \mathcal{M}$ a linear mapping. Let $\{0\} \overset{\circ}{\subset} U$, where U is a closed and bounded subset of $\mathcal{M}$. Suppose that

$$\ell U \overset{\circ}{\subset} U. \tag{2.1}$$

Then

(a) The resolvent $(E - \ell)^{-1}$ exists, at least on U and both ℓ and $(E - \ell)^{-1}$ are continuous.

(b) Either $\Lambda(\ell) = \phi$ or $|\Gamma(\ell)| < 1$.

(c) If U is balanced then $\| \bullet \|_U$ is a norm and $\| \ell \|_U < 1$. Furthermore, for all $n \in N$ with

$$\ell^{n+1}(U) \overset{\circ}{\subset} \ell^n(U),$$

we have $\| \ell \|_{\ell^n(U)} < 1$.

(d) If (2.1) is replaced upon $\ell U \overset{\circ}{\subset} \kappa U$ for some $\kappa \in ₵$, then the conclusions (a)-(c) of this theorem remain valid by replacing ℓ by ℓ / κ.

Proof:

(a) To demonstrate the existence of $(E - \ell)^{-1}$ on U suppose that p is a nonnull element of U such that $(E - \ell)p = 0$. Since $\{0\} \overset{\circ}{\subset} U$, $q = \sigma p \in U$ for all sufficiently small scalars σ. Since U is closed and bounded there exists a scalar $\bar{\sigma}$ such that $\bar{q} = \bar{\sigma} p \in \partial U$. Then

$$\ell \bar{q} = (E - (E - \ell))\bar{q} = \bar{q} \in \partial U.$$

This is a contradiction since the left member here is in the interior of U. Thus $(E - \ell)$ annihilates no nontrivial element of U, and so, $(E - \ell)^{-1}$ exists on U. Since ℓ is bounded on U, it is continuous there. Similarly $(E - \ell)^{-1}$ is continuous.

(b) If $\ell \equiv 0$ then $\Lambda(\ell) = \{0\}$. Suppose $\ell \neq 0$ and that $\Lambda(\ell) \neq \phi$. Then let $\lambda \in \Lambda(\ell) \subseteq \mathbb{C}$ be an eigenvalue (which we may suppose is nonnull) and let v be a corresponding eigenvector so normalized that $v \in U$. Consider the set $\Psi := \{\psi \in \mathbb{C} \mid \psi v \in U\}$. Ψ is closed since U is closed. Thus $|\psi^*| := \max\limits_{\psi \in \Psi} |\psi|$ exists and moreover $\psi^* \in \Psi$. Thus $\psi^* v \in U$ and so by hypothesis $\ell(\psi^* v) \overset{\circ}{\subset} U$, i.e.

$$\ell(\psi^* v) = \psi^* \ell v \overset{\circ}{\subset} U.$$

Then there exists a $\sigma \in \mathbb{R}$ with $\sigma > 1$ so that

$$\sigma \psi^* \lambda c \in \partial U \subset U.$$

Then by the definition of the set Ψ, we note that $\tilde{\psi} := \sigma \psi^* \lambda \in \overline{\Psi}$. But then

$$|\psi^*| \geq |\tilde{\psi}| = |\sigma \psi^* \lambda| = \sigma |\psi^*| \, |\lambda|.$$

Thus $\quad |\lambda| \leq \frac{1}{\sigma} < 1.$

Since λ is arbitrary, $|\Lambda(\ell)| \leq q < 1$ with $q = \frac{1}{\sigma}$.

(c) Since U is balanced the Minkowski functional $\|y\|_U$ is a norm. Moreover, $\|v\|_U \leq 1$ for $v \in U$ but not otherwise. For an arbitrary $v \in U$, $\{\ell v\} \overset{\circ}{\subset} U$. Thus $\exists t, q \in \mathbb{R}$ with $t \leq q < 1$ such that $\ell v \in tU$ and $\|\ell v\|_U \leq q < 1$. Combining we have

$$\| \ell \|_U = \sup_{\| v \|_U \leq 1} \| \ell v \|_U = \sup_{v \in U} \| \ell v \|_U < 1,$$

the first assertion of (c).

Clearly for arbitrary continuous linear ℓ we have $\{0\} \overset{\circ}{\subset} \ell^n(U)$, and moreover, that $\ell^n(U)$ is balanced. Thus the second assertion of (c) follows along the lines of the first.

(d) This is evident since $\ell U \overset{\circ}{\subset} \kappa U$ implies $\frac{\ell}{\kappa} U \overset{\circ}{\subset} U$. ■

Remark:

1. From conclusion (b) of this lemma we may infer that

$$r(\ell) := \sup_{\lambda \in \Lambda(\ell)} |\lambda| = q < 1.$$

Then from the extremal property of $r(\ell)$ with respect to the class of operator norms, there exists a norm (in general unknown) such that

$$\| \ell \| \leq q + \varepsilon < 1$$

for arbitrary and sufficiently small ε. Thus ℓ is a contraction operator (in a norm which is unknown).

2. Without the hypothesis of balance for U, the Minkowski functional is not a norm. However, the property $\| \ell \|_U \leq q < 1$ prevails. Loss of the norm property appears to result in the loss of contractivity. However, uniqueness is achieved by alternative means as shall be seen in the fixed-point theorems to follow.

And now the Theorem: We shall make use of $\mathbb{L}(\mathcal{M}, \mathcal{M})$, the set of linear operators taking $\mathcal{M}$ into itself. Corresponding to a set Y we consider a subset $\mathcal{K}(Y) \subseteq \mathbb{L}$, i.e.

$$\mathcal{K}(Y) = \{k(y) \in \mathbb{L}(\mathcal{M}, \mathcal{M}) \mid y \in Y\}.$$

Thus, Y is a parametrization of the set of operators $\mathcal{K}$.

Theorem 1: Let $\mathcal{M}$ be a complex Banach space and let $\mathcal{M} \supseteq Z \supseteq Y$, where Y is a nonempty, convex, closed, and bounded subset of $\mathcal{M}$. Let $f\colon Z \to \mathcal{M}$ be a compact mapping with the following property. To each $z \in Z$ and to each arbitrary but fixed $g \in \mathcal{M}$ there exists a compact set of compact linear operators $\mathscr{K}(Y) \subseteq \mathbb{L}(\mathcal{M}, \mathcal{M})$ so that the following conditions prevail:

(i) $$\bigwedge_{y \in Y} f(y) \in g + \mathscr{K}(Y)(y - z)$$

(ii) $$\bigwedge_{y_i, y_2 \in Y} f(y_1) - f(y_2) \in \mathscr{K}(Y)(y_1 - y_2)$$

(iii) There exists a set $\mathscr{L}(Y)$ of linear operators so that $\mathscr{K}(Y) \subset \mathscr{L}(Y)$ and

$$g + \mathscr{L}(Y)(Y - Z) =: F(Y) \overset{\circ}{\subset} Y. \tag{2.2}$$

Then

(a) There exists exactly one fixed-point $\hat{y}$ of f and moreover $\hat{y} \in F(Y)$.

(b) For each compact $\ell \in \mathscr{L}(Y)$ either $\Lambda(\ell) = \phi$ or $|\Lambda(\ell)| \leq q < 1$ and the resolvent $(E - \mathscr{K})^{-1}$ exists on $Y - \hat{y}$.

Proof:

(a) existence

From (i) and (iii) we have $f\colon f(Y) \subseteq Y$. Indeed, the hypothesis of the theorem then allows us to apply the Schauder fixed-point theorem from which we conclude the existence of a fixed-point $\hat{y} \in f(Y)$.

Uniqueness

To each compact $\ell \in \mathscr{L}(Y)$ we associate the affine mapping $h(y)$

$$h(y) := g + \ell(y - z)\colon Z \to \mathcal{M}.$$

From (iii)

$$h(Y) = g + \ell(Y - z) \subseteq F(Y) \overset{\circ}{\subset} Y. \tag{2.3}$$

Thus, by the Schauder fixed-point theorem h has a fixed-point $y^* = y^*(\ell) \in Y$, i.e.,

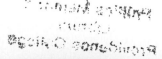

$$h(y^*) = g + \ell(y^* - z) = y^*. \tag{2.4}$$

Let $U := Y - y^*$ so that $\{0\} \overset{\circ}{\subset} U$. Now subtracting (2.4) from (2.3), we get

$$\ell(Y - y^*) = \ell(Y - z) - \ell(y^* - z) \overset{\circ}{\subset} Y - y^*,$$

i.e.,

$$\ell U \overset{\circ}{\subset} U. \tag{2.5}$$

Now from Theorem 1(a) we have that $(E - \ell)^{-1}$ exists. Suppose y_1 and y_2 are fixed-points of f. Then from (ii) we have

$$y_1 - y_2 = f(y_1) - f(y_2) \in \mathcal{K}(y_1 - y_2).$$

Then there exists a mapping $\ell \in \mathcal{K}$ such that

$$y_1 - y_2 = \ell(y_1 - y_2),$$

i.e., $(E - \ell)(y_1 - y_2) = 0$. Since $(E - \ell)^{-1}$ exists, $y_1 - y_2 = 0$.

(b) Using (2.5) and Lemma 1(a) and (b), we have that either $\Lambda(\ell) = \phi$ or $|\Lambda(\ell)| < 1$ and the resolvent exits on $U = Y - y^*$. ■

We now comment on this Theorem 1.

Remarks:

1. Referring to the remark following Lemma 1 we see that for each $k \in \mathcal{K}$ there exists a norm $\| \bullet \|_k$ (not explicitly known) with respect to which k is contractive.

2. Typically the set $\mathcal{K}(Y)$ of linear operators is taken to be the Frechet derivative f' of f. For example, take $y \in Y := A + Bx$ and $\mathcal{K}(Y)\bullet = \int (A + Bt)\bullet dt$ where A and B are intervals of $\mathbb{IPR}$. Apply $\mathcal{K}(Y)$ to $Y - z$. The result $\mathcal{K}(Y)(Y - x) = \int Y(Y - z)dt = \int (A + Bt)((A - a) + (B - b)t)dt$ is always a compact set. However, this compactness property is not conveyed if the set is only represented by means of its boundaries (e.g. interval polynomials). However, the compactness of $\mathcal{K}$ is a theoretical requirement. Indeed, in practice $\mathcal{K}$ is replaced by a superset $\mathcal{L}(Y)$ which is typically neither compact nor required to

be. In the situation of the example discussed here we have

$$\begin{aligned}\mathcal{K}(Y)(Y-z) &\subseteq \mathcal{L}(Y)(Y-z)\\ &:= A(A-a)\int dt + (A(B-b)+B(A-a))\int t\,dt\\ &\qquad + B(B-b)\int t^2 dt,\end{aligned}$$

which, while simple in form for computation, is in general not a compact set.

3. The typical application of the theorem is to provide an inclusion of the solution of explicit differential equations and integral equations of the second kind. As a by-product, a proof of existence and uniqueness is also established. For implicit problems, generalizations of this theorem are required, and some of these are developed in [12].

4. Referring to [12], we note that compactness in the theorem above may be replaced by α-condensing using the Krasnoselski-Darbo fixed-point theorem.

3. PRACTICAL USE OF THE FIXED POINT THEOREMS

In all modified fixed point theorems the conditions for existence and inclusion are coupled with the requirements (2.2):

$$F(Y) \overset{\circ}{\subset} Y. \tag{3.1}$$

In practice f is usually contracting. However, we do not know this until (3.1) has been verified (mostly on a computer). If F (corresponding to f as displayed in the theorem above) satisfies (3.3) relative to a set $Y \subset D \subset \mathcal{M}$, then the iteration process

$$y_{i+1} := f(y_i) \text{ with } y_0 \in Y \tag{3.2}$$

is convergent on Y. It similarly follows that

$$Y_{i+1} := F(Y_i) \text{ with } Y_0 = \tilde{y} \tag{3.3}$$

converges in the power set $\mathbb{P}\mathcal{M}$, also in a neighborhood of Y. Hence it is reasonable to compute such a Y by the following method:

$$Y_0 := \tilde{y}$$

$$\underline{\text{repeat}}\quad Y_{i+1} := F(Y_i)\ \underline{\text{until}}\ Y_{i+1} \overset{\circ}{\subset} Y_i \tag{3.4}$$

$$Y := Y_{i+1}$$

where $\tilde{y}$ is a cleverly (but in principal an arbitrarily) chosen approximation of the fixed point. E.g., $\tilde{y}$ could be determined by a computing process similar to (3.2). If $Y_{i+1} \overset{\circ}{\subset} Y_i$ occurs then the iteration process (3.4) is a posteriori mathematically validated.

In the following we consider the abstract fixed point problem

$$y = f(y).$$

If no uniqueness is desired it suffices to put theoretically $F = f$. But if uniqueness is of interest then in the case of $f \in C'_M$, F can for example be defined by a set of linearizations of f:

$$F(Y) = f(\tilde{y}) + f'(Y)(Y - \tilde{y}),$$

where $\tilde{y}$ is an approximate solution contained in Y.

The next question is how to compute (3.4) practically and economically on a computer.

4. FUNCTIONAL ARITHMETIC AND ROUNDINGS

For simplicity, we restrict ourselves to functional equations in a Hilbert space $\mathcal{M}$ of functions over a given domain X, with a generating system $\{\phi_i\}_{i=0}^{\infty}$. Some of the purely mathematical assumptions such as completeness of $\mathcal{M}$ are actually superfluous but simplify the development.

We call $\mathcal{M}$ the theoretical space. The subspace $[\phi_i]_{i=0}^{N}$, is called a screen of $\mathcal{M}$ and is denoted by $S_N(\mathcal{M})$. The mapping S_N: $\mathcal{M} \to S_N(\mathcal{M})$, which project an element $y \in \mathcal{M}$ into the subspace $S_N(\mathcal{M})$, is called a reduction or rounding operator. This terminology calls to mind not only the extensive analogy with the rounding in floating point numbers but also the fact that S_N possesses more properties than a customary projection. S_N is a semimorphism [14], which is closely related to homomorphisms.

For implementation on a computer we need an isomorphism

$i := S_N(\mathcal{M}) \to R_N(\mathcal{M})$.

In the most typical case, $R_N(\mathcal{M}) = \mathbb{R}^{N+1}$, and i maps the vector $\sum_{i=0}^{N} c_i \phi_i$ into its practical realization, the $(N+1)$-tuple $(c_0,\ldots,c_N)$. In the case of spline bases $R_N(\mathcal{M})$ is itself a function space, for example $(C_X^p)^{N+1}$. Therefore, the mapping $R_N := iS_N$: $\mathcal{M} \to R_N(\mathcal{M})$ is the rounding operator associated with S_N. Let, for example, $\phi_N := (\varphi_1,\ldots,\varphi_N)$ be the vector of the basis elements of $S_N(\mathcal{M})$. Then any $y \in \mathcal{M}$ has the representation

$$S_N y = \phi_N * R_N y \tag{4.1}$$

where $*$ denotes the scalar product in $S_N\mathcal{M} \times R_N\mathcal{M}$. Along the lines in [14] the semimorphism S_N induces in $S_N(\mathcal{M})$ a semimorphic algebraic structure corresponding to $\mathcal{M}$. Let $(\mathcal{M}, +, -, \bullet, /, \int)$ be such an algebraic structure. Then the semimorphic structure $(S_N(\mathcal{M}),$ $\boxplus, \boxminus, \boxdot, \boxed{/}, \boxed{\int})$ is induced by

$$\text{(RS)} \qquad \bigwedge_{\circ \in \{+,-,\bullet,/,\int\}} \ \bigwedge_{y,z \in S_N(\mathcal{M})} y \boxed{\circ} z := S_N(y \circ z)$$

(The operator $\int$ is to read either as a monadic operator $y \int z := \int z$ or as $y \int z := \int_0^y z$, $y \int z := \int_y^x z$, etc.). Hence, if an arithmetic operator f: $\mathcal{M}^n \to \mathcal{M}$ is given, i.e. f is defined by the algebraic operations in $\mathcal{M}$, then $S_N f|_{(S_N(\mathcal{M}))^n}$: $S_N(\mathcal{M})^n \to S_N(\mathcal{M})$ is easy to realize by replacing all operations $\circ$ by the corresponding semimorphic algebraic operations $\boxed{\circ}$ in $S_N(\mathcal{M})$. We sometimes abbreviate $S_N f$ also by $\boxed{f}$.

In bounded domains X we take polynomials as specific bases: Chebyshev polynomials, Legendre polynomials, Bernstein polynomials, Beta polynomials (for random problems) and Spline bases; in infinite and semiinfinite intervals we typically take exponential polynomials, Fourier and exponential Spline bases.

The corresponding round operators are mostly problem dependent. To introduce the reader to the general complex situation, we stretch our procedure in the simplest case involving linear (problem independent) rounding operators. Furthermore, for simplicity, we assume that corresponding to N there exists an $M \geq N$ such that for all $\circ \in \{+,-,\bullet,/,\int\}$ the image of $\circ: S_N(\mathcal{M}) \times S_N(\mathcal{M}) \to S_N(\mathcal{M})$ lies in $S_M(\mathcal{M})$.

Now, let the following representations be given

$$y = \phi_M * a = \phi_N * R_N y \text{ for } y \in S_N(\mathcal{M}),$$

$$z = \phi_M * b = \phi_M * R_M z \text{ for } z \in S_M(\mathcal{M}).$$

Furthermore, let the rounding operator S_N: $\mathcal{M} \to S_N(\mathcal{M})$ be linear with $S_N \varphi_k = \varphi_k$ for $0 \leq k \leq N$. Then an element $z \in S_M(\mathcal{M})$ with the property

$$z = \phi_M * b = \sum_{i=0}^{M} \varphi_i b_i = \sum_{i=0}^{N} \varphi_i b_i + \sum_{i=N+1}^{M} \varphi_i b_i$$

satisfies

$$S_M z = S_M(\phi_M) * b = \sum_{i=0}^{N} \varphi_i b_i + \sum_{i=N+1}^{M} S_N(\varphi_i) b_i.$$

Consequently the matrix $S_N(\phi_M)$ has the form

$$S_N(\phi_M) = (\mathrm{E}, \Psi_{NM}), \tag{4.2}$$

where E is the $(N+1) \times (N+1)$ identity matrix and Ψ_{NM} is an $(N+1) \times (M+N)$-matrix. Ψ_{NM} is characterized by

$$S_N(\varphi_j) = \sum_{j=0}^{N} \Psi_{ij} \varphi_i, \text{ for all } N+1 \leq j \leq M.$$

Thus, for the practical representation of a linear rounding operator it suffices to compute a rounding table in which (in principle) the elements of Ψ_{NM} are listed. As an example we present a table (Table 4.7) of Chebyshev roundings for polynomial basis taken from [12]. We choose as basis the following multiples of the monomials $\phi_j = 2^{j-1} x^j, j = 0,1,\ldots$. We list $S_N(\phi_j)$ and $\sigma_j = \| (E - S_N) \phi_j \|$ for $1 \leq N \leq 8$, $1 \leq j \leq 9$ and $j \geq N$, and for the range $X = [-1, 1]$.

Table 4.1

Chebyshev Rounding of Monomials

j	ϕ_j	$S_7(\phi_j)$, $S_6(\phi_j)$	σ_j	$S_5(\phi_j)$, $S_4(\phi_j)$	σ_j	$S_3(\phi_j)$,$S_2(\phi_j)$	σ_j	$S_1(\phi_j)$, $S_0(\phi_j)$	σ_j
0	1	1	0	1	0	1	0	1	0
2	$2x^2$	$2x^2$	0	$2x^2$	0	$2x^2$	0	1	1
4	$8x^4$	$8x^4$	0	$8x^4$	0	$8x^2-1$	1	3	5
6	$32x^6$	$32x^6$	0	$48x^4-19x^2+1$	1	$30x^2-5$	7	10	22
8	$128x^8$	$256x^6-160x^4+32x^2-1$	1	$224x^4-112x^2+7$	9	$112x^2-21$	37	35	93

j	ϕ_j	$S_8(\phi_j)$, $S_7(\phi_j)$	σ_j	$S_6(\phi_j)$, $S_5(\phi_j)$	σ_j	$S_4(\phi_j)$, $S_3(\phi_j)$	σ_j	$S_2(\phi_j)$, $S_1(\phi_j)$	σ_j	$S_0(\phi_j)$,	σ_j
1	x	x	0	x	0	x	0	x	0	0	1
3	$4x^3$	$4x^3$	0	$4x^3$	0	$4x^3$	0	$3x$	1	0	4
5	$16x^5$	$16x^5$	0	$16x^5$	0	$20x^3-5x$	1	$10x$	6	0	16
7	$64x^7$	$64x^7$	0	$112x^5-56x^3+7x$	1	$84x^3-28x$	8	$35x$	29	0	64
9	$256x^9$	$576x^7-432x^5+120x^3-9x$	1	$576x^5-384x^3+54x$	10	$336x^3-126x$	46	$126x$	130	0	256

From Table 4.1 we can read for example $S_\psi x^6 = S_5 x^6 = \dfrac{48x^4-18x^2+1}{32}$; hence $\Psi_{6,0} = \dfrac{1}{32}$, $\Psi_{6,1} = 0$, $\Psi_{6,2} = -\dfrac{18}{32}$, $\Psi_{6,3} = 0$ and $\Psi_{6,4} = \dfrac{48}{32}$.

The approximation methods and results following from the above material are merely a unification of known results in the recent literature. For example, the Spline bases lead to classical discretization, multiple shooting or finite element methods. But our aim of developing error determining and controlling E-methods requires that we introduce some additional tools.

Therefore we need the powerset $\mathcal{PM}$ of $\mathcal{M}$ and a corresponding embedding of the theoretical problem, functions, operators and operations into this space $\mathcal{PM}$. The subset $[\phi_i]_{1=0}^{N} := \{\sum_{i=0}^{N} \phi_i A_i \mid A_i \in I\mathbb{C}\} \subset \mathcal{PM}$ may be denoted by $IS_N(\mathcal{PM})$. The corresponding mapping IS_N: $\mathcal{PM} \to IS_N(\mathcal{PM})$ is called a directed reduction or directed rounding and maps an interval function $Y \subset \mathcal{PM}$ into the subset $IS_N(\mathcal{PM})$. IS_N is a directed semimorphism.

Analogously, for realization on computers, we need an isomorphism ι: $IS_N(\mathcal{PM}) \to IR_N(\mathcal{PM})$, where in our (special) case we can put $IR_N(\mathcal{PM}) = (I\mathbb{C})^{N+1}$. The mapping $IR_N = \iota IS_N = \mathscr{PM} \rightarrow IR_N(\mathscr{PM})$ is the directed rounding operator associated with IS_N for practical realizations of $IS_N(\mathscr{PM})$.

Corresponding to (4.1), for a $Y \epsilon \mathscr{PM}$ we have the representation

$$IS_N Y = \phi_N * IR_N Y \tag{4.4}$$

where $*$ denotes the scalar product in $IS_N(\mathscr{PM}) \times IR_N(\mathscr{PM})$.

Furthermore, the directed semimorphism IS_N induces a directed semimorphic algebraic structure in $IS_N(\mathscr{PM})$ corresponding to that of $\mathscr{PM}$ as follows: $\mathscr{M}$ induces an algebraic structure in $\mathscr{PM}$, $(\mathscr{PM}, +\ -, \cdot, /, \int)$ and through

$$(IRS) \quad \bigwedge_{\circ \epsilon \{+,-,\cdot,/,\int\}} \quad \bigwedge_{Y,z \epsilon IS_N(\mathscr{PM})} Y(\circ)z := IS_N(Y \circ z),$$

IS_N induces the directed semimorphic structure $(IS_N(\mathscr{PM}), (+), (-), (\cdot), (/), (\int))$. The same remarks which were made regarding f and $\boxed{f}$ are valid here for the corresponding extension F: $(\mathscr{PM})^n \rightarrow \mathscr{PM}$ and its realization

$$IS_N f = (f): (IS_N(\mathscr{M})^n \rightarrow IS_N(\mathscr{M}).$$

Just as in the earlier case, we assume IS_N to be a sublinear directed rounding operator, i.e., for $\alpha, \beta \epsilon I\mathbb{C}$, we have

$$IS_N(\alpha Y + \beta Z) \subseteq \alpha IS_N(Y) + \beta IS_N(Z). \tag{4.5}$$

Additionally we assume for simplicity that all operations $\circ \epsilon \{+,-,\cdot,/,\int\}$ maps $IS_N(\mathscr{PM})$ into $IS_M(\mathscr{PM})$ with an $M \geq N$. Thus for a $Z = \phi_M * B = \sum_{i=0}^{M} \varphi_i B_i \epsilon IS_M(\mathscr{PM})$ we have $S_N Z = S_N(\phi_M) * B \subseteq \sum_{i=0}^{N} \varphi_i B_i + \sum_{i=N+1}^{M} IS_N(\varphi_i) B_i$. The interval matrix $IS_N(\phi M)$ then has the similar structure

$$IS_N(\phi_M) = (\mathrm{E}, I\Phi_{NM}) \tag{4.6}$$

and

$$\varphi_i \in IS_N(\varphi_i) = \sum_{j=0}^{N} \varphi_j I\Psi_{ij} \text{ for all } N+1 \leq i \leq M. \tag{4.7}$$

Here also, it suffices to compute a rounding table. Those bases for which the matrix $I\Psi_{NM}$ can be computed easily using Ψ_{NM} are especially advantageous. In Table 4.1 there is a column for σ_i which gives $\varphi_i \in IS_N(\varphi_i) = S_N\varphi_i + \frac{[-1,1]\sigma_i}{2^{i-1}}$; in our example $\phi_i = x^6$ we find that for $\sigma_6 = 1$

$$IS_4x^6 = IS_5x^6 = \frac{48x^4 - 18x^2 + 1}{32} + \frac{[-1,1]}{32}$$

and thus

$$I\Phi_{6,0} = \frac{[0,2]}{32} \text{ and } I\Phi_{6,j} = \Psi_{6,i} \text{ for } j \geq 1.$$

5. ALGORITHMIC EXECUTION OF ITERATIONS

Again, consider the fixed point problem $y = f(y)$ in the space $\mathcal{M} = C[-1,1]$. For simplicity we dispense with the verification of uniqueness for nonlinear f and assume $f \in C'_{\mathcal{M}}$ so that an iteration process

$$y_{k+1} := f(y_k) \text{ in } \mathcal{M}$$

seems to be successful.

The isomorphic iteration in $R_N(\mathcal{M})$ then reads (for i see 4.)

$$v_{k+1} = \iota y_{k+1} = \iota S_N f(y_k) = g(\iota y_k) = g(v_k). \tag{5.1}$$

Using (5.1), let an approximation $\tilde{y} \in S_N(\mathcal{M})$ resp. $\tilde{v} = \iota\tilde{y} \in R_N(\mathcal{M})$ be computed. Starting with $Y_0 = \tilde{y}$ we continue iterating in $\mathcal{PM}$:

$$Y_{k+1} := f(Y_k) \text{ until } Y_{k+1} \overset{\circ}{\subset} Y_k \text{ occurs.}$$

The isomorphic iteration $IR_N(\mathcal{M})$ then reads

$$V_{k+1} := \iota V_{k+1} = \iota IS_N f(Y_k) := G(\iota Y_k) = G(V_k) \tag{5.2}$$

until $V_{k+1} = \iota Y_{k+1} \overset{\circ}{\subset} \iota Y_k = V_k$ occurs.

The construction of the operators g resp. G from f is, in case of arithmetical operators, more or less difficult dependent on the chosen basis ϕ_N. Generally it is necessary to construct auxiliary operators which realize the arithmetic operations $\iota\boxplus$, $\iota\boxminus$, $\iota\boxdot$, $\iota\boxslash$, $\iota\boxed{\int}$ in $R_N(\mathcal{M})$ and $\iota(+)$, $\iota(-)$, $\iota(\bullet)$, $\iota(/)$ and $\iota(\int)$ in $IR_N(\mathcal{PM})$, isomorphic in each case to those of $S_N(\mathcal{M})$ and $IS_N(\mathcal{PM})$, resp.

Let, for example, $\boxdot$ be the polynomial product in $S_N(\mathcal{M})$ (basis ϕ_N is the polynomial basis), then $\iota\boxdot$ represents a tensor $\mathcal{P} = (P_{i,j,k})$. With $y = \phi_N * b$ and $z = \phi_N * c$ we have

$$\iota(y \boxdot z) = \iota((\phi_N * b) \boxdot (\phi_N \bullet c)) = \phi_N * (bPc)$$

where naturally bPc is a vector and bP and Pc are matrices.

For the isomorphic image in $R_N(\mathcal{M})$ we have

$$\iota(y)\iota\boxdot\iota(z) = \iota(y)P\iota(z) \tag{5.3}$$

where $\iota\boxdot$ is realized by a tensor multiplication.

Correspondingly all operations $\iota\boxdot$ are realized by tensors of degree three. Due to the rules of tensor calculation the operator $g = \iota f$ is a nested sequence of vectors, matrices and tensor operations.

For example, let the linear operator $f(y) = a(x) \boxdot y(x) \boxplus b(x)$ and the corresponding tensors $P := \iota \boxdot$, $+ := \iota \boxplus$ and the vectors $\iota a = \alpha, ib = \beta$ and $\iota y = V$ be given.

Then we have

$$g(v) = \iota(f(y)) = \alpha P v + \beta = \Gamma v + \beta$$

with the matrix $\alpha P =: \Gamma$, hence g is a linear mapping $\mathbb{R}^{N+1} \rightarrow \mathbb{R}^{N+1}$:

The iteration corresponding to (5.1) now reads

$$v_{k=1} := \Gamma v_k + \beta = g(v_k), \tag{5.4}$$

and the one corresponding to (5.2) is

$$V_{k+1} := I\Gamma V_k + I\beta = g(V_k) \tag{5.5}$$

iterated in $IR_N(\mathcal{M})$ (where $I\Gamma =$ integral matrix, $I\beta =$ interval vector).

In order to conclude $Y_{k+1} \overset{\circ}{\subset} Y_k$ in the space $IS_N(\mathcal{M})$ from the verification of $V_{k+1} \overset{\circ}{\subset} V_k$ in the space $IR_N(\mathcal{M})$, it is necessary to verify for the chosen basis that for every two elements $U,V \in IR_N(\mathcal{M})$ satisfying $U \overset{\circ}{\subset} V$ one has $\phi_N * U \overset{\circ}{\subset} \phi_N * V$.

This verification as well as the construction of G requires certain properties of appropriate bases ϕ_N.

6. APPLICATIONS TO DIFFERENTIAL AND INTEGRAL EQUATIONS

We remark that in general the basis and the type of rounding must be adapted to the given problem in order to assure effective computation.

For example, the basis must be sufficiently consistent with the given boundary condition $Ry = 0$ to allow the boundary value problem to be transformable into a related integral equation. Such transformations are necessary to assure compactness of the iterating operator, insofar this property is required.

For example, let us consider the boundary value problem

$$y'' = f(x,y,y',y'') \text{ with } Ry = 0. \tag{6.1}$$

Here, for example, Ry may be nonlinear and may contain very general boundary conditions such as

$$Ry = \begin{pmatrix} y(0) - y(1/2) \\ y'(1/3) \cdot y(1/4) + \int_0^1 y(t)dt - 1 \end{pmatrix}.$$

Nearly all functional equations occurring in practice can be transformed into algebraic functional equations in the space $(\mathcal{M}, +, -, \bullet, /, \int)$ as required in 4. The following example may illustrate this obvious fact:

with the substitutions ($\mathcal{Y}_p$ Bessel's function)

$$w(x) := \mathcal{Y}_p(e^x) \text{ and } v(x) := e^{y(x)}$$

the transcendental differential equation

$$y'' = f(y) := \mathcal{Y}_p(e^x)y' + e^y$$

is transformed into the system

$$\begin{pmatrix} y'' \\ v'' \\ w'' \end{pmatrix} = f(y,v,w) = \begin{pmatrix} w \bullet y' + v \\ 2v' \\ (p^2 - v)w \end{pmatrix}$$

The transformation of (6.1) into an integral equation is achievable in different ways, of which the most natural appears to be the following:

By setting $z := y''$ and introducing a function a (compatible to the basis) with $a'' = 0$, we arrive at the equations

$$\begin{aligned} z &= f(x,\, a(x) + \int_0^x \int_0^x z\,dx^2,\, a'(x) + \int_0^x z\,dx,\, z) =: f(z) \\ R(a &+ \int_0^x \int_0^x z\,dx^2) = 0. \end{aligned} \tag{6.2}$$

This shows that we may suppose that our problem takes the form

$$z = f(z) \text{ in } (\mathcal{M}, +, -, \bullet, /, \int), \tag{6.3}$$

where f is a compact operator (although more generally the α-condensing property would also suffice). Summarizing, we formulate an E-method. The practical execution on a computer due to Section 4 is guaranteed by the isomorphism $IR_N(\mathcal{PM}) = \iota IS_N(\mathcal{PM})$. Let be given an approximation $\tilde{z}$ in $S_N(\mathcal{M})$ then the algorithm reads:

(1) Compute $U := W_0 := IS_N(f(\tilde{z}) - \tilde{z})$

(2) <u>repeat</u> $W := U{*}{*}\varepsilon$

$U := IS_N(W_0 + f'((W + \tilde{z}) \cup \tilde{z})W)$

<u>until</u> $U \overset{\circ}{\subset} W$ <u>or</u> limit

(3) We then have: existence, uniqueness and containment (inclusion) of the solution z of the problem (6.3):

$$z(x) \in \tilde{z}(x) + U(x) \text{ for } x \in [-1,1]$$

and for the solution $y(x)$ of problem (6.1) we have:

$$\begin{aligned} y''(x) &\in \tilde{z}(x) + U(x) \\ y'(x) &\in a'(x) + \int_0^x (\tilde{z}(t) + U(t)dt \qquad \text{for all } x \in [-1,1]. \\ y(x) &\in a + \int_0^x \int_0^t (\tilde{z}(t) + U(t)dt^2 \end{aligned}$$

The notation $U{*}{*}\varepsilon$ is defined as follows:

$$U{*}{*}\varepsilon := U + [-\varepsilon, \varepsilon]\frac{d(U)}{\| U \|},$$

a so-called artificial blowing up (i.e., enlargement) of the segment function U. This is done to achieve a faster convergence and to achieve $U \overset{\circ}{\subset} W$.

The short notation "<u>or</u> limit" denotes an exit in the form of termination of the loop in (2). Criteria for this could be: $d(U)$ too large or, time-cost-limit-overflow, etc. Such criteria are necessary and natural, since due to the point of view of E-methods for (6.3) we do not know a priori that the iteration converges, nor do we have any other guarantee of success. Experience shows that it is more economical for an investigator to conduct several tests based on E-methods (with different arrangements) than it is for him to try to solve the problem theoretically, a process which in the worst case could fail after an expenditure of weeks.

In principle these E-methods are also applicable to a large class of problems, including: systems of equations, partial differential equations, integral equations, free boundary value problems, eigenvalue problems, problems of cybernetics, of reliability analysis and studies of parameters.

7. SOME EXAMPLES

We consider some examples illustrating the efficiency of E-methods compared to other recent methods (although an E-method delivers significantly more information such as existence, uniqueness, containment). The complexity of our examples is limited by the size of contemporary computers. Hence, all examples were computed on a microprocessor (!) ZILOG Z80 (with a 12 decimal digit mantissa) of the Institute für Angewandte Mathematik der Universität Karlsruhe. This mini computer possesses all important basic routines which are described in [3], [11], [14], [19] and in the present volume. Notice that this mini computer has only about 20 KBYTE storage units free available for the programs to be executed.

Note, that in all examples the inclusion $Y(x)$ for the solution $y(x)$ is computed as a segment polynomial of the form $Y(x) := \sum_{i=0}^{12} Y_i x^i$. Although we have actually determined bounds $Y(x)$ for $y(x)$ which are valid continuously for all $x \in [-1,1]$, our tables list some results for only a few points.

Example 1:

Elliptical Integral: $y(x) = \int_{-1}^{x} \frac{dx}{\sqrt{x^3-56x^2+896x-4096}}$

$Y_0 = 1.4824223378^{9}_{8}E-02$

$Y_1 = 1.562^{50000001}_{49000000}\ E-02$ $\quad Y_5 = 3.8414727063^{9}_{6}E-07$ $\quad Y_9 = 3.50279666^{202}_{199}E-11$

$Y_2 = 8.54492187^{502}_{498}E-04$ $\quad Y_6 = 3.5208661733^{4}_{1}E-08$ $\quad Y_{10} = 3.6930867885^{3}_{0}E-12$

$Y_3 = 5.7856241863^{5}_{2}E-05$ $\quad Y_7 = 3.3953912382^{3}_{0}E-09$ $\quad Y_{11} = 4.1568672357^{8}_{4}E-13$

$Y_4 = 4.4927000999^{3}_{0}E-06$ $\quad Y_8 = 3.4007983875^{9}_{6}E-10$ $\quad Y_{12} = 4.5125810622^{6}_{2}E-14$

x	$Y(x)$	x	$Y(x)$	x	$Y(x)$
-0.6	$5.744897554^{9}_{6}E-03$	0.0	$1.4824223379^{9}_{8}E-02$	0.6	$2.4519951384^{9}_{5}E-02$
-0.2	$1.173294728^{51}_{49}E-02$	0.2	$1.7983873230^{9}_{7}E-02$	1.0	$3.13664875^{401}_{397}E-02$

Example 2:

Boundary Value Problem: $2y' - y = 0,\ y(-1) = 1$

exact solution $\quad y(x) = \exp\left(\frac{x+1}{2}\right)$

$Y_0 = 1.648721270^{71}_{68}E+00$

$Y_1 = 8.243606353^{51}_{42}E-01$ $\quad Y_5 = 4.293544975^{24}_{19}E-04$ $\quad Y_9 = 8.873700933^{3}_{1}E-09$

$Y_2 = 2.060901588^{38}_{35}E-01$ $\quad Y_6 = 3.57795414^{604}_{599}E-05$ $\quad Y_{10} = 4.436850466^{63}_{58}E-10$

$Y_3 = 3.434835980^{63}_{59}E-02$ $\quad Y_7 = 2.555681678^{22}_{19}E-06$ $\quad Y_{11} = 2.029434175^{71}_{68}E-11$

$Y_4 = 4.293544975^{79}_{74}E-03$ $\quad Y_8 = 1.5973010488^{8}_{7}E-07$ $\quad Y_{12} = 8.455975732^{1}_{0}E-13$

x	$Y(x)$	x	$Y(x)$	x	$Y(x)$
-0.8	$1.105170918^{10}_{05}E+00$	0.0	$1.648721270^{71}_{68}E+00$	0.8	$2.459603111^{17}_{12}E+00$
-0.6	$1.221402758^{18}_{13}E+00$	0.2	$1.822118800^{41}_{36}E+00$	1.0	$2.718281828^{48}_{42}E+00$
-0.4	$1.349858807^{59}_{55}E+00$	0.4	$2.013752707^{49}_{44}E+00$		
0.2	$1.491824697^{66}_{62}E+00$	0.6	$2.225540928^{51}_{46}E+00$		$e = 2.7182818284\ 590 \in Y(1.0)$

Example 3:

Boundary Value Problem in Comparison with Multiple Shooting Method:

$4y'' - y = 0.5 + 0.5x, \quad y(-1) = 0, \quad y(1) = 1$

x	$Y(x)$	$x = 0.0$		x	$Y(x)$
-0.8	0.704674068^{86}_{75}	0.3868188839^{71}_{69}	$Y(0.0)$	0.2	0.48348014891^{8}_{5}
-0.6	0.1426409088^{61}_{57}	Shooting method with stepwidth		0.4	0.590898524736^{6}_{2}
-0.4	0.21824367622^{4}_{0}	0.386819	$h = 0.1$	0.6	0.71141096008^{5}_{1}
-0.2	0.29903320048^{6}_{2}	0.38681888	$h = 0.01$	0.8	0.84696338169^{4}_{0}
				1.0	1.00000000001 0.999999999998

Example 4:

Transcendental Integral: $\int_{-1}^{x} \frac{\sqrt{t+47}}{100+t^2}\, e^{\frac{t^2-x^2}{(100+x^2)(100+t^2)}}\, dt$

i	Y_i	i	Y_i
0	$6.918443366^{40}_{39}E-2$	7	$2.03618523340^{6}_{5}E-9$
1	$7.00000000001E-2$ 6.99999999999	8	$6.070316697^{70}_{69}E-12$
2	$1.0489871193^{8}_{9}E-3$	9	$-1.1341540062^{5}_{4}E-11$
3	$2.3330903790^{9}_{8}E-3$	10	$-5.1998532183^{8}_{7}E-14$
4	$-5.772122425^{50}_{49}E-7$	11	$7.0378100823^{1}_{0}E-14$
5	$-4.6786104839^{7}_{6}E-7$	12	$4.0658578766^{6}_{5}E-16$
6	$-4.4971956624^{6}_{5}E-10$		

$Y(0.0)$	$Y(0.5)$	$Y(1.0)$
0.06918443366^{40}_{39}	0.10447579340^{5}_{4}	0.14046568640^{4}_{3}

Example 5:

Double Integral: $\int_{-1}^{x} \int_{-1}^{t} \dfrac{d\tau dt}{4\sqrt{(104.8576-t^4)(104.8576-\tau^4)^3}}$

i	Y_i	i	Y_i	i	y_i
0	$4.7805602^{6}_{3}E-3$	5	$4.5538772^{7}_{4}E-6$	9	$1.4617144^{77}_{69}E-8$
1	$9.550449^{24}_{19}E-3$	6	$6.062978^{31}_{27}E-6$	10	$2.2264934^{3}_{1}E-8$
2	$4.7683715^{9}_{6}E-3$	7	$3.497838^{11}_{08}E-10$	11	$2.990652^{90}_{88}E-10$
3	$2.2041223^{7}_{5}E-11$	8	$7.26768^{303}_{299}E-10$	12	$5.1650195^{7}_{4}E-10$
4	$6.877356^{62}_{58}E-11$				

$Y(0.0)$	$Y(0.5)$	$Y(1.0)$
0.0047805602^{6}_{3}	0.0107481148^{8}_{3}	$0.0119110036^{81}_{7\$}$

Example 6:

Extended Boundary Value Problem:

$(32 + 2x)y'' - (13 + x)y' - y = 0;$

$7y(-1) + 8y(1) - 15y'(-1) - 17y'(1) = 0;\ 4y(0) + 128y'(0) = 16$

Exact Solution: $y(x) = \exp\left(\frac{x}{2}\right) \Big/ \sqrt{x+16},\ y(1) = \sqrt{e/17}$

$Y(0.0)$	$Y(0.5)$	$Y(1.0)$
0.25000000000^{1}_{0}	0.31610520599^{2}_{0}	0.399873643^{902}_{898}

Example 7:

Boundary Value Problem with Singularity:

$(0.64 - x^2)y'' + 18xy' - 90.000000538y = 0.734 + 2x,\ y(-1) = 1,\ y(1) = 2$

i	Y_i	i	Y_i
0	$-7.2484643415^{7}_{2}E-3$		
1	$-2.3809830107^{5}_{2}E-2$	7	$1.8163822862^{8}_{5}E-1$
2	$6.3779847938^{6}_{3}E-2$	8	$2.433008291^{60}_{58}E-1$
3	$7.439901548^{65}_{56}E-2$	9	$2.3650813116^{6}_{2}E-3$
4	$4.6506139568^{9}_{6}E-1$	10	$8.447947705^{41}_{37}E-3$
5	$2.4412177268^{6}_{2}E-1$	11	$1.96057592^{401}_{397}E-10$
6	$7.2665844378^{6}_{3}E-1$	12	$6.936636546^{12}_{06}E-11$

x	$Y(x)$	x	$Y(x)$
-0.9	0.5703954185^{4}_{2}	0.1	-0.0088675552203^{9}_{0}
-0.8	0.3159702240^{45}_{38}	0.2	-0.007992360187^{2}_{0}
-0.7	0.16962535138^{9}_{5}	0.3	-0.001696303771^{9}_{5}
-0.6	0.08793456336^{9}_{7}	0.4	$0.016039819^{4001}_{3996}$
-0.5	0.043586507936^{9}_{0}	0.5	0.05656452580^{50}_{41}
-0.4	0.01995740361^{41}_{36}	0.6	0.14011506726^{3}_{0}
-0.3	0.00730523604^{23}_{19}	0.7	0.30121459611^{9}_{5}
-0.2	0.000180275516^{6}_{3}	0.8	0.596580044^{402}_{396}
-0.1	-0.0042593060402^{9}_{0}	0.9	0.1163900426^{6}_{4}
0.0	-0.0072484643415^{7}_{2}		1

REFERENCES

[1] Adams, E. und Spreuer, H. (1975). Konvergente numerische Schrankenkonstruktion mit Splinefunktionen für nichtlineare gewöhnliche bzw. lineare parabolische Randwertaufgaben. Lecture Notes in Computer Science 29, 1180126, Berlin-Heidelberg-New York: Springer.

[2] Alefeld, G., Herzberger, J. (1974). Einführung in die Intervallrechnung. Mannheim-Wien-Zürich: Bibliographisches Institut.

[3] Bohlender, G., Grüner, K. (1982). Gesichtspunkte zur Realisierung einer optimalen Arithmetik; in diesem Band.

[4] Collatz, L. (1952). *Aufgaben monotoner Art. Arch. Math.* 3, 366-376.

[5] Collatz, L. (1964). *Funktionalanalysis und Numerische Mathematik,* Berlin-Göttingen-Heidelberg: Springer.

[6] Collatz, L. (1981). Anwendung von Monotoniesätzen zur Einschliessung der Lösung von Gleichungen. Jahrbuch Überblicke Mathematik 1981, Bibliographisches Institut Mannheim.

[7] Kaucher, E. (1977). Ein Fixpunktsatz für Iterationsverfahren mit isoton gerundeter Iterationsfunktion in bedingt vollständigen Verbänden; ZAMM 57, T284-T286.

[8] Kaucher, E. (1977). Algebraische Erweiterungen der Intervallrechnung unter Erhaltung der Ordnungs - und Verbandsstrukturen, Computing, Supplementum 1, 65-79.

[9] Kaucher, E., Rump, S. M. (1980). Small Bounds for the Solution of Systems of Linear Equations; Computing, Supplementum 2, 157-164.

[10] Kaucher, E., Rump, S. M. (1980). Generalized Iteration Methods for Bounds of the Solution of Fixed Point Operator-Equations: Computing 24, 131-137.

[11] Kaucher, E., Rump, S. M. (1982). E-Methods for Fixed Point Equations $f(x) = x$, Computing 28, 31-46.

[12] Kaucher, E. (1981). Neuere Methoden zur Einschliessung der Lösung von Gleichungen mit automatischem Existenz- und Eindeutigkeitsnachweis; Interne Mitteilung am Institut für Angewandte Mathematik der Universität Karlsruhe.

[13] Kaucher, E. (1982). Lösung von Funktionalgleichungen mit garantierten und genauen Schranken, im Tagungsband: Wissenschaftliches Rechnen und Programmiersprachen; Hrsg.: U. Kulisch/Ch. Ullrich in Berichte 10, Teubner Verlag.

[14] Kulisch, U., Miranker, W. L. (1981). Computer Arithmetic in Theory and Practice, Academic Press.

[15] Spreuer, H. (1975). Konvergente numerische Schranken für partielle Randwertaufgaben von monotoner Art. Lecture Notes in Computer Science 29, 298-305, Berlin-Heidelberg-New York: Springer.

[16] Talbot, T. D. (1968). Guaranteed Error Bounds for Computed Solutions of Non-linear Two-Point Value Problems. MRC TEchnical Summary Report no. 875, University of Wisconsin, Madison.

[17] Walter, W. (1970). Differential and integral Inequalities. Berlin-Heidelberg-New York: Springer.

[18] Bohlender, G., Kaucher, E., Klatte, R., Kulisch, U., Miranker, W. L., Ullrich, Ch., Wolff von Gudenberg, J. (1980). FORTRAN for Contemporary Numerical Computation. RC 8348, IBM Research Center, Yorktown Heights, NY. 66 Seiten. Auch in Computing 26 (1981) erschienen.

[19] Kulisch, U., et al. (1982). Wissenschaftliches Rechnen und Programmiersprachen; Berichte des German Chapter of the ACM, 10, Teubner Verlag.

[20] Acker, A. (1981). How to approximate the Solution of Certain Free Boundary Problems for the Laplace Equation by using the Contraction Principle. *J. Appl. Math Phys.* 32.

[21] Kaucher, E., Miranker, W. L. Numerics with Guaranteed Accuracy for Function Space Problems, RC 9789, IBM Research Center, Yorktown Heights, NY.

ULTRA-ARITHMETIC: THE DIGITAL COMPUTER SET IN FUNCTION SPACE

W. L. Miranker

IBM Thomas J. Watson Research Center
Yorktown Heights, New York 10598

Methods of numerical analysis require a computational environment in which a variety of data types and associated arithmetic operations are available as computer primitives. (These types include reals and complex numbers, intervals of these as well as vectors and matrices over all of them.) In the same sense we propose that the techniques of numerical analysis which deal with problems set in function spaces (e.g., integral equations, boundary value problems, etc.) and their associated numerical methods (e.g., finite elements, collocation methods, spectral methods, etc.) should be supplied with a computational environment in which data types corresponding to functions and operations corresponding to operations on functions are available. This is the point of view of ultra-arithmetic. We introduce the data types, roundings as well as digital versions of ultra-arithmetic operations (including differentiation, integration and compositions of functions). An interval-arithmetic for function data types is also discussed. Examples and error analyses are given.

1. INTRODUCTION

Architectural concepts for computers frequently evolve from the nature of data structures and data types which occur in computation as well as from the nature of the processing to be done with these structures and types. Contemporary scientific computation employs a variety of such data types such as reals (actually floating point numbers), vectors, matrices and complex versions of these as well as intervals over all of these. (See [3] for a comprehensive treatment of floating point arithmetic.) A well taken point is that the methods of numerical analysis have generated many of these data structures and types as well as processing requirements associated with them. Numerical analysis itself finds its procedures in turn evolving from the body of mathematical methodology, and in a sense, is the bridge between that methodology and scientific computation.

Taken in this way but viewed from the vantage of the digital computer, mathematical methodology amounts to a limited set of operations consisting more or less to numerical algebra applied to the limited collection of data types enumerated above. This is a limitation of approach which unnecessarily curtails both the power of mathematics and the power of digital computation. The operations and constructs of mathematics which can be implemented

ISBN 0-12-428660-7

directly in digital computers are far greater in number than those which are currently implemented. In a sense the digital computer can be made to appear as a deeper and somewhat more accurate image of mathematical constructs and operations than it now is. This is the viewpoint of ultra-arithmetic where the structures, data types and operations corresponding to functions are developed for direct digital implementation.

Table Contrasting Number and Function Data Types

Numbers $a \in \mathbb{R}$	Functions $f \in L^2$
Representation as a decimal expansion	Representation as a generalized Fourier series
$a = \sum_{i=M}^{\infty} \frac{a_i}{10^i}$	$f = \sum_{j=-\infty}^{\infty} a_j \phi_j$
"basis" for the expansion	basis for the expansion
$e_i = 10^{-i}, \; i = M, M+1, \ldots$	$\phi_j = e^{ijx}, \; j = 0, \pm 1, \ldots$
"coefficients"	coefficients
$a_i \in \{0, 1, \ldots, 9\}$	$a_j = \frac{1}{\sqrt{2\pi}} \langle f, \phi_j \rangle = \frac{1}{\sqrt{2\pi}} \int_{-\pi}^{\pi} f \phi_j dx$
"normalized basis" and "coefficients"	"normalized basis" and "coefficients"
$\tilde{e}_i = 1$ $\tilde{a}_i \in \{0, 10^{-i}, \ldots, 9 \times 10^{-i}\}$	$\tilde{\phi}_j = \frac{1}{\sqrt{2\pi}} \phi_j$ $\tilde{a}_j = a_j$ and $\lvert \langle e^{ijx}, f \rangle \rvert \le const \frac{\lVert f^{(p)} \rVert}{j^p}$.
The $\tilde{a}_i$. decay exponentially with i .	The $\tilde{a}_j$ decay like the pth power provided f is periodic and p-times differentiable.
Rounded number	Rounded function
$S_N a := \sum_{i=M}^{N} \tilde{a}_i \tilde{e}_i$	$S_N f := \sum_{j=-N}^{N} a_j \phi_j$
S_N is some rounding operator.	The rounding operator S_N corresponds to truncation.
data type	data type
$a_M a_{M-1} \ldots a_0 \bullet a_{-1} a_{-2} \ldots a_{-N}$,	$(a_{-N}, a_{-N+1}, \ldots, a_0, \ldots, a_N)$,
a finite string of integers.	a finite string of reals.
Rounding error estimate	Rounding error estimate
$\lVert a - S_N a \rVert \le .5 \times 10^{-N}$	$\lVert (I - S_N) f \rVert \le const \; \lVert f^{(p)} \rVert / N^{p-1/2}$
This rounding error estimate follows from the exponential decay of the coefficient $\tilde{a}_i$.	This rounding error estimate follows from the power-like decay of the a_j.

A digital computer designed or equipped with ultra-arithmetic will be a far more congenial tool for scientific computation. Indeed problems associated with functions will be solvable on computers just as now we solve algebraic problems. Moreover, the considerably enlarged set of structures, data types and operations is likely to make for the generation of more far reaching concepts of computer architecture and in particular, concepts of parallelism.

To get a feeling for this point of view let us contrast notions and methodology for approximating and representing real numbers by means of a decimal expansion on the one hand and functions by means of a Fourier expansion on the other hand. We do this in the table above.

We begin in section 2 with a review of ultra-arithmetic, that is the specification of function data types, of basic rounded operations (i.e., computer versions of the operations $+, -, \times, /, \frac{d}{dx}, \int^x$ applied to functions), and of algorithms for these operations. Samples of error estimates are also given.

In section 3, we consider three basic problems of numerical analysis whose solutions are functions from the point of view of ultra-arithmetic. The three problems are i) a Fredholm integral equation, ii) a boundary value problem for a second order differential equation and iii) the composition problem, $f(g) = h$. The composition problem: find the function g being given the functions f and h is the basis of ultra-numerical methods for non-linear problems. The point of view of ultra-arithmetic is to develop the numerical treatment of these problems in such a manner that their numerical solution may be reduced to computer operations in a direct sense. Thus we do not take as our objective here the customary criteria of numerical analysis which include studying methods with one or another superior computational feature, for example. (However, we do believe that superior computation in every sense will quickly follow.) Instead we shall treat simple numerical methods for approximating problems whose solutions are functions and show what such methods require in the design and specification of operators which are then intended to be implemented directly in computers. It will appear that certain special operations occur, and these are likely independent of the underlying algorithmic choices (e.g., choice of basis, choice of numerical method, etc). These include some more ramified operations of matrix-vector algebra such as linear system solvers, as we shall see. A complete list of operators is not yet at hand, not to say a primitive list. We view the treatment here simply as a study on the way to compiling this list.

Finally in section 4 we review notions of an ultra-interval arithmetic for functions, in particular, an ultra-arithmetic for intervals of polynomials. This treatment is a step toward equipping the approach of ultra-arithmetic with the facility of computer error specification, i.e., computation with guarantees.

2. A REVIEW OF ULTRA-ARITHMETIC

We consider functions $f \epsilon L^2[-1, 1]$ with norm $\|f\|$. Let $\{\phi_j\}_{j=0}^{\infty}$ be an orthonormal basis in L^2. We call the projection S_N a rounding operator where

$$S_N\colon \sum_{j=0}^{\infty} f_j \phi_j(x) \rightarrow \sum_{j=0}^{N} f_j \phi_j(x) := S_N f(x).$$

Here $f_j = (f, \phi_j)$ so that the rounding operator S_N maps the generalized Fourier series for $f(x)$ into its N-th partial sum. (In fact rounding operators need not be chosen so restrictively.) The rounding error is $(I - S_N)f$, and the norm of this error is characterized in terms of smoothness properties of $f(x)$.

Given functions $f, g, \ldots \epsilon L^2$, their corresponding ultra-arithmetic data types are $S_N f$, $S_N g$, ... or equivalently the vectors $\boldsymbol{f} := (f_0, \ldots, f_N)^T$, $\boldsymbol{g} := (g_0, \ldots, g_N)^T$, ... (We write $S_N f \sim \boldsymbol{f}$, $S_N g \sim \boldsymbol{g}$, ...) The basic ultra-arithmetic operations (addition, subtraction, multiplication, reciprocation, differentiation and integration) corresponding to these data types are given in the following list:

$$
\begin{aligned}
f \pm g &\rightarrow S_N f \pm S_N g \\
fg &\rightarrow S_N(S_N f \cdot S_N g) \\
fh = 1 &\rightarrow \min_{h_N \in S_N(L^2)} \| 1 - h_N S_N f \| \\
\frac{d}{dx} f &\rightarrow S_N\left[\frac{dS_N f}{dx}\right] \\
\int^x f &\rightarrow S_N \int^x S_N f.
\end{aligned}
\tag{2.1}
$$

The development including algorithms and error analyses for these operations is given in [2] where superior properties of a polynomial basis were noted. For this reason we will henceforth restrict our attention to the basis $\phi_j(x) = T_j(x)$, the j-th Chebyshev polynomial, viz,

$$T_n(x) = \cos\ n\theta, \quad \theta = \cos^{-1} x$$

$$T_n(x) = 2x\ T_{n-1}(x) - T_{n-2}(x), \ n \geq 2$$

with

$$T_0(x) = 1 \quad \text{and} \quad T_1(x) = x.$$

This is an unnormalized basis so that for $f \epsilon L^2$ we have

$$f(x) = \sum_{j=0}^{\infty} f_j T_j(x)$$

with

$$f_0 = \frac{1}{\pi} \int_{-1}^{1} f(x) \frac{dx}{\sqrt{1-x^2}}$$

$$f_j = \frac{2}{\pi} \int_{-1}^{1} f(x) T_j(x) \frac{dx}{\sqrt{1-x^2}}, \quad j \geq 1.$$

Let $\Phi(x) = [T_0(x), \ldots, T_N(x)]^T$ so that

$$S_N f(x) = \sum_{j=0}^{N} f_j T_j(x) = \boldsymbol{f}^T \Phi. \tag{2.2}$$

Using the notation

$$\| g \|_\mu = \left(\int_{-1}^{1} | g(x) |^2 \frac{dx}{\sqrt{1-x^2}} \right)^{1/2},$$

we note that the rounding error admits of the following bound

$$\| (I - S_N) f \|_\mu \leq O(N^{-p} \log N) \tag{2.3}$$

whenever $f \epsilon C^p[-1, 1]$ (see [1]). If $g \epsilon C^p[-1, 1]$ as well, we note by way of example, that ultra-multiplication admits of the following error bound

$$\| fg - S_N[S_N f \cdot S_N g] \|_\mu \leq O(N^{-p} \log N). \tag{2.4}$$

These results may be found in [2] from which (also by way of example) we reproduce the following algorithm for ultra-multiplication.

Example: Algorithm for Multiplication

Let h be the data type corresponding to ultra-multiplication of $\boldsymbol{f}$ and $\boldsymbol{g}$, i.e.,

$$fg \rightarrow S_N[S_N f \cdot S_N g] \sim \boldsymbol{h}.$$

Then

$$\boldsymbol{h} = \frac{1}{2}(U + V + W)\boldsymbol{f} \tag{2.5}$$

where

$$U = \begin{bmatrix} g_0 & & & & O \\ g_1 & \cdot & & & \\ \cdot & & \cdot & & \\ \cdot & & & \cdot & \\ \cdot & & & & \cdot \\ g_N & \cdot & \cdot & g_1 & g_0 \end{bmatrix}, \quad W = \begin{bmatrix} g_0 & g_1 & \cdot & \cdot & \cdot & g_N \\ g_1 & & & & \cdot & \\ \cdot & & & \cdot & & \\ \cdot & & \cdot & & & \\ \cdot & \cdot & & & & O \\ g_N & & & & & \end{bmatrix} \tag{2.6}$$

and $V = U^T$.

Finally we introduce the rounding operator $S_{N,N}$ for functions of two variables

$$S_{N,N}k(x,y) = \sum_{i,j=0}^{N} k_{ij}T_i(x)T_j(y)$$

$$k_{ij} = \left(\frac{\sigma_{ij}}{\pi}\right)^2 \int_{-1}^{1}\int_{-1}^{1} k(x,y)T_i(x)T_j(y)\frac{dx}{\sqrt{1-x^2}}\frac{dx}{\sqrt{1-x^2}} \tag{2.7}$$

$$\sigma_{ij} = \begin{cases} 1, & i=j=0, \\ \sqrt{2}, & i+j=1, \\ 2, & i,j \geq 1. \end{cases}$$

If $k \epsilon C^p$ (meaning that all mixed partial derivatives of order p are continuous) then

$$\| (I-S_{N,N})k \|_{\mu} \leq O(N^{-p} \log \ N),$$

where the notation here is clear from the context.

Since confusion will not result we will hereafter omit the subscript μ on norms.

Error estimates analogous to (2.3) and (2.4) for all of the ultra-arithmetic operations in (2.1) and for a variety of choices of bases may be found in [2]. Algorithms analogous to (2.5) and (2.6) for these operations may also be found in [2].

3. APPLICATIONS OF ULTRA-ARITHMETIC†

A Fredholm Integral Equation

Our first application of ultra-arithmetic is to the Fredholm integral equation for the unknown function $v(x)$.

$$\mathscr{L}v := v(x) - \int_{-1}^{1} k(x,y)v(y)dy = f(x). \tag{3.1}$$

We seek a rounded version of this problem, that is a problem for the data type $\boldsymbol{v} \sim S_N v$, the round of v itself. A rounded version of (3.1) may be derived most directly by successively applying various ultra-arithmetic operation in the list (2.1), viz,

† The material in this section was developed jointly with L. Brieger, cf. [1] for further details.

$$S_N \mathscr{L} S_N v := S_N v(x) - S_N \int_{-1}^{1} S_N[(S_{N,N} k(x, y)) S_N v(y)]dy = S_N f(x).$$

This approach involves committing a "rounding" error for each such operation. To reduce the number of such errors, we proceed instead by defining a new ultra-arithmetic operator specific to the integral equation. Our point of view is that an algorithm for executing this operator is to be explicitly implemented into the computing environment (such as the algorithm for multiplication displayed in section 2.)

We replace (3.1) by

$$\mathscr{L} S_N v := S_N v(x) - \int_{-1}^{1} S_{N,N} k(x, y) S_N v(y) dy = S_N f(x) \tag{3.2}$$

which we refer to as the rounded problem. Considering the definition of S_N in section 2, we see that (3.2) is Fredholm's method (a relatively primitive numerical method) for approximating (3.1). We have already discussed our reasons for using such a primitive numerical method in the Introduction.

We write $S_N f = \boldsymbol{f}^T \Phi(x)$ and $S_N v = \boldsymbol{v}^T \Phi(x)$ (see (2.2)). Correspondingly using (2.7) we write

$$S_{N,N} k(x, y) = \Phi^T(x) K^T \Phi(y)$$

where $K = [k_{ij}]$ is a matrix.

Using these notations along with the Gramian

$$\Gamma := \left[\int_{-1}^{1} T_i(y) T_j(y) dy \right]$$

in (3.2), we find

$$\mathscr{L}_N S_N v := \boldsymbol{v}^T \Phi(x) - \boldsymbol{v}^T \Gamma^T K \Phi(x) = \boldsymbol{f}^T \Phi(x).$$

From this we obtain the ultra-arithmetic operator which we seek, given implicitly by the following system of linear algebraic equations.

$$(I - K^T \Gamma) \boldsymbol{v} = \boldsymbol{f}. \tag{3.3}$$

Here I is the $N \times N$ identity matrix.

Error Analysis

Let $u(x)$ denote the solution of (3.1) and let $S_N v(x)$ denote the solution to (3.2). (Of course, $S_N v \sim \mathbf{v}$ or $S_N v = \mathbf{v}^T \Phi$ where $\mathbf{v}$, the solution of (3.3) is what is actually computed.) By combining (3.1) and (3.2) the following equation for the error

$$\varepsilon(x) = u(x) - S_N v(x)$$

may be derived.

$$\varepsilon(x) = -\int_{-1}^{1} k(x, y)\epsilon(y)dy + F(x) \tag{3.4}$$

with

$$F(x) = (I - S_N)f(x) + \int_{-1}^{1} (I - S_{N,N})k(x, y)S_N v(y)dy. \tag{3.5}$$

Multiplying (3.4) by $\epsilon(x)$ and integrating with respect to x we get

$$\begin{aligned} \|\varepsilon\|^2 &= -\int_{-1}^{1}\int_{-1}^{1} \varepsilon(x)k(x, y)\epsilon(y)dydx + \int_{-1}^{1} \varepsilon(x)F(x)dx \\ &\le \|\varepsilon\|^2 \|k\| + \|\varepsilon\| \|F\| \end{aligned} \tag{3.6}$$

where

$$\|k\|^2 := \int_{-1}^{1}\int_{-1}^{1} k^2(x, y)dxdy.$$

Utilizing the properties $f \epsilon C^p$ and $k \epsilon C^p$ we will show that $\|F\| = O(N^{-p} \log N)$. Then by assuming that $\|k\| < 1$ we obtain the following error estimate

$$\|\varepsilon\| \le \frac{1}{1 - \|k\|} O(N^{-p} \log N).$$

Now $\| F \|$ is bounded by the norm of the two summands in (3.5). For the first such term we have

$$\| (I-S_N)f \| = O(N^{-p} \log N)$$

since $f \epsilon C^p$. For the second summand we have

$$\| \int_{-1}^{1} (I-S_{N,N})kS_N v dy \|^2 \leq \int_{-1}^{1} \left(\int_{-1}^{1} [(I-S_{N,N})k(x,y)]^2 dy \int_{-1}^{1} [S_N v(y)]^2 dy \right) dx$$

$$= \| S_N v \|^2 \| (I-S_{N,N})k \|^2.$$

Since $k \epsilon C^p$ the second factor here is $O(N^{-p} \log N)$. Indeed the entire right member here is of this order since $\| S_N v \|$ is the norm of a polynomial. Combining these observations yields the desired estimate for $\| F \|$ and completes the error analysis.

A Boundary Value Problem

Our second application of ultra-arithmetic concerns the boundary value problem

$$\begin{aligned} \mathscr{L}v &:= v''(x) - v(x) = f(x), \quad x \epsilon (-1, 1), \\ v(-1) &= c_0, \\ v(1) &= c_1. \end{aligned} \tag{3.7}$$

We write the boundary conditions as

$$\mathscr{B}v = \boldsymbol{c}$$

where $\boldsymbol{c} = (c_0, c_1)^T$ and $\mathscr{B}$ is an appropriate operator.

We proceed by replacing the boundary value problem by a rounded version for the data type $S_N v = \boldsymbol{v}^T \Phi$:

$$\begin{aligned} &S_N \mathscr{L} S_N v = (S_N v)'' - S_N v = S_N f \\ &\left. \begin{aligned} S_N v(-1) &= c_0 \\ S_N v(-1) &= c_1 \end{aligned} \right\} \quad \mathscr{B} S_N v = \boldsymbol{c}. \end{aligned} \tag{3.8}$$

Here we have used the fact that $\mathrm{sp}\{T_j\}_{j=0}^{N}$ (the span of the Chebyshev polynomials $T_0, \ldots, T_n$) is invariant under differentiation.

The problem (3.8) consists of $N + 3$ equations in $N + 1$ unknowns:

$$\begin{aligned} (D^2 - I)\mathbf{v} &= \mathbf{f}, \\ B\mathbf{v} &= \mathbf{c}. \end{aligned} \tag{3.9}$$

Here D is the lower triangular $(N + 1)\times(N + 1)$ matrix relating Chebyshev polynomials and their derivatives: $\Phi'(x) = D^T\Phi(x)$. In particular,

$$D^T = \begin{bmatrix} 0 & & & & & & \\ 1 & 0 & & & & & \\ 0 & 4 & 0 & & & & \\ 3 & 0 & 6 & 0 & & & \\ 0 & 8 & 8 & 8 & 0 & & \\ 5 & 0 & 10 & 0 & 10 & 0 & \\ 0 & 12 & 12 & 12 & 12 & 12 & 0 \\ & & . & . & . & & \end{bmatrix}$$

The matrix B in (3.9) is given by

$$B = \begin{bmatrix} T_0(-1) & T_1(-1) & \ldots & T_N(-1) \\ T_0(1) & T_1(1) & \ldots & T_N(1) \end{bmatrix}$$

$$= \begin{bmatrix} 1 & -1 & \ldots & (-1)^N \\ 1 & 1 & \ldots & 1 \end{bmatrix}$$

We now consider two approaches for solving the overdetermined problem (3.9): i) a method of "decrease of precision" ii) a method of constraint least squares.

i) decrease of precision

In this method we discard the last two equations in the first member of (3.9) (equivalently (3.8)). The result is the problem

$$\begin{aligned} &S_{N-2}\mathscr{L}S_N v = S_{N-2}f, \\ &\mathscr{B}S_N v = c, \end{aligned} \tag{3.10}$$

or equivalently

$$\begin{bmatrix} (D^2-I)_{N-1} \\ B \end{bmatrix} \boldsymbol{v} = \begin{bmatrix} \boldsymbol{f}_{N-1} \\ \boldsymbol{c} \end{bmatrix}. \tag{3.11}$$

Here the subscript $N-1$ indicates that the last two rows (entries) of the subscripted matrix (vector) have been discarded.

Note that except for its last two rows the matrix of the system (3.11) is upper triangular. Thus, (3.11) is easily solved by backward substitution.

ii) constrained least squares

In this approach we determine $\boldsymbol{v}$ as the solution of the following constrained minimization problem:

$$\begin{aligned} &\min_{\boldsymbol{v}} \|(D^2-I)\boldsymbol{v}-\boldsymbol{f}\| \\ &B\boldsymbol{v} = \boldsymbol{c} \end{aligned} \tag{3.12}$$

(equivalently $\min\limits_{v(x)} \| S_N \mathscr{L}S_N v - S_N f \|$ subject to $\mathscr{B}S_N v = \boldsymbol{c}$).

Error Analysis Let $u(x)$ be the solution to (3.7) and let $S_N v(x)$ be the solution of the rounded problem (3.8) obtained by method (i) or (ii) as the case may be. Let $\varepsilon(x) = u(x) - S_N v(x)$ be the error. We now derive estimates for $\|\varepsilon\|$ in the two cases. We assume that $f(x) \epsilon C^p$.

Case i) By combining (3.7) and (3.10) we find

$$\mathscr{L}\varepsilon = F(x),$$

$$\mathscr{B}\varepsilon = 0,$$

where

$$F(x) = (I - S_{N-2})f - (I - S_{N-2})\mathscr{L} S_N v.$$

We observe that $\| F(x) \| = O((N-2)^{-p} \log N)$ since $f \in C^p$ and $\mathscr{L} S_N v$ is a polynomial. Indeed this estimate may be made uniform with respect to all problems corresponding to say all f in the unit ball in H^{p+1} (the $(p+1)$-st Sobolev space) as an appeal to the Lax-Milgram lemma shows. Thus

$$\varepsilon'' - \varepsilon = \delta, \quad \delta = O((W-2)^{-p} \log N),$$

$$\varepsilon(0) = \varepsilon(1) = 0.$$

By multiplying the differential equation by ε and integrating by parts, several error estimates may be derived, viz,

$$\| \varepsilon \|, \ \| \varepsilon \|_1, \ \| \varepsilon \|_\infty \leq O\big((N-2)^{-p} \log N\big).$$

Here $\| \varepsilon \|_1$ denotes the norm in H^1, while $\| \varepsilon \|_\infty$ is the sup-norm.

Case ii) The following error equation may be derived:

$$\mathscr{L}\varepsilon = S_N \mathscr{L}(I - S_N)u + S_N f - S_N \mathscr{L} S_N v$$

$$\mathscr{B}\varepsilon = 0.$$

The term $S_N \mathscr{L}(I - S_N)u = O(N^{-p} \log N)$ since $f \in C^p$. The term $S_N f - S_N \mathscr{L} S_N v$ is the quantity minimized (see (3.12)f). To estimate the norm of the latter term consider the function

$$\phi := u + ax + b$$

where a and b are determined by solving the pair of linear equations $\mathscr{B} S_N \phi = \mathbf{c}$. Determining a and b in this manner makes this function ϕ admissible to the minimization problem. Then insertion of ϕ into the expression to be minimized will provide an upper bound for the latter. In particular

$$\begin{aligned}\min \| S_N \mathscr{L} S_N v - S_N f \| &\le \| S_N \mathscr{L} S_N (u + ax + b) - S_N f \| \\ &= \| -S_N \mathscr{L}(I - S_N) u + S_N \mathscr{L} S_N (ax + b) \| \\ &\le O(N^{-p} \log N) + \| S_N \mathscr{L} S_N (ax + b) \| .\end{aligned}$$

Now

$$\begin{aligned}0 &= \mathscr{B} S_N \phi - \boldsymbol{c} \\ &= \mathscr{B}[u + (S_N - I)u + S_N(ax + b)] - \boldsymbol{c} \\ &= \mathscr{B}[(S_N - I)u + S_N(ax + b)].\end{aligned}$$

Thus

$$S_N (ax + b) = \mathscr{B}(S_N - I)u = O(N^{-p} \log N).$$

Combining these estimates, we find for the error equation that

$$\begin{aligned}\mathscr{L}\varepsilon &= O(N^{-p} \log N), \\ \mathscr{B}\varepsilon &= 0.\end{aligned}$$

Thus

$$\varepsilon = O(N^{-p} \log N).$$

The Non-linear Equation

Our third application is to the composition problem. That is to solve the non-linear equation

$$f(g) = h \quad \text{or} \quad f \circ g = h \tag{3.13}$$

for $g(x)$ the functions $f(x)$ and $h(x)$ being given. The rounded version of this problem is taken to be

$$S_N[S_N f \circ S_N g] = S_N h. \tag{3.14}$$

Thus we are given the data types $S_N f$ and $S_N h$ (equivalently $\boldsymbol{f}$ and $\boldsymbol{h}$), and we seek the solution $S_N g$ (equivalently $\boldsymbol{g}$), of (3.14). (3.14) corresponds to a system of $N+1$ non-linear equations for the components of $\boldsymbol{g}$. This system which is to be specified below may be solved by one or another iteration method and the latter in turn requires an appropriate starting vector. The starting vector will be determined from the following variant of (3.14),

$$C_N[S_N f \circ S_N g] = S_N h, \tag{3.15}$$

which is much easier to solve for $\boldsymbol{g}$. The operator C_N is to be specified.

(3.15) depends on a monomial representation of the data types which is given by replacing Φ by Z where

$$Z = (1, x, \ldots, x^N)^T.$$

Notice that if for any data type

$$S_N g = \boldsymbol{g}^T \Phi = \boldsymbol{n}^T Z,$$

then

$$\boldsymbol{g}^T \mathcal{T} = \boldsymbol{n}^T \quad \text{or} \quad \mathcal{T} Z = \Phi, \tag{3.16}$$

where $\mathcal{T}$ is the following lower triangular $(N+1) \times (N+1)$-matrix.

$$\mathcal{T} = \begin{bmatrix} t(0,0) & & & & & & & & \\ 0 & t(1,0) & & & & & & & \\ t(2,1) & 0 & t(2,0) & & & & & & \\ 0 & t(3,1) & 0 & t(3,0) & & & & & \\ t(4,2) & 0 & & & \cdot & & & & \\ 0 & \cdot & & & & \cdot & & & \\ \cdot & \cdot & & & & & \cdot & & \\ \cdot & \cdot & & & & & & \cdot & \\ \cdot & \cdot & & & & & & & \\ t(N, \frac{N}{2}) & 0 & \cdot & \cdot & \cdot & \cdot & \cdot & \cdot & t(N,0) \end{bmatrix}$$

in the case that N is even. When N is odd the last row of $\mathcal{T}$ *is*

$$\left(0,\ t(N, \frac{N-1}{2}),\ 0, \ldots, t(N,0)\right).$$

Here

$$t(n, k) = (-1)^k \sum_{j=k}^{[\frac{n}{2}]} \binom{n}{2j} \binom{k}{j}.$$

We now specify the operator C_N. Let $f(x) \epsilon L^2$. $C_N f$ transforms the Chebyshev expansion of $f(x)$ into a monomial expansion and truncates the latter to degree N. Thus the range of C_N is $\boldsymbol{m}^T Z$ for all N-vectors $\boldsymbol{m}$. (For analytic f, $C_N f$ is just the finite Taylor series.)

Now consider the following data types

$$\begin{aligned} S_N f(x) &= \boldsymbol{f}^T \Phi = \boldsymbol{m}^T Z \\ S_N h(x) &= \boldsymbol{h}^T \Phi = \boldsymbol{w}^T Z \\ S_N g(x) &= \boldsymbol{g}^T \Phi = \boldsymbol{n}^T Z. \end{aligned} \tag{3.17}$$

Then

$$S_N f \circ S_N g = \boldsymbol{u}_{N^2}^T \Phi_{N^2} = \boldsymbol{p}_{N^2}^T Z_{N^2}. \tag{3.18}$$

Here the subscript N^2 is used to indicate that the subscripted quantities are N^2-vectors. Indeed the components of $\boldsymbol{u}_{N^2}$ and $\boldsymbol{p}_{N^2}$ respectively, are functions of the pairs of N-vectors $(\boldsymbol{f}, \boldsymbol{g})$ and $(\boldsymbol{m}, \boldsymbol{n})$ respectively:

$$u_j = u_j(\boldsymbol{f}, \boldsymbol{g}),$$

$$p_j = p_j(\boldsymbol{m}, \boldsymbol{n}), \quad j = 0, 1, \ldots, N^2.$$

The equation (3.14) yields

$$u_j(\boldsymbol{f}, \boldsymbol{g}) = h_j, \quad j = 0, \ldots, N, \tag{3.19}$$

while (3.15) yields

$$p_j(\boldsymbol{m}, \boldsymbol{n}) = w_j, \quad j = 0, \ldots, N. \tag{3.20}$$

Our claim that (3.20) is easier to solve than (3.19) is based on the following assertion concerning the form of the equations (3.19):

$$\begin{aligned} p_0 &= \sum_{k=0}^{N} m_k n_0^k \\ p_j &= p_0(n_0, \ldots, n_j), \quad j \geq 1, \end{aligned} \tag{3.21}$$

where moreover for $j \geq 1$, p_j is linear in n_j. Thus solving (3.20) requires solving one polynomial equation followed by N divisions.

Having solved (3.20) for $\boldsymbol{n}$ we use (3.16) to write

$$\boldsymbol{g}^{(0)} = (\mathscr{T}^T)^{-1} \boldsymbol{n}. \tag{3.22}$$

Of course this $\boldsymbol{g}^{(0)}$ does not in general solve (3.19), but we use it as a starting value for an iterative method of solution.

Since it is useful to understand the nature of the equations generated by (3.14) and by (3.15), we display tables of the functions $u_j(\boldsymbol{f}, \boldsymbol{g})$, $j \leq N = 1, 2, 3$ and $p_j(\boldsymbol{m}, \boldsymbol{n})$, $j \leq N = 1, 2, 3, 4$.

Table 3.1†

N	j	p_j
1	0	$m_1n_0 + m_0$
	1	m_1n_1
2	0	$m_2n_0^2 + m_1n_0 + m_0$
	1	$n_1(2n_0m_2 + m_1)$
	2	$n_2(2n_0m_2 + m_1) + n_1^2m_2$
3	0	$m_3n_0^3 + m_2n_0^2 + m_1n_0 + m_0$
	1	$n_1(3n_0^2m_3 + 2n_0m_2 + m_1)$
	2	$n_2(3n_0^2m_3 + 2n_0m_2 + m_1) + n_1^2(3n_0m_3 + m_2)$
	3	$n_3(3n_0^2m_3 + 2n_0m_2 + m_1) + 2n_2n_1(3n_0m_3 + m_2) + n_1^3m_3$
4	0	$m_4n_0^4 + m_3n_0^3 + m_3n_0^2 + m_1n_0 + m_0$
	1	$n_1(4n_0^3m_4 + 3n_0^2m_3 + 2n_0m_2 + m_1)$
	2	$n_2(4n_0^3m_4 + 3n_0^2m_3 + 2n_0m_2 + m_1) + n_1^2(6n_0^2m_4 + 3n_0m_3 + m_4)$
	3	$n_3(4n_0^3m_4 + 3n_0m_3 + 2n_0m_2 + m_1) + 2n_2n_1(6n_0^2m_4 + 3n_0m_3 + m_2)$ $+ n_1^3(4n_0m_4 + m_3)$
	4	$n_4(4n_0^3m_4 + 3n_0m_3 + 2n_0m_2 + m_1) + 2n_3n_1(6n_0^2m_4 + 3n_0m_3 + m_2)$ $+ n_2^2(6n_0^2m_4 + 3n_0m_3 + m_2) + 3n_2n_1^2(4n_0m_4 + m_3) + n_1m_4$

† This table was prepared by D. Yun employing the SCRATCHPAD system.

Table 3.2†

N	j	u_j
1	0	$g_0 f_1 + f_0$
	1	$g_1 f_1$
2	0	$g_2^2 f_2 + g_1^2 f_2 + 2g_0^2 f_2 + g_0 f_1 - f_2 + f_0$
	1	$g_1(2g_2 f_2 + 4g_0 f_2 + f_1)$
	2	$4g_2 g_0 f_2 + g_2 f_1 + g_1^2 f_2$
3	0	$g_3^2(6g_0 f_3 + f_2) + 6g_3 g_2 g_1 f_3 + g_2^2(6g_0 f_3 + f_2) + 3g_2 g_1 f_3$
		$+ g_1^2(6g_0 f_3 + f_2) + 2g_0^2(2g_0 f_3 + f_2) - g_0(3f_3 - f_1) - f_2 + f_0$
	1	$6g_3^2 g_1 f_3 + g_3(3g_2^2 f_3 + 12g_2 g_0 f_3 + 2g_2 f_2 + 3g_1^2 f_3)$
		$+ 6g_2^2 g_1 f_3 + 2g_2 g_1(6g_0 f_3 + f_2)$
		$+ 3g_1^3 f_3 + g_1(12g_0^2 f_3 + g_1(12g_0^2 f_3 + 4g_0 f_2 - 3f_3 + f_1)$
	2	$6g_3^2 g_2 f_3 + 2g_3 g_1(3g_2 f_3 + 6g_0 f_3 + 2f_2)$
		$+ 3g_2^3 f_3 + g_2(6g_1^2 f_3 + 12g_0^2 f_3 + 4g_0 f_2 - 3f_3 + f_1)$
		$+ g_1^2(6g_0 f_3 + f_2)$
	3	$3g_3^3 f_3 + g_3(6g_2^2 f_3 + 6g_1^2 f_3 + 12g_0^2 f_3 + 4g_0 f_2 - 3f_3 + f_1)$
		$+ 3g_2^2 g_1 f_3 + 2g_2 g_1(6g_0 f_3 + f_2) + g_1^3 f_3$

† This table was prepared by D. Yun employing the SCRATCHPAD system.

In the remainder of this section we will i) demonstrate the assertion concerning the form of (3.21), ii) estimate the quality of $\mathbf{g}^{(0)}$ in (3.22) as a starting value and iii) provide an error estimate for the difference of the solution to (3.13) and its rounded version (3.14).

i) Proof of the structure of (3.21).

Referring to (3.17), we have

$$S_N f \circ S_N g = \sum_{j=0}^{N} m_j [S_N g]^j$$

$$:= \sum_{j=0}^{N} m_j \sum_{k=0}^{jN} v_{jk} x^k$$

which defines the v_{jk}, $k = 0, 1, \ldots, jN$, $j = 0, 1, \ldots, N$. Notice that $v_{ok} = \delta_{ok}$, the Kronecker delta. Then

$$C_N[S_N f \circ S_N g] = \sum_{j=0}^{N} m_j \sum_{k=0}^{N} v_{jk} x^k.$$

Now using (3.18) we have

$$p_n = \sum_{j=0}^{N} m_j v_{jk}, \quad k = 0, \ldots, N. \tag{3.23}$$

Notice that

$$\left[\sum_{1=0}^{N} n_i \, x^i \right]^j = \sum_{k=0}^{jN} v_{jk} x^k. \tag{3.24}$$

Then setting $x = 0$, we get $n_0^j = v_{j0}$. Then from (3.23) we have

$$p_k = \sum_{j=0}^{N} m_j \, n_0^j.$$

Since $v_{0k} = \delta_{0k}$, we also have from (3.23) that

$$p_k = \sum_{j=1}^{N} m_j v_{jk}, \quad k \geq 1.$$

Thus the desired properties of the p_k will follow from these properties being possessed by the $v_{j,k}$ $k \geq 1$, $j \geq 1$. Notice that (3.24) implies that

$$v_{jk} = v_{jk}(n_0, \ldots, n_k) \tag{3.25}$$

while setting $j = 1$ in (3.24) gives $v_{1k} = n_k$, $k \geq 1$.

There remains to show that v_{jk} being linear in n_k implies that $v_{j+1,k}$ is linear in n_k. From (3.24) $v_{j+1,k}$ is the coefficient of x^k in

$$\left[\sum_{i=0}^{N} n_i\, x^i\right]^{j+1} = \sum_{i=0}^{N} n_i\, x^i \left[\sum_{i=0}^{N} n_i\, x^i\right]^{j}$$

$$= \sum_{i=0}^{N} n_i\, x^i \sum_{k=0}^{jN} v_{jk} x^k.$$

The coefficient in question is then the following convolution

$$\sum_{m=0}^{k} n_m v_{j,k-m}.$$

Using (3.25) we see that only the first and last terms in this sum depend on n_k, viz,

$$n_0 v_{j,k} + n_k v_{j,0}.$$

Then appealing to the inductive hypothesis concludes the proof.

ii) Estimate of quality of $\mathbf{g}^{(0)}$

$\mathbf{g}^{(0)}$ denotes the solution to (3.15). Let $\mathbf{g}^*$ denote the solution to (3.14). Subtracting (3.14) and (3.15) gives

$$S_N[S_N f \circ S_N g^* - S_N f \circ S_N g^{(0)}] = (C_N - S_N)[S_N f \circ S_N g^{(0)}]. \tag{3.26}$$

Now C_N is the identity operator on polynomials of degree N. Then

$$(C_N - S_N)\phi \;=\; C_N(I - S_N)\phi. \tag{3.27}$$

We seek an estimate of the norm of this expression, but we are only interested in the case in which ϕ is a polynomial of degree N^2. In this case $(I - S_N)\phi$ is also such a polynomial. Then in the term $C_N(I - S_N)\phi$ we may therefore replace C_N by $C_N|_{N^2}$, the restriction of C_N to the linear manifold of polynomials of degree N^2. But $C_N|_{N^2}$ is then a finite dimensional operator and so all of its norms are equivalent. In particular for the norm,

$$\| \sum_{i=0}^{N^2} a_i x^i \|_*^2 := \sum_{i=0}^{N^2} a_i^2,$$

we have $\| C_N |_{N^2} \|_* \leq 1$ since $C_N |_{N^2}$ annihilates all coefficients of a polynomial beyond the Nth, leaving the remaining coefficients invariant. Then

$$\begin{aligned} \| C_N(I-S_N)\phi \| &\leq const \| C_N |_{N^2} \|_* \| (I-S_n)\phi \| \\ &\leq O(N^{-p} \log N), \ \forall p>0. \end{aligned}$$

Since $S_Nf \circ S_Ng^{(0)}$ is a polynomial of degree N^2, we may combine (3.26) and (3.27) to write

$$S_N[S_Nf \circ S_Ng^k - S_Nf \circ S_Ng^{(0)}] = O(N^{-p} \log N), \ \forall p>0.$$

For the term S_N in front of the bracket here, write $I + (S_N - I)$. Note that the summand $S_N - I$ operating on the bracket is $O(N^{-p} \log N)$, $\forall p>0$ since the bracket is a polynomial. Thus we get

$$S_Nf \circ S_Ng^* - S_Nf \circ S_Ng^{(0)} = O(N^{-p} \log N), \ \forall p>0.$$

Then employing the mean value theorem,

$$\min_{x \in (-1,1)} |(S_Nf)'| (S_Ng^* - S_Ng^{(0)})(x) = O(N^{-p} \log N), \ \forall p>0.$$

If the minimum here is positive (which we assume), we have the following estimate for the quality of the starting iterate.

$$\| S_Ng^* - S_Ng^{(0)} \|_\infty = O(N^{-p} \log N), \ \forall p>0.$$

iii) The error estimate

Recall that g^* denotes the solution of (3.14). Then combining (3.13) and (3.14) we obtain

$$\begin{aligned} f(g) - f(S_Ng^*) &= (I-S_N)[h - f \circ S_Ng^* - S_Nf \circ S_Ng^*] \\ &= O(N^{-p} \log N). \end{aligned}$$

The second line follows if $h(x) \in C^p$, which we assume, since the remaining two terms in the bracket here are polynomials. Now using the mean value (and assuming that $\min_{x \in (-1,1)} |f'(x)|$ is positive), we obtain the following error estimate.

$$\| g - S_N g^* \|_\infty = O(N^{-p} \log N).$$

4. THE ARITHMETIC OF INTERVALS OF POLYNOMIALS

Background Material on Polynomials

The polynomial $p(x) = a_0 + \dots + a_k x^k$ is said to be of degree k, and we write $\partial p = k$. Non-zero constants are polynomials of degree zero, and the polynomial which is identically zero is assigned the degree -1. We write $\mathscr{P}$ for the set of all polynomials with real coefficients, and put

$$\mathscr{P}_n = \{p \in \mathscr{P} : \partial p = n\}.$$

Note that since we do not require the coefficients of the highest power appearing in a polynomial to be non-zero we have $\mathscr{P}_k \subset \mathscr{P}_n$, when $k = -1, 0, \dots, n-1$.

If $q_j \in \mathscr{P}_n$, $j = 0, 1, \dots, n$ are linearly independent then every $p \in \mathscr{P}_n$ can be written in the form

$$p(x) = a_0 q_0(x) + \dots + a_n q_n(x),$$

i.e., $q_0, \dots, q_n$ form a basis for $\mathscr{P}_n$.

A particular basis which we will make explicit use of is furnished by the Bernstein polynomials

$$q_j(x) = \binom{n}{j} x^j (1-x)^{n-j}, \; j = 0, \dots, n.$$

That these polynomials form a basis for $\mathscr{P}_n$ is obvious from the following identities.

$$x^j = \sum_{i=j}^{n} \frac{\binom{i}{j}}{\binom{n}{j}} \binom{n}{i} x^i (1-x)^{n-i},\ j = 0, \ldots, n.$$

Moreover, the functions

$$\phi_j^{(k)}(x) := \binom{k}{j} x^j (1-x)^{k-j},\ j = 0, \ldots, k$$

form a basis for $\mathscr{P}_k$, and so if $p_n \epsilon \mathscr{P}_n$ and $n \leq k$ we may write

$$p_n(x) = \sum_{j=0}^{k} b_j^{(k)} \phi_j^{(k)}(x).$$

We will also make use of the following two lemmas. (For proofs of these Lemmas and for other details concerning the material presented in this section, cf. [3]).

Lemma 1. If $k \geq n$ and

$$p(x) = \sum_{j=0}^{k} b_i \phi_j^{(k)}(x),$$

then $p \epsilon \mathscr{P}_n$ if, and only if,

$$\Delta^i b_0 = 0,\ i = n+1, \ldots, k.$$

Here $\Delta b_j := b_{j+1} - b_j$ and $\Delta^i := \Delta(\Delta^{i-1})$, i.e., Δ is the forward difference operator.

Lemma 2. If

$$p(x) = \sum_{j=0}^{k} b_j^{(k)} \phi_j^{(k)}(x),$$

then

$$\min_{0 \leq j \leq k} b_j^{(k)} \leq \min_{x \epsilon I} p(x) \leq \max_{x \epsilon I} p(x) \leq \max_{0 \leq j \leq k} b_j^{(k)}.$$

Intervals of Polynomials

Let k, ℓ, m be integers not less than -1, and I be the interval $[0,1]$. Given $p_1 \in \mathscr{P}_k$, $p_2 \in \mathscr{P}_\ell$ we define

$$[p_1, p_2](k, \ell; m) := \{p \in \mathscr{P}_m: p_1(x) \leq p(x) < p_2(x), x \in I\}.$$

$[p_1, p_2](k, \ell; m)$ is an *interval of polynomials of index* $(k, \ell; m)$ *with endpoints* p_1, p_2. If ∞ appears in place of k, ℓ, or m, respectively, that means that $\mathscr{P}_k$, $\mathscr{P}_\ell$ or $\mathscr{P}_m$ are to be replaced by $\mathscr{P}$, respectively. We shall generally be in a situation in which $(k, \ell; m)$ is fixed in a discussion and will then suppress the index in our notation for intervals of polynomials. Note, (i) if p_1, p_2 satisfy $p_2(x) < p_1(x)$ for some $x \in I$ then the interval $[p_1, p_2]$ is empty, and (ii) the (finite) index of a given interval is not unique.

For example, let

$$\|f\| = \max_{0 \leq x \leq 1} |f(x)|$$

for $f \in C(I)$. Then the unit ball in $\mathscr{P}_n$, $\{p \in \mathscr{P}_n: \|p\| \leq 1\} = \mathscr{B}_n$, is $[-1, 1](k, \ell; n)$ for any $k, \ell \geq 0$.

A related notation is that of a *polynomial with interval coefficients.* If $\boldsymbol{a}:(a_0, ..., a_n)$ and $\boldsymbol{b}:(b_0, ..., b_n)$ are given vectors of real numbers we say that $\boldsymbol{a} \leq \boldsymbol{b}$ if $a_j \leq b_j$, $j = 0, ..., n$. Suppose $\boldsymbol{a} \leq \boldsymbol{b}$, and $q_0, ..., q_n$ is a basis for $\mathscr{P}_n$.

$$p_n^{\mathscr{I}}(x) := \sum_{i=0}^{n} [a_i, b_i]q_i = \{\sum_{i=0}^{n} c_i q_i: \boldsymbol{a} \leq \boldsymbol{c} \leq \boldsymbol{b}\}$$

is a polynomial with interval coefficients. The graph of $p_n^{\mathscr{I}}$ on I is

$$G(p_n^{\mathscr{I}}) = \{(x, p_n^{\mathscr{I}}(x)): 0 \leq x \leq 1\}.$$

After these preliminaries we turn next to our main concern, a study of the arithmetic of intervals of polynomials. While polynomials with interval coefficients are "graphically" related to intervals of polynomials, we prefer to deal exclusively with intervals of polynomials in what follows as a more appropriate model for the applications of these concepts that we have in mind later.

Addition and Subtraction of Intervals of Polynomials

Let $n \geq -1$ be a fixed integer. We write $[p, q]$ as shorthand for $[p, q](n, n; n)$. Given $[p_1\ p_2]$ and $[q_1, q_2]$ we denote the sum set by

$$S = \{p + q\colon p \in [p_1, p_2], q \in [q_1, q_2]\}.$$

Clearly, $S \subset [p_1 + q_1, p_2 + q_2]$. Moreover, if $S \subset [r, s]$ then $[p_1 + q_1, p_2 + q_2] \subset [r, s]$, since $r(x) \leq p_1(x) + q_1(x)$ and $p_2(x) + q_2(x) \leq s(x)$, all $x \in I$. In short, $[p_1 + q_1, p_2 + q_2]$ is the smallest interval containing S, and we define

$$[p_1, p_2] + [q_1, q_2] = [p_1 + q_1, p_2 + q_2].$$

In the same way we define

$$[p_1, p_2] - [q_1, q_2] = [p_1 - q_2, p_2 - q_1].$$

Multiplication of Intervals of Polynomials

Again let $n \geq -1$ be a fixed integer, and write $[p, q]$ in short for $[p, q](n, n; n)$. Given $[p_1, p_2]$ and $[q_1, q_2]$ we denote the product set by

$$P = \{pq\colon p \in [p_1, p_2], q \in [q_1, q_2]\}.$$

It is easy to see that $P \subset [r_1, r_2](n, n; 2n)$ if, and only if,

$$r_2(x) \geq M(x),\ r_1(x) \leq m(x),\ x \in I$$

where

$$M(x) = \max\ (p_1(x)q_1(x),\ p_1(x)q_2(x),\ p_2(x)q_1(x),\ p_2(x)q_2(x)),$$

and

$$m(x) = \min\,(p_1(x)q_1(x),\ p_1(x)q_2(x),\ p_2(x)q_1(x),\ p_2(x)q_2(x)),$$

and $r_1,\ r_2 \in \mathscr{P}_n$.

Let $r_2^*(x)$, $r_1^*(x)$ be the best uniform approximations out of $\mathscr{P}_n$ on I to $M(x)$ and $m(x)$, respectively, producing errors $E_n(M)$ and $E_n(m)$ respectively. Then $\bar{r}_2(x) = r_2^*(x) + E_n(M)$ and $\bar{r}_1(x) = r_1^*(x) - E_n(m)$ are the best one-sided uniform approximations out of $\mathscr{P}_n$ on I to $M(x)$ and $m(x)$ from above and below. There is no interval $[r_1, r_2](n, n; 2n)$ satisfying

$$P \subset [r_1, r_2](n, n; 2n) \subset [\bar{r}_1, \bar{r}_2](n, n; 2n)$$

for the generic product set P. We shall seek an interval $[r_1, r_2](n, n; 2n)$ containing P, whose endpoints are more readily attainable than $\bar{r}_1$ and $\bar{r}_2$. To this end we write the polynomials p_1q_1, p_1q_2, p_2q_1, p_2q_2 in Bernstein form, say,

$$p_1q_1 = \sum_{j=0}^{2n} a_j\phi_j^{(2n)}$$

$$p_1q_2 = \sum_{j=0}^{2n} b_j\phi_j^{(2n)}$$

$$p_2q_1 = \sum_{j=0}^{2n} c_j\phi_j^{(2n)}$$

$$p_2q_2 = \sum_{j=0}^{2n} d_j\phi_j^{(2n)}.$$

Now put

$$u_j = \max\,(a_j, b_j, c_j, d_j),\ j = 0, \ldots, 2n$$

$$v_j = \min\,(a_j, b_j, c_j, d_j),\ j = 0, \ldots, 2n.$$

Let

$$u(x) = \sum_{j=0}^{2n} u_j \phi_j^{(2n)}(x)$$

$$v(x) = \sum_{j=0}^{2n} v_j \phi_j^{(2n)}(x).$$

$v(x)$ and $u(x)$ furnish bounds for $m(x)$ and $M(x)$. Indeed

$$v(x) \le m(x) \le M(x) \le u(x).$$

Bounds for the discrepancies $u(x)-M(x)$, and $m(x)-v(x)$ are given next. They make us of the notion of modulus of continuity ω which we recall. Suppose $f(x)$ is defined for $x \epsilon I$. Given $\delta > 0$ we have

$$\omega(f;\ \delta) := \sup_{\substack{x_1, x_2 \epsilon I \\ |x_1 - x_2| \le \delta}} |f(x_1) - f(x_2)|.$$

The bounds are stated in terms of the quantities

$$A_i(2n) = \sum_{j=1}^{2n} (j-1)^2 |f_{i,j}|,\ i = 1, 2, 3, 4$$

where

$$f_i = \sum_{j=0}^{2n} f_{i,j} x^j,\ i = 1, 2, 3, 4$$

and $f_1 = p_1 q_1$, $f_2 = p_1 q_2$, $f_3 = p_2 q_1$, $f_4 = p_2 q_2$. The bounds are given by the following theorem.

Theorem 3. For $x \epsilon I$

$$0 \le u(x) - M(x) \le \frac{3}{2} \max_{1 \le i \le 4} \omega\left(f_i;\ (2n)^{-1/2}\right) + \max_{1 \le i \le 4} \frac{A_i(2n)}{2n}.$$

$$0 \le m(x)-v(x) \le \frac{3}{2} \max_{1 \le i \le 4} \omega\left(f_i;\ (2n)^{-1/2}\right) + \max_{i \le i \le 4} \frac{A_i(2n)}{2n}.$$

■

Next we seek to approximate $u(x)$ from above by a polynomial of degree at most n. We seek

$$p(x) = \sum_{j=0}^{2n} p_i \phi_i^{(2n)}(x) \tag{4.1}$$

such that (i) $p \epsilon \mathscr{P}_n$ (ii) $p_i \ge u_i$, $i = 0, \ldots, 2n$ and (iii) $\max\ (p_i - u_i)$, $i = 0, \ldots, 2n$ is minimum for all choices of p which satisfy (i) and (ii). Recalling Lemma 1, we have the following linear programming problem, with variables $p_0, \ldots, p_{2n}$, and ρ. Minimize ρ, subject to the constraints

$$p_i \ge u_i, \qquad i = 0, \ldots, 2n,$$

$$u_i \ge p_i - \rho, \quad i = 0, \ldots, 2n,$$

$$\Delta^i p_0 = 0, \qquad i = n+1, \ldots, 2n.$$

The inequalities here are feasible since

$$p_i = c = \max_{i=0,\ldots,2n} u_i,\ i = 0, \ldots, 2n$$

and

$$\rho = c - \min_{i=0,\ldots,2n} u_i$$

satisfy them. Let $p^*(x)$ be the polynomial corresponding to a solution $\rho^*, p_0^*, \ldots, p_{2n}^*$ of the linear programming problem via (4.1). Put $r_2(x) = p(x)$, and let $r_1(x)$ be determined in an analogous fashion as a polynomial approximation of degree n to $v(x)$ from below. It may be established that $P \subset [r_1, r_2]\ (n, n; 2n)$.

Bounds for $r_2(x) - M(x)$ and for $m(x) - r_1(x)$ are stated in the following theorem.

Theorem 4. For $x \epsilon I$ each of $r_2(x) - M(x)$ and $m(x) - r_1(x)$ is bounded by

$$\frac{3}{2}\max_{1\le i\le 4}\omega\left(f_i;\,(2n)^{-\frac{1}{2}}\right)+\max_{1\le i\le 4}\frac{A_i(2n)}{2n}+\frac{1}{\sqrt{2\pi}}\left(\frac{e}{2}\right)^n\|E^{(n+1)}\|$$

with

$$\|E^{(n+1)}\|\le\left(\sum_{j=n+1}^{2n}|\Delta^j u_0|\binom{2n}{j}\right)n^{-(n+1)}\frac{4n^2(4n^2-1)(4n^2-4)\ldots(4n^2-n^2)}{(2n+1)(2n-1)\ldots3.1}.$$

Here Δ is the usual forward difference operator (cf. Lemma 2).

■

Division of Polynomial Intervals

We view division as reciprocation followed by multiplication. Since we have just finished our discussion of multiplication we restrict our attention to reciprocation. Namely, consider $[p_1, p_2](n, n; n)$ subject to the condition

$$0 < b \le p_1(x),\ x\epsilon I$$

for some fixed b. We denote the reciprocal set by

$$P^{-1} = \{p^{-1}\colon p\epsilon[p_1, p_2]\},$$

and we seek $q_1, q_2\epsilon\mathscr{P}_n$ such that

$$q_1(x) \le \frac{1}{p_2(x)} \le \frac{1}{p_1(x)} \le q_2(x),\ x\epsilon I.$$

Let us first give our attention to the upper bound. We observe that $q_2(x) = b^{-1}$ is a trivial upper bound, and indeed, sometimes b^{-1} is the best uniform approximation from above to $p_1^{-1}(x)$ out of $\mathscr{P}_n$. For example; if $p_1(x) = 2-T_n(x)$ then $b = 1$ and by virtue of the optimal min-max approximation property of T_n, we see that 1 is the best uniform approximation to $(2-T_n(x))^{-1}$ from above out of $\mathscr{P}_n$. To obtain another upper bound, we will choose a $q_2\epsilon\mathscr{P}_n$ satisfying $1 \le p_1(x)q_2(x)$ on I. Let

$$p_1(x) = \sum_{j=0}^{n} a_j \phi_j^{(n)}(x)$$

and

$$q_2(x) = \sum_{j=0}^{n} t_j \phi_j^{(n)}(x).$$

Then

$$p_1(x) q_2(x) = \sum_{j=0}^{2n} c_j \phi_j^{(2n)}(x)$$

where

$$c_j = \sum_{k=0}^{j} a_{j-k} \frac{\binom{n}{j-k}\binom{n}{k}}{\binom{2n}{j}} t_k,\ j = 0, \ldots, 2n \tag{4.2}$$

with the assumption that $a_i = t_i = 0,\ i > n$. We write (4.2) as $\boldsymbol{c} = B\boldsymbol{t}$, where B is the $(2n+1)\times(n+1)$ matrix of the transformation. We want to choose $\boldsymbol{t}$ so that $\boldsymbol{c} \geq \boldsymbol{1}$, for this would insure that $1 \leq p_1(x)q_2(x),\ x \epsilon I$. Moreover, we would like to minimize $\max\,(c_j - 1)$, $j = 0, \ldots, 2n$ among all choices of $\boldsymbol{t}$, as well. Thus we solve the linear programming problem, with variables $\boldsymbol{t}$, r, of minimizing r subject to the constraints

$$B\boldsymbol{t} - \boldsymbol{1} \geq \boldsymbol{O}$$

$$\boldsymbol{r} - B\boldsymbol{t} + \boldsymbol{1} \geq \boldsymbol{O}$$

(where $\boldsymbol{r} = (r, \ldots, r)$ with similar conventions for $\boldsymbol{1}$ and $\boldsymbol{O}$). Let $\boldsymbol{t} = \boldsymbol{t}^*$ and $r = r^*$ be a solution of the linear programming problem. If q_2 is now chosen to be the polynomial with coefficients $\boldsymbol{t}^*$ then $q_2(x) \geq p_1^{-1}(x),\ x \epsilon I$ and

$$\max_{x \epsilon I} \left[q_2(x) - \frac{1}{p_1(x)} \right] \leq \frac{\max_{x \epsilon I} [p_1(x)q_2(x) - 1]}{b} \leq \frac{r^*}{b},$$

in view of Lemma 2.

Integration

We suppose the basic interval I to be $[-1, 1]$. Suppose we are given a polynomial interval $[p_1, p_2](n, n; n)$. If $p \epsilon [p_1, p_2]$ then

$$\int_{-1}^{x} p_1(x)dx \le \int_{-1}^{x} p(x)dx \le \int_{-1}^{x} p_2(x)dx.$$

We seek, $q_1, q_2 \epsilon \mathscr{P}_n$ such that

$$\int_{-1}^{x} p_2 \le q_2, \int_{-1}^{x} p_1 \ge q_1, x \epsilon I.$$

Let

$$p_2(x) = \sum_{j=0}^{n}{}' A_j T_j(x),$$

where Σ' means the term $j = 0$ is halved and $T_j(x)$ is the Chebyshev polynomial of degree j.

Using the relations

$$\int_{-1}^{x} T_j(t)dt = \begin{cases} T_1 + 1, \ j = 0, \\ \frac{1}{2}\left(\frac{T_2}{2} + \frac{1}{2}\right) - \frac{1}{2}, \ j = 1, \\ \frac{1}{2}\left(\frac{T_{j+1}}{j+1} - \frac{T_{j-1}}{j-1}\right) + \frac{(-1)^{j+1}}{j^2 - 1}, \ j = 2, 3, \ldots, \end{cases}$$

we find that

$$r_2 = \int_{-1}^{x} p_2(x) = \sum_{k=0}^{n+1}{}' B_k T_k(x),$$

where

$$\frac{1}{2} B_0 = \frac{A_0}{2} - \frac{A_1}{4} + \sum_{j=2}^{n} (-1)^{j-1} \frac{A_j}{j^2 - 1},$$

$$B_k = \frac{1}{2}\left(\frac{A_{k-1} - A_{k+1}}{k}\right), \ k = 1, 2, \ldots,$$

with $A_{n+1} = A_{n+2} = 0$. Note that

$$B_{n+1} = \frac{A_n}{2(n+1)}.$$

It is clear that the best uniform approximation to $r_2(x)$ by a polynomial of degree at most n on I is given by

$$\sum_{k=0}^{n}{}' B_k T_k(x) = r_2(x) - B_{n+1} T_{n+1}(x),$$

and also that

$$\max_{x \in I} | r_2(x) - \sum_{k=0}^{n}{}' B_k T_k(x) | \leq | B_{n+1} | .$$

Thus, if we choose

$$q_2(x) = | B_{n+1} | + \sum_{k=0}^{n}{}' B_k T_k(x)$$

then $q_2(x) \geq r_2(x)$ and

$$q_2(x) - r_2(x) \leq 2 | B_{n+1} |,$$

is an error bound. q_1 is obtained in an entirely analogous fashion.

Differentiation

Again take $I = [-1, 1]$ and consider the interval $[p_1, p_2](n, n;,n)$. If $p \epsilon [p_1, p_2]$ then

$$| p(x) - \frac{p_1 + p_2}{2} | \leq \frac{p_2 - p_1}{2}, \ x \epsilon I.$$

Suppose

$$\max \frac{p_2 - p_1}{2} = M,$$

then, the Theorem of A. Markov (cf. Rivlin [5r) tells us that

$$-Mn^2 \leq p'(x) - \frac{p'_1 + p'_2}{2} \leq Mn^2,$$

i.e., $p' \epsilon [-Mn^2 + \frac{p'_1 + p'_2}{2}, Mn^2 + \frac{p'_1 + p'_2}{2}](n, n; n)$.

This crude bound can sometimes be improved.

REFERENCES

[1] Brieger, L., Miranker, W. L. (1982). "Three Applications of Ultra-arithmetic", IBM Research Center Report, RC9235.

[2] Epstein, C., Miranker, W. L., Rivlin, T. J. (1982). "Ultra-Arithmetic, Part I: Function Data Types", *Math. and Computers in Simulation,* XXIV 1-18.

[3] Epstein, C., Miranker, W. L., Rivlin, T. J. (1982). "Ultra-Arithmetic, Part II: Intervals of Polynomials", *Math. and Computers in Simulation,* XXIV 19-29.

[4] Kulisch, U. W. and Miranker, W. L. (1981). *Computer Arithmetic in Theory and Practice,* Academic Press, New York.

[5] Rivlin, T. J. (1970). "Bounds on a Polynomial", J. of Research of the N.B.S. - B. Math. Sci., **74B,** 47-54.

A FORTRAN EXTENSION FOR SCIENTIFIC COMPUTATION

C. P. Ullrich

Institute for Applied Mathematics
University of Karlsruhe
Karlsruhe, West Germany

In this paper we propose an extension of FORTRAN which satisfies contemporary requirements of numerical computation. The fundamental data types in computation are the integers, the real and complex numbers, the real segments (intervals) and complex segments as well as vectors and matrices defined over all of these. In our extended language, operands and operators for all these types are accepted as primitives in expressions. We give the description of language extension including the additional basic external and intrinsic functions for the new data types in a new powerful form of easily traceable syntax diagrams.

CONTENTS

1. MOTIVATION

The following explanations concerning the language extension are based on an IBM Research Report, which was prepared in March 1980 at the "Physik-Zentrum Bad Honnef", West Germany. In the meantime this report has also been published in Computing [2].

ISBN 0-12-428660-7

In that report, the authors intended to illustrate that higher programming languages do not satisfy current requirements of scientific computation. Moreover, a proposal for necessary extensions for a well known programming language was outlined. For this enterprise, the programming language FORTRAN was selected because of the dominant use of this language in technical scientific computation. Other programming languages seem to be better adapted to this purpose than FORTRAN, for example BASIC or APL (with their matrix-vector-package), PASCAL or PL/1 (due to their structured data types), ALGOL 68 or ADA (due to their operator concepts).

In contrast, FORTRAN [9] has few advantages. All numeric spaces, which must be available over the real and complex numbers according to column 3 of the following table (see [7], [8]), have to be realized by means of the data structure array.

1		2		3	4
		$\mathbb{R}$	$\supset$	R	$+ \; - \; * \; /$
					$\times$
		$V\mathbb{R}$	$\supset$	VR	$+ \; -$
					$\times$
		$M\mathbb{R}$	$\supset$	MR	$+ \; - \; *$
$\mathbb{P}\mathbb{R}$	$\supset$	$\mathbb{I}\mathbb{R}$	$\supset$	$\mathbb{I}R$	$+ \; - \; * \; /$
					$\times$
$\mathbb{P}V\mathbb{R}$	$\supset$	$\mathbb{I}V\mathbb{R}$	$\supset$	$\mathbb{I}VR$	$+ \; -$
					$\times$
$\mathbb{P}M\mathbb{R}$	$\supset$	$\mathbb{I}V\mathbb{R}$	$\supset$	$\mathbb{I}MR$	$+ \; - \; *$
		$\mathbb{C}$	$\supset$	C	$+ \; - \; * \; /$
					$\times$
		$V\mathbb{C}$	$\supset$	VC	$+ \; -$
					$\times$
		$M\mathbb{C}$	$\supset$	MC	$+ \; - \; *$
$\mathbb{P}\mathbb{C}$	$\supset$	$\mathbb{I}\mathbb{C}$	$\supset$	$\mathbb{I}C$	$+ \; - \; * \; /$
					$\times$
$\mathbb{P}V\mathbb{C}$	$\supset$	$\mathbb{I}V\mathbb{C}$	$\supset$	$\mathbb{I}VC$	$+ \; -$
					$\times$
$\mathbb{P}M\mathbb{C}$	$\supset$	$\mathbb{I}M\mathbb{C}$	$\supset$	$\mathbb{I}MC$	$+ \; - \; *$

The operations in these spaces, according to column 4, invariably require cumbersome subroutine calls. Therefore, a FORTRAN extension for scientific computation must provide all operations listed in column 4 as operators, and must have corresponding facilities for writing expressions in these data types. Thus, the computer implementation of the resulting extension of FORTRAN has many strong points, such as shorter and therefore more reliable programs, shorter compilation time, shorter running time in matrix and vector computations because roundings are avoided, error analysis by the computer, and so on.

Before we commence with our discussion of the language extension in detail we remark that the proposed extension is intended to serve as a model for language designs but also as a working plan for implementations. The authors themselves have not yet tried to implement it.

2. NOTATION OF THE LANGUAGE EXTENSION

We give an exact and clear description of the language extension from the point of view of syntax by means of so-called syntax diagrams. We employ this notation not only in its basic form [6] but also in a recently developed collective or selective form which contains what we denote by $\mathscr{T}$-diagrams. The compact form of these diagrams makes them especially easy to read and comprehend.

a) Simple notation: We consider, for example, the following diagram for a logical expression

F81: L EXPRESSION

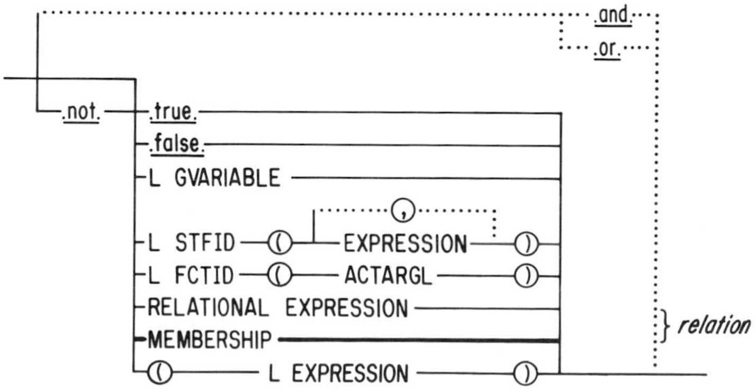

and explain how to read and interpret it:

- Syntax variables are written in upper case letters
- Terminals may be characters, sequences of characters or word symbols. Characters are enclosed in circles; sequences of characters in ovals; and word symbols are underlined.
- Solid lines are to be traversed from left to right and from top to bottom. Dotted lines are to be traversed oppositely.

b) A short notation for diagrams with identical structure ($\mathcal{T}$-diagrams)

F44: $\mathcal{T}$ PRIMARY (PRIMARY OF TYPE $\mathcal{T}$) $\mathcal{T}\in\{$I,R,D,C,S,CS$\}$

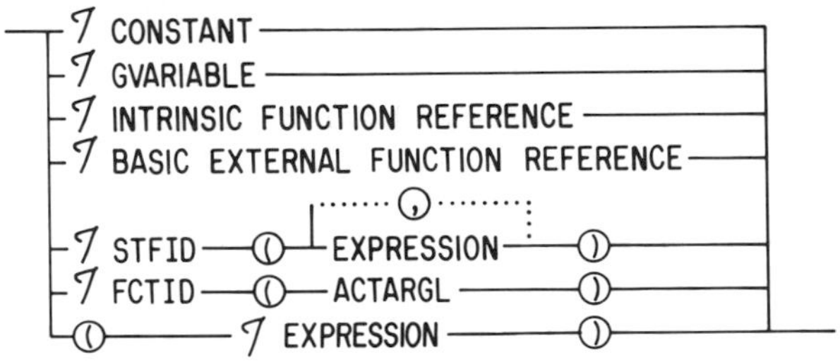

Identically structured diagrams, which only differ in type are represented by a so-called $\mathcal{T}$-diagram and thus determine a $\mathcal{T}$-syntax variable. For $\mathcal{T}\in\{I,$R,D,C,S,CS$\}$ the above diagram includes six diagrams in simple notation; for example, in the case $\mathcal{T} = I$, we have

F44: I PRIMARY (PRIMARY OF TYPE I)

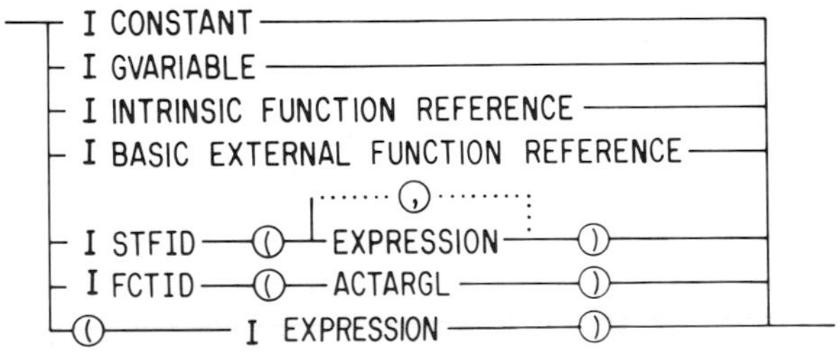

c) Collective diagrams: The $\mathcal{T}$-syntax variable can be applied as a tool for the efficient shortening of diagrams. In the following diagram:

F 30: EXPRESSION

—— $\mathcal{T}$ EXPRESSION ——

$$\mathcal{T} \epsilon \left\{ \begin{array}{l} I, R, D, C, \underline{S}, \underline{CS}, L, \\ \underline{VI}, \underline{VR}, \underline{VD}, \underline{VC}, \underline{VS}, \underline{VCS}, \underline{VL}, \\ \underline{MI}, \underline{MR}, \underline{MD}, \underline{MC}, \underline{MS}, \underline{MCS}, \underline{ML}, \\ \underline{TI}, \underline{TR}, \underline{TD}, \underline{TC}, \underline{TS}, \underline{TCS}, \underline{TL} \end{array} \right\}$$

$\underbrace{\qquad\qquad\qquad}_{\text{"COLLECTOR"}}$

the $\mathcal{T}$-syntax variable $\mathcal{T}$ EXPRESSION in connection with the corresponding "collector" constitutes the collection of 28 analogous diagram lines, which are only distinguished by the type substituted for $\mathcal{T}$.

```
┬─ I  EXPRESSION ┐
├─ R  EXPRESSION ┤
├─ D  EXPRESSION ┤
├─ S  EXPRESSION ┤
├─ CS EXPRESSION ┤
├─ L  EXPRESSION ┤
├─ VI EXPRESSION ┤
├─ VR EXPRESSION ┤
┊                ┊
└─ TL EXPRESSION ┴──
```

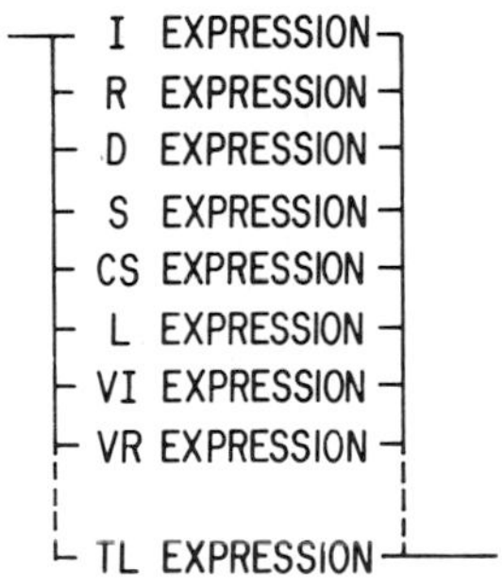

The depicted short notation can be applied repeatedly in a diagram. For example, the first line of the diagram:

F 22: ASSIGNMENT STATEMENT

$$\mathcal{R}, \mathcal{S} \epsilon \{I, R, D\}$$
$$\mathcal{T}_1 \epsilon \quad \{C, \underline{S}, \underline{CS}, L\}$$
$$\mathcal{T}_2 \epsilon \{I, R, D, C, S, CS, L\}$$

```
┬─ R       GVARIABLE ─────(=)───── S   EXPRESSION ─┐
├─ T₁      GVARIABLE ─────(=)───── T₁  EXPRESSION ─┤
├─ VT₂     GVARID ────────(=)──── VT₂  EXPRESSION ─┤
├─ MT₂     GVARID ────────(=)──── MT₂  EXPRESSION ─┤
├─ TT₂     GVARID ────────(=)──── TT₂  EXPRESSION ─┤
└─ assign ── LABEL ─ to ───────── I GVARID ────────┘
```

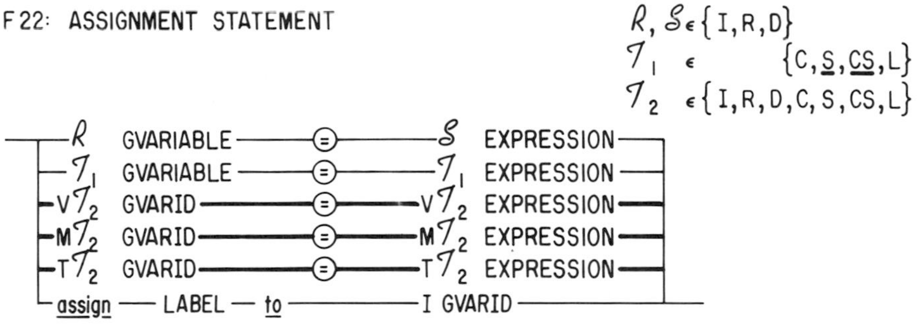

means the following diagram part in simple notation:

```
┬─ I GVARIABLE ┐
├─ R GVARIABLE ┤                  ┬─ I EXPRESSION ┐
└─ D GVARIABLE ┴──(=)──           ├─ R EXPRESSION ┤
                                  └─ D EXPRESSION ┴──
```

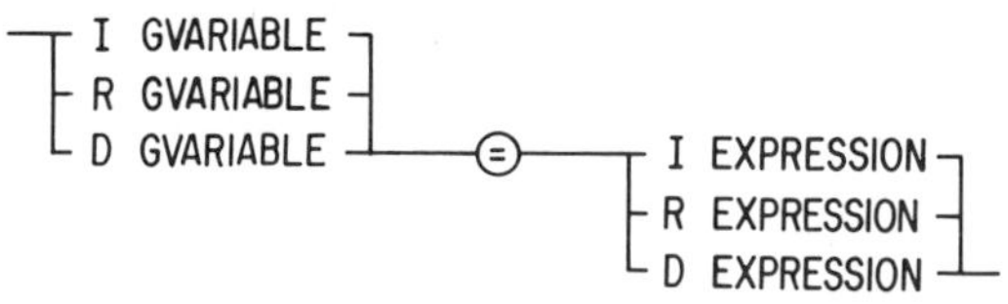

Thus, during the passage through a diagram each incidence of $\mathcal{T}$ must be replaced by a type chosen out of the corresponding collector. This rule can lead to different results if we consider a diagram including a recursion. Hence, by the diagram part:

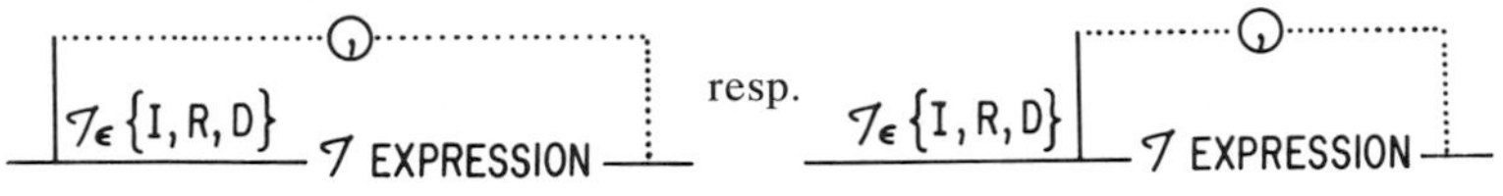

a list of expressions with possibly different types, or, resp., the same type is produced.

d) Selective $\mathcal{T}$-diagrams: This notation allows one to collect semantically analogous but syntactically different concepts of the language into one $\mathcal{T}$-diagram. We gather, for example, the following diagrams of the constants

F45: DSQ (DIGIT SEQUENCE)
I CONSTANT (I CN)

F46: R CONSTANT

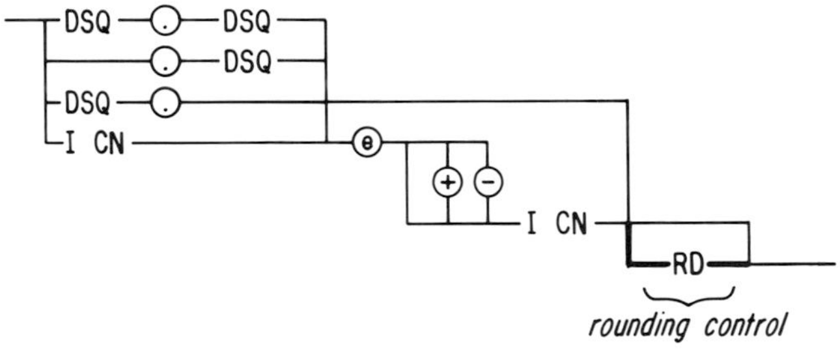

F47: D CONSTANT

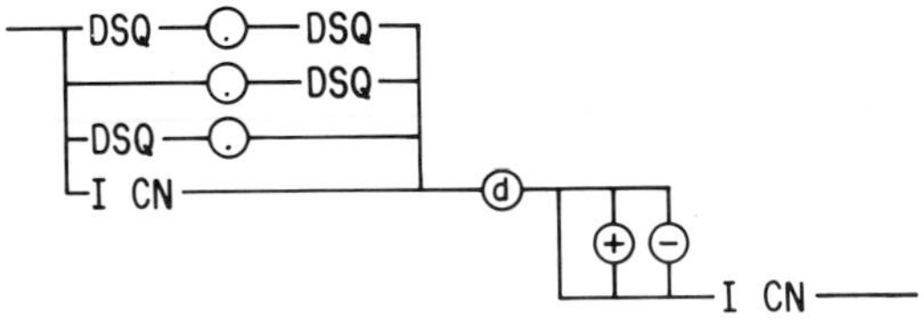

and so on, in a so-called selective $\mathcal{T}$-diagram:

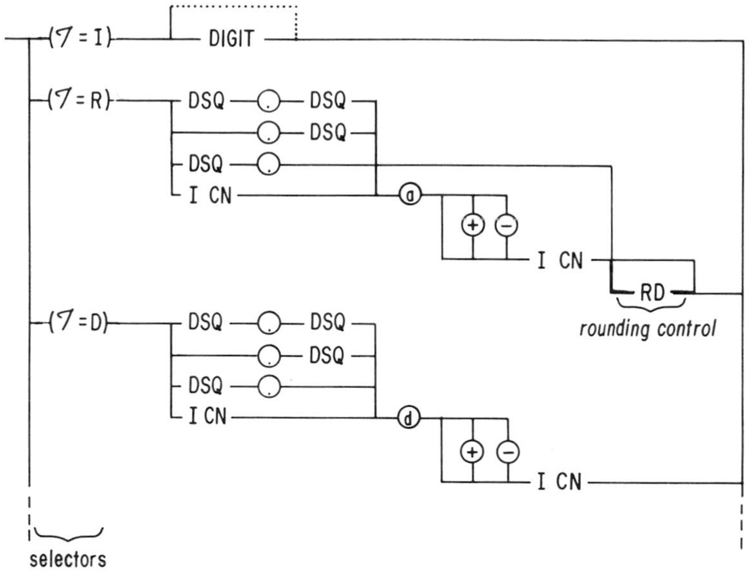

in these diagrams a selector (viz ($\mathscr{T} = I$), ($\mathscr{T} = R$), ($\mathscr{T} = D$), etc.) works as a barrier which permits passage only if the diagram is traversed for the types listed in the selector.

3. SYNTAX AND SEMANTICS OF THE EXTENSION

A complete and formal description of the extended language is given in [2]. In the following section we discuss the most important points of the extension in a more informal way. Nevertheless, we use syntax diagrams for the introduction of new language concepts, which are identified by bold lines or underlined types in $\mathscr{T}$-diagrams. Thus elimination of the bold line parts in all diagrams as well as the underlined types in $\mathscr{T}$-diagrams leaves the syntax of standard FORTRAN ([9]).

3.1 Character set

The additional characters ∇, $\triangle$, [,] are used. Since the characters ∇, $\triangle$ are unusual, ∇ and $\triangle$ may be represented by < and >, respectively.

The complete character set is displayed by the following diagram:

F94: CHARACTER

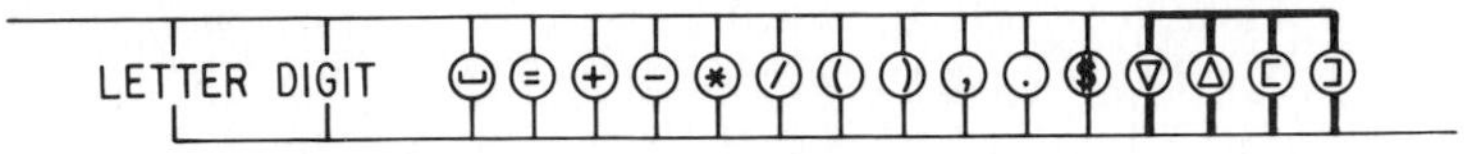

3.2 Data types

In addition to the computer representable integers I we consider five basic data types which are denoted by R,D,C,S and CS. These six types may be called integer, real, double precision, complex, segment and complex segment. The type segment resp. complex segment denotes the computer representable set of real resp. complex intervals (segments). Both types are not complemented with double precision counterparts.

Building on these six basic data types $\Phi \in \{I,R,D,C,S,CS\}$, we consider the following structured data types:

The sets $V\Phi$ of vectors with components in Φ,

$M\Phi$ of matrices with components in Φ,

$T\Phi$ of tensors with components in Φ.

A comparison of these data types with the spaces in column 3 of the table above yields the following correspondence:

spaces	$\mathbb{IR}$	$\mathbb{IVR}$	$\mathbb{IMR}$	$\mathbb{IC}$	$\mathbb{IVC}$	$\mathbb{IMC}$
data types	S	VS	MS	CS	VCS	MCS

Some of these corresponding spaces are formally different according to their construction, but the general theory of computer arithmetic shows that they are isomorphic and thus no confusion should result.

The following table displays all data structures of the extended language:

I	(VI)	(MI)	(TI)
R	VR	MR	(TR)
D	(VD)	(MD)	(TD)
C	VC	MC	(TC)
S	VS	MS	(TS)
CS	VCS	MCS	(TCS)
L	(VL)	(ML)	(TL)

The first column lists all basic data types including L for the type logical and the following columns display the vector, matrix and tensor types built on these. The language provides an arithmetic in the cases of the nonparenthesized data structures while variables, functions and expressions of a simple form are allowed for all data structures.

The declaration of the new data types (framed part of the table) must be realized by using new type identifiers:

F8: $\mathcal{T}$ TPID ($\mathcal{T}$ TYPE IDENTIFIER, TYPE IDENTIFIER FOR TYPE $\mathcal{T}$) $\mathcal{T} \in \{I, R, D, C, \underline{S}, \underline{CS}, L\}$

```
──┬─(𝒯 = I )── integer ──────────┐
  ├─(𝒯 = R )── real ─────────────┤
  ├─(𝒯 = D )── double precision ─┤
  ├─(𝒯 = C )── complex ──────────┤
  ├─(𝒯 = S )── segment ──────────┤
  ├─(𝒯 = CS)── complex segment ──┤
  └─(𝒯 = L )── logical ──────────┴──
```

Consequently, in connection with the declaration of arrays for these types, all needed data types are available.

Example: The declarations

complex c1, c2, c3

segment a, b

segment v(10), m(10,10)

complex segment w(100)

establish the complex variables c1, c2, c3, the real segments (intervals) a,b, the vector *v* with 10 segment components, the 10×10-matrix *m* with segment components and the vector *w* with 100 complex segment components.

Note that a complex segment entity is counted as four and a segment entity as two logically consecutive storage units, resp.

3.3 Constants

For the new data types S and CS constants are provided by analogy to their mathematical notation:

F49: S CONSTANT

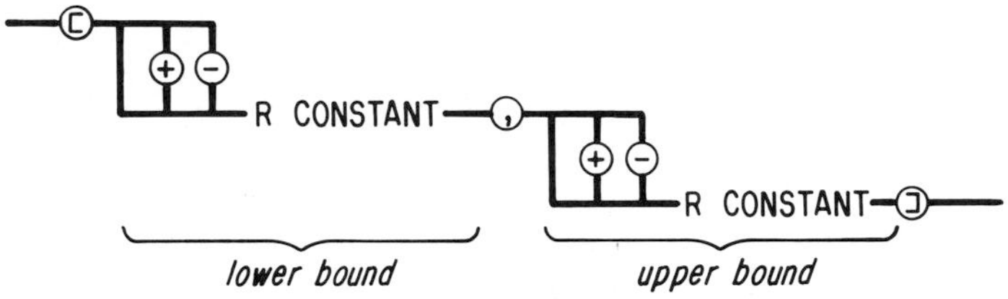

F50: CS CONSTANT

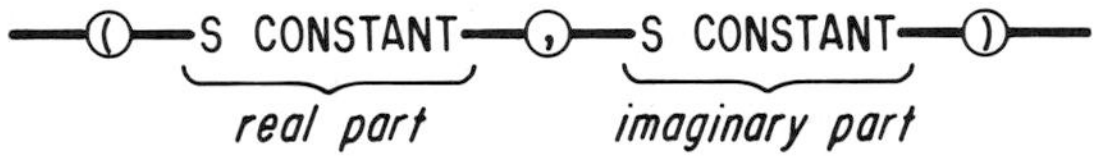

The definition of the segment constant and hence of the complex segment constant is based on the real constant. The correct rounding of a segment constant, i.e. the rounding of its bounds, requires the rounding control for real constants. For that purpose the sign ∇, resp. Δ, is appended to the real constant, if it is to be rounded downwardly or, resp., upwardly.

F46: R CONSTANT

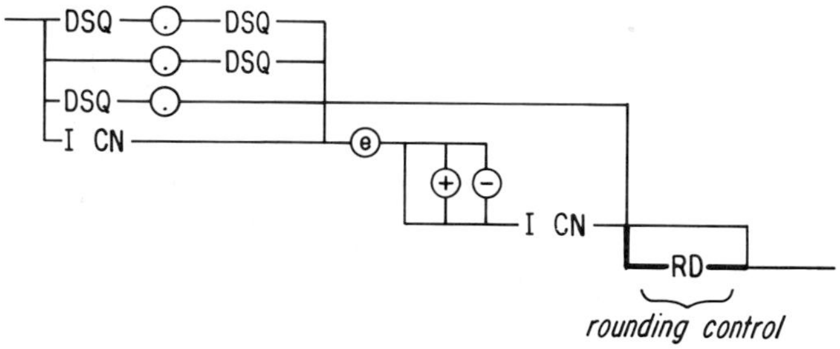

F45: DSQ (DIGIT SEQUENCE)
I CONSTANT (I CN)

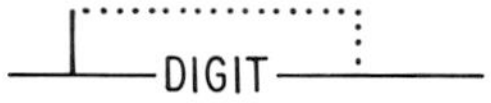

F80: RD (ROUNDING)

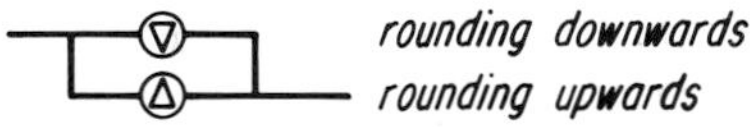

It the rounding control is missing, the implicit rounding □ is applied.

We observe that the rounding operates on the signless real constant. Therefore, the rounding control for a constant giving the bounds of a segment constant is to be determined corresponding to the sign of the bounds.

	no sign	+	−
lower bound	▽	▽	△
upper bound	△	△	▽

Examples:

$5.37625e - 10\ \triangledown$

$[-1.823\ e + 2\ \triangle, -2.376\ e - 2\ \triangledown]$

$([+37.5\ \triangledown,\ 61.47\ \triangle],\ [-1.592\ \triangle,\ 0.13\ \triangle])$

3.4 Input/Output

In addition to a correct representation of the constants used in expressions rounded computation also requires a correct conversion of the values read at input and written at output. That conversion is affected exclusively by an extension of the format statement, while the

F17: FIELD DESCRIPTOR

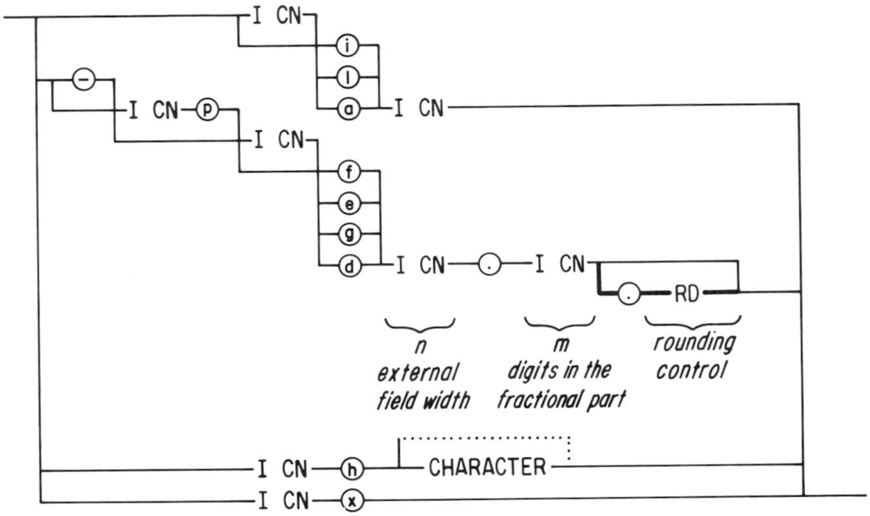

representation of the input and output data remains unchanged. Again, the rounding control ∇, and $\triangle$, effects the downwardly, resp. upwardly, directed rounding, if the rounding control is missing the implicit rounding $\square$ is applied.

For reading and writing values of the new data types, additional field descriptors are not necessary. We only note that two consecutive real field descriptors (the first one with rounding control ∇, the second one with rounding control $\triangle$) are required for a real segment value and analogously, four consecutive real field descriptors are required for a complex segment value.

Examples:

> dimension a(100)
>
> segment sa, sb, vs(10)
>
> complex segment cs1, cs2
>
> 100 format (1h⊔,f10.5.△,10e12.5.▽,g15.8)
>
> write (6,100) x, (a(i),i = 1,10), y
>
> 150 format (1h⊔,12(1h[,e12.5.▽,1h,,e12.5.△,1h]))
>
> write (6,150) sa, sb, vs
>
> 200 format (1h⊔,2(2h(⊔,2(1h[,e12.5.▽,1h,,e12.5.△,1h]),2h⊔)))
>
> write (6,200) cs1, cs2

3.5 Expressions

The following diagram gives an overview of the admissible expressions:

F30: EXPRESSION

── 𝒯 EXPRESSION ──

$\mathcal{T} \in \{$ I, R, D, C, S̲, C̲S̲, L,
V̲I̲,V̲R̲,V̲D̲,V̲C̲,V̲S̲,V̲C̲S̲,V̲L̲,
M̲I̲,M̲R̲,M̲D̲,M̲C̲,M̲S̲,M̲C̲S̲,M̲L̲,
T̲I̲,T̲R̲,T̲D̲,T̲C̲,T̲S̲,T̲C̲S̲,T̲L̲ $\}$

In the cases of vectors and matrices over the data types integer, double precision, and logical and of the tensors over all basic data types, only very simple expressions are allowed:

F31: 𝒯 EXPRESSION (EXPRESSION OF TYPE 𝒯) $\mathcal{T} \in \{$ VI, VD,   VL,
MI, MD,   ML,
TI,TR,TD,TC,TS,TCS,TL $\}$

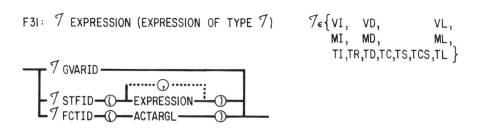

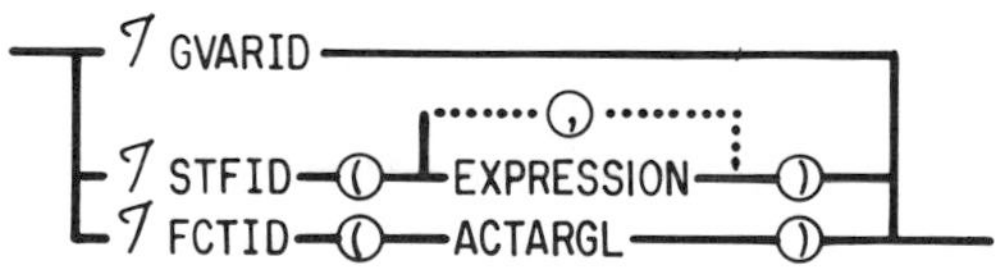

Such expressions consist of one operand, either a variable or a function call.

In the remaining cases, expressions can occur having a great variety of operations, which are added to the operations available in standard FORTRAN. We present all operations in the two following tables. For each operator we list the permitted types of operands and the result type, for example, the 2nd line in the 1st subdivision of the second table has the meaning, that the dyadic operators $**\nabla$, $**\triangle$ can be applied to two real operands, but also to a first operand of type real and a second operand of type integer. In both cases the result type is real. The rounding control ∇, resp. $\triangle$, behind the operator sign $**$ means that the result value has to be computed from the exact result by the downwardly, resp. upwardly, directed rounding.

In addition to the well known operators $+, -, *, /$ and $**$, there is a new operator .ch., resp. is., which evaluates the convex hull, resp. the intersection, of the two interval operands. Moreover, the new relation operator .in. delivers the value true. if the value of the first operand is an element of the second operand (corresponding segment expression) and it delivers the value .false. otherwise.

Monadic operators (inversion): $\mathscr{T} \in \{S,CS,VR,VC,VS,VCS,MR,MC,MS,MCS\}$

result type	operator	operand of type
$\mathscr{T}$	+,−	$\mathscr{T}$

Dyadic operators:

result type	operator	operands of type
R	+▽,−▽,*▽,/▽,+△,−△,*△,/△ **▽,**△	R×R R×I,R×R
C	+▽,−▽,*▽,/▽,+△,−△,*△,/△ **▽,**△	R×C,C×R,C×C C×I
S	+,−,*,/,.ch., .is. **	S×I,I×S,S×S S×I
CS	+,−,*,/,.ch., .is. **	S×CS,CS×S,CS×CS CS×I
VR	+,−,+▽,−▽,+△,−△ *,*▽,*△	VR×VR R×VR,MR×VR
VC	+,−,+▽,−▽,+△,−△ *,*▽,*△	VC×VC C×VC,MC×VC
VS	+,−,.ch., .is. *	VS×VS S×VS,MS×VS
VCS	+,−,.ch., .is. *	VCS×VCS CS×VCS,MCS×VCS
MR	+,−,+▽,−▽,+△,−△ *,*▽,*△	MR×MR R×MR,MR×MR
MC	+,−,+▽,−▽,+△,−△ *,*▽,*△	MC×MC C×MC,MC×MC
MS	+,−,.ch., .is. *	MS×MS S×MS,MS×MS
MCS	+,−,.ch., .is. *	MCS×MCS CS×MCS,MCS×MCS
L	.lt.,.le.,.eq.,.ne., .gt.,.ge. .in.	$\mathscr{T}\times\mathscr{T}$ $\mathscr{T}\in$ {C,S,CS,VI,VR,VD,VC,VS,VCS, MI,MR,MD,MC,MS,MCS, TI,TR,TD,TC,TS,TCS] I,R,D×S,C×CS VI,VR,VD×VS,VC×VCS MI,MR,MD×MS,MC×MCS TI,TR,TD×TS,TC×TCS

A correct and detailed definition of the syntax of the corresponding expressions is given by the following diagrams.

F32: I EXPRESSION

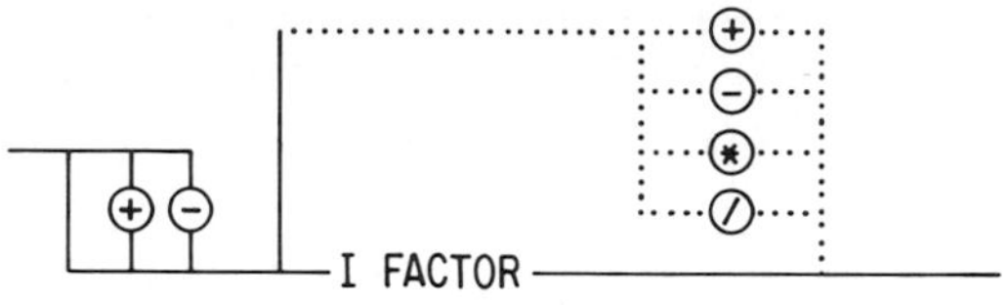

F33: R EXPRESSION

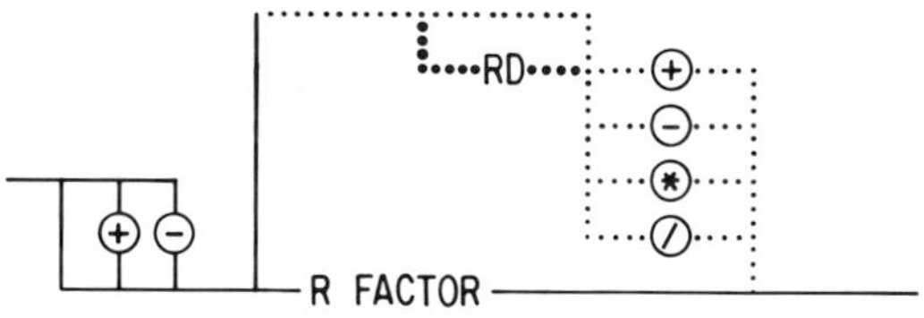

F34: D EXPRESSION

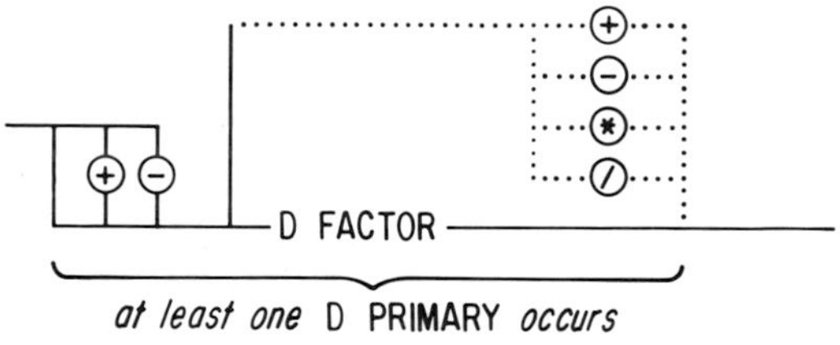

F35: C EXPRESSION

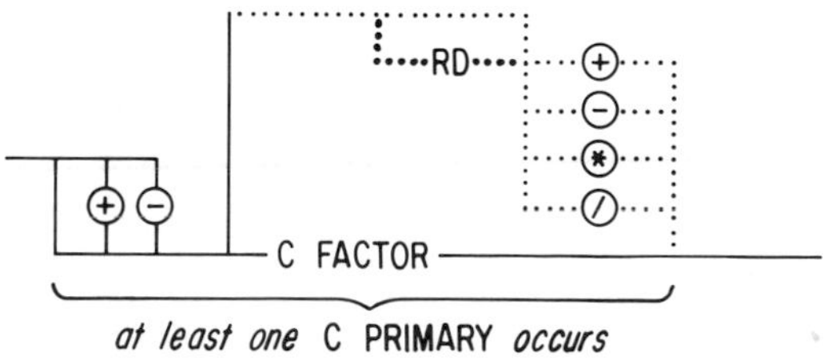

F36: τ EXPRESSION $\tau \in \{S, CS\}$

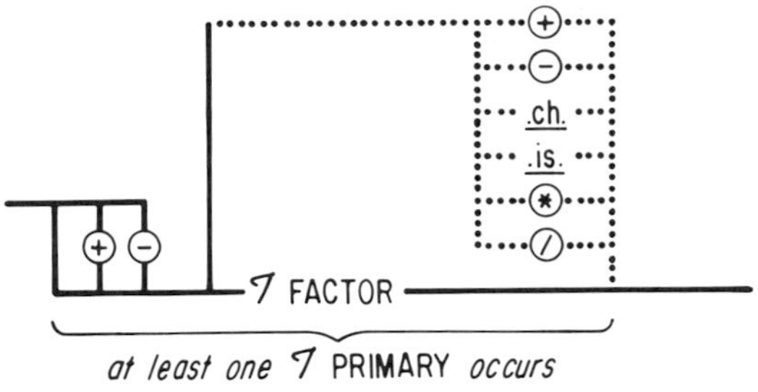

F38: I FACTOR

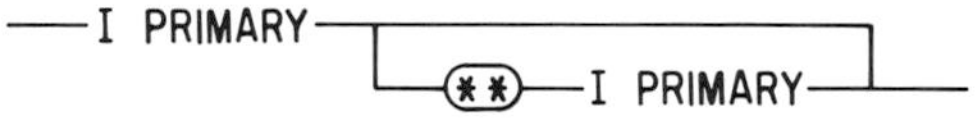

F39: R FACTOR

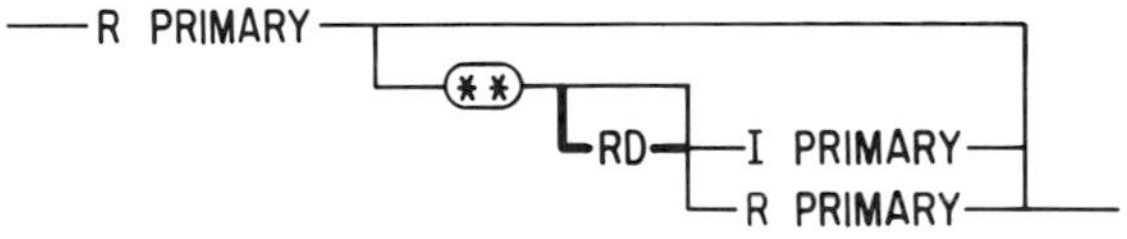

F40: D FACTOR

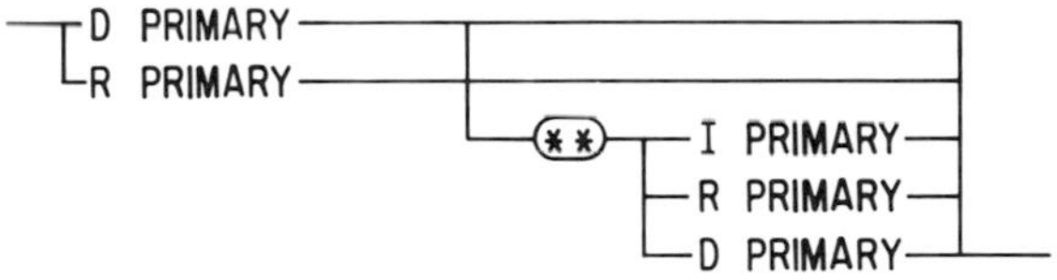

F41: C FACTOR

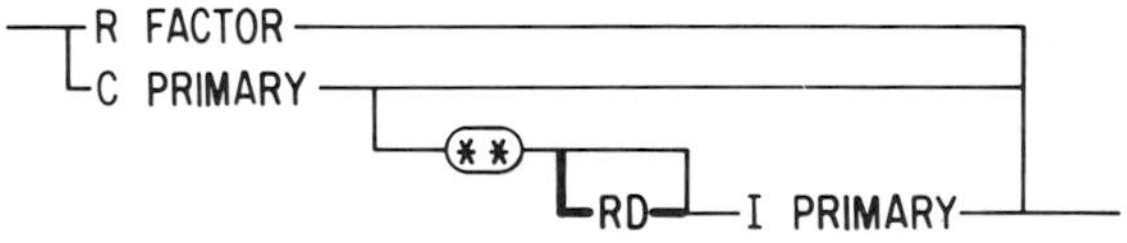

F42: τ FACTOR $\tau \in \{S, CS\}$

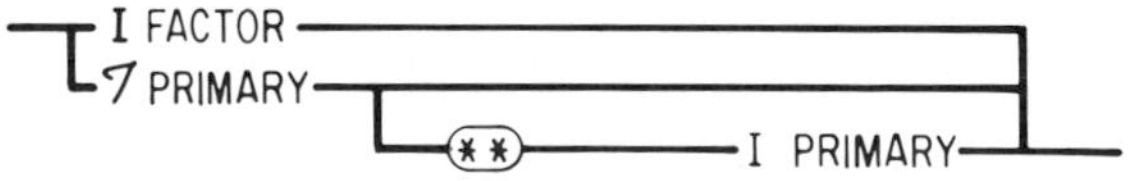

F44: 𝒯 PRIMARY (PRIMARY OF TYPE 𝒯) $\mathcal{T} \in \{I, R, D, C, \underline{S}, \underline{CS}\}$

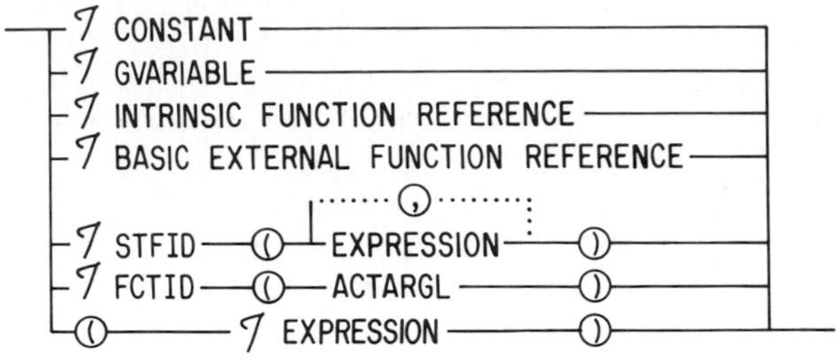

F63: V𝒯 EXPRESSION $\mathcal{T} \in \{R, C\}$

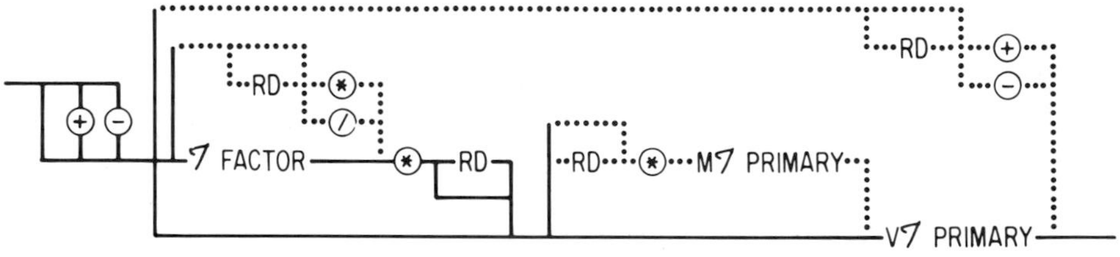

F65: V𝒯 EXPRESSION $\mathcal{T} \in \{S, CS\}$

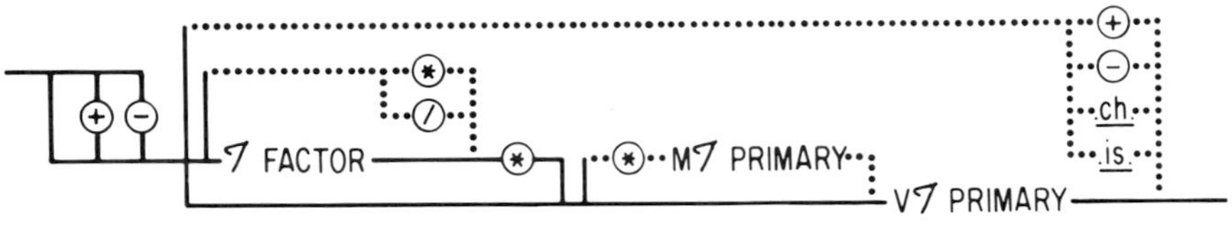

F67: M𝒯 EXPRESSION $\mathcal{T} \in \{R, C\}$

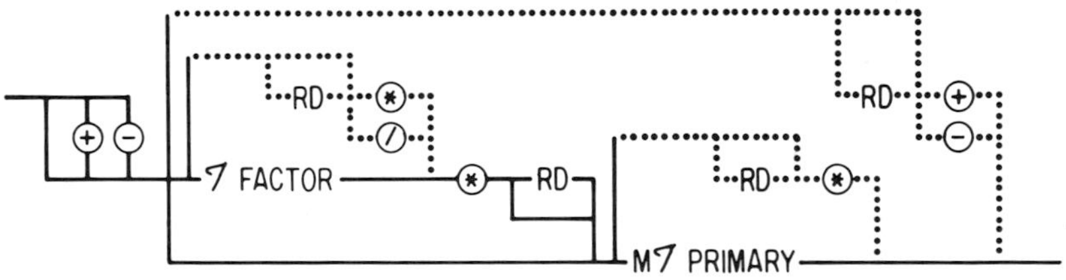

F69: M𝒯 EXPRESSION $\mathcal{T} \in \{S, CS\}$

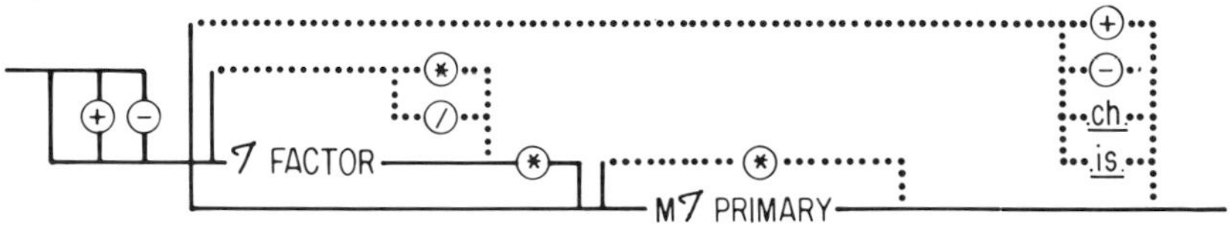

F71: $\mathcal{T}$ PRIMARY (PRIMARY OF TYPE $\mathcal{T}$) $\mathcal{T} \in$ {VR, VC, VS, VCS, MR, MC, MS, MCS}

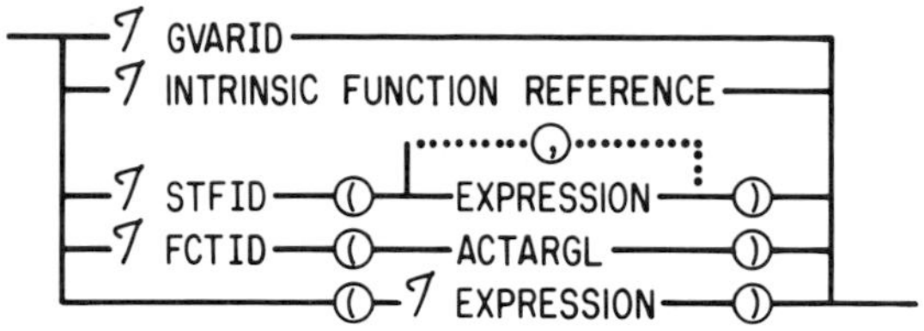

F81: L EXPRESSION

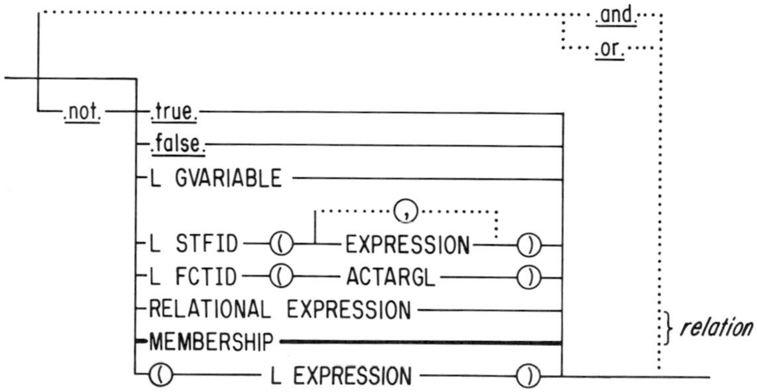

F82: RELATIONAL EXPRESSION $\mathcal{T} \in$ { I, R, D, C, S, CS, VI, VR, VD, VC, VS, VCS, MI, MR, MD, MC, MS, MCS, TI, TR, TD, TC, TS, TCS }

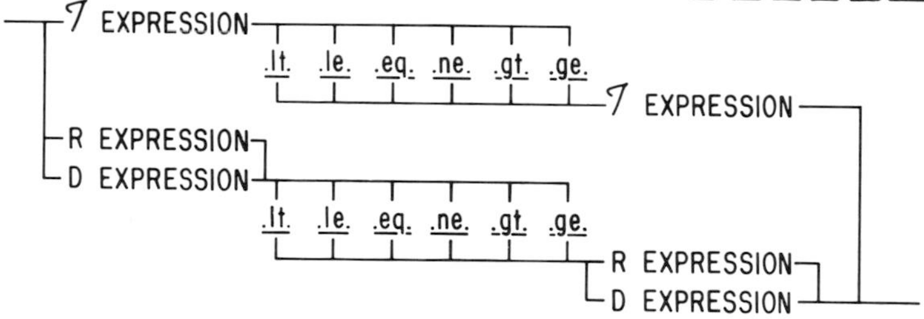

F83: MEMBERSHIP $\mathcal{T} \in$ {I, R, D}

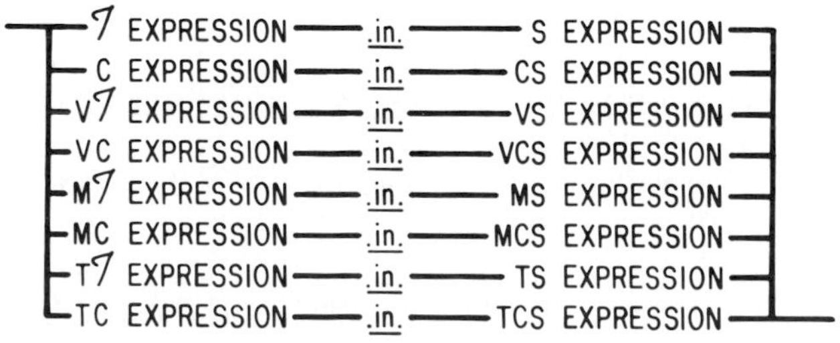

3.6 Functions

The extension of the concept for expressions to the type segment resp. complex segment and to vectors, matrices, tensors over all basic data types leads to the need for functions with arbitrary result type. We obtain functions either as statement functions (F12) or as external functions (F3).

The declaration of a statement function is applied nearly unchanged in the case of vector, matrix and tensor functions:

F12: STFUNCTION (STATEMENT FUNCTION)

$\mathcal{R}, \mathcal{S} \in \{I,R,D\}$
$\mathcal{T}_1 \in \{C,\underline{S},\underline{CS},L\}$
$\mathcal{T}_2 \in \{I,R,D,C,S,CS,L\}$

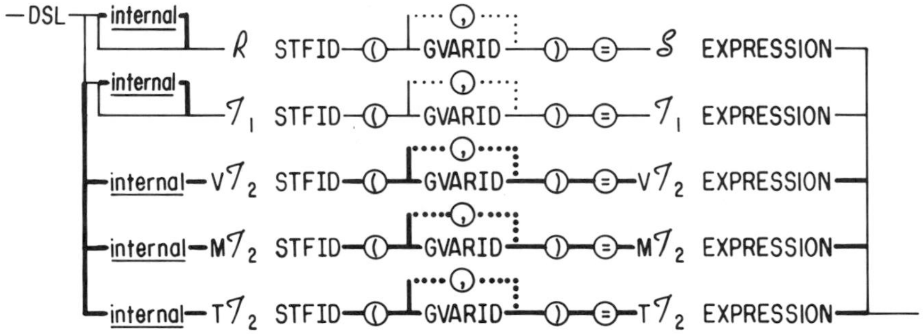

The new word symbol internal simplifies the syntactical analysis of this element and prevents ambiguities. (Otherwise, confusion with the assignment statement for array elements would occur.) The declaration of an external function, which produces a vector, matrix or tensor value, commences with a specification of the result type.

F3: SUBPROGRAM

$\mathcal{T} \in \{I, R, D, C, \underline{S}, \underline{CS}, L\}$

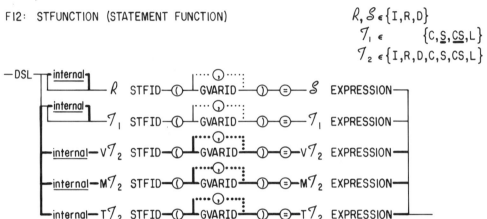

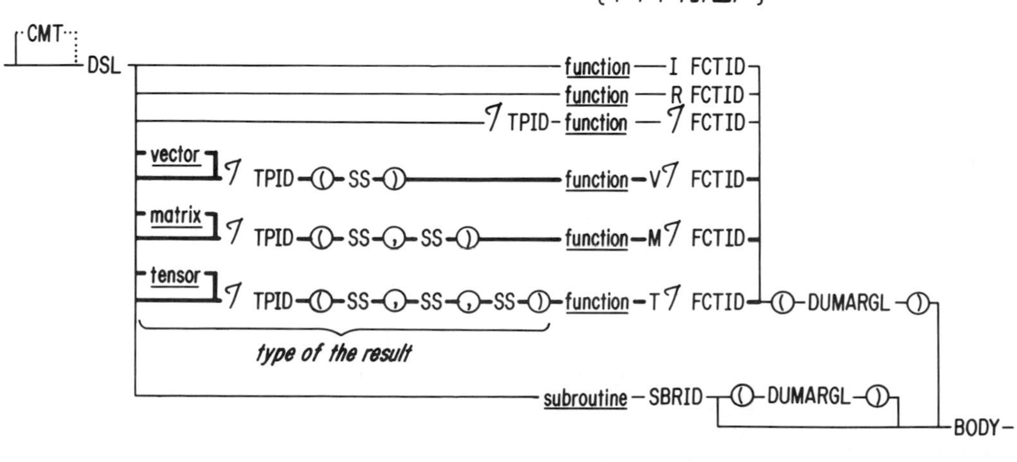

Constants and variables of type integer can both be inserted as index bounds, where variables must be taken from the dummy argument list. When the function is called, such arguments are used for computing the length of the result value before computing the function.

Example:

vector real (n) function eigvec (n,a,y)

dimension y(n), a(n,n)

do 1 i=1,n

... .

In the using program unit type and kind of the result value of a function are specified by a specification statement:

F7: SPECIFICATION (SPECIFICATION STATEMENT) $\tau \in \{$I,R,D,C,S,CS,L$\}$

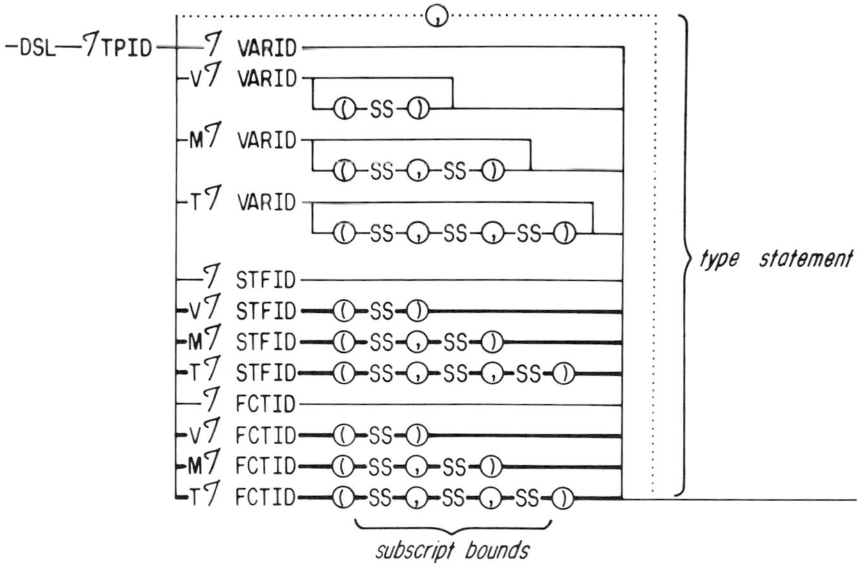

Some functions which are useful for the new data types are introduced as additional basic functions. The following table presents 56 new intrinsic functions which are ordered according to their result type. In line 1 of the 2nd subdivision of the table, for example, we find five different intrinsic functions whose identifier contains the word real. The complete identifier is taken by appending real to the corresponding result type S, VR, MR, VS and MS resp. (i.e., sreal, v(r)real, m(r)real, vsreal and ms real resp.). This rule for the identifiers does not apply to functions with result type <u>real</u>. Their identifiers are signldwn, singlup, inf and sup. (Another way for the identification of these functions would be the use of overloaded identifiers and the distinction according to the argument and result types.)

R	C	S	CS	VR	MR	VC	MC	VS	MS	VCS	MCS	result type / identifier
D												singldwn
D												singlup
		CS		VC	MC			VCS	MCS			real
		CS		VC	MC			VCS	MCS			imag
S	CS			VS	MS	VCS	MCS					inf
S	CS			VS	MS	VCS	MCS					sup
			S,S			VR,VR	MR,MR			VS,VS	MS,MS	complx
		R,R	C,C					VR,VR	MR,MR	VC,VC	MC,MC	sgmt
				I	I	I	I	I	I	I	I	null
					I		I		I		I	id
			CS			VC	MC			VCS	MCS	conj
					MR				MS			transp
							MC				MCS	herm

The permitted argument list of a function is given by the enumeration of the argument types. In the following list we state the meanings of all the identifiers from which function identifiers can be composed:

singldwn, singlup, resp.: conversion from double precision to real (rounded downwardly resp. upwardly)

real, imag: real resp. imaginary part

inf, sup: infinum resp. supremum

complex: composes a complex value from two real values

sgmt: composes a segment from two values

null: vector resp. matrix of zeros

id: identity matrix

conj: takes the conjugate of the complex value

transp: transposes the matrix

herm: gives the conjugate transpose of the matrix

Analogously, we describe 22 new external functions:

random: produces an integer chosen at random from the interval specified by two integer expressions

abs: largest absolute value in the segment expression

dot, dotdwn, dotup resp.: scalar product of two vector expressions with implicit rounding, rounded downwardly or rounded upwardly, resp.

exp, log, log10, sin, cos, tanh, atan, sqrt: segment, resp, complex segment, versions of the corresponding real functions. (These functions have to be implemented optimally, i.e., they have to compute the smallest segment containing the exact value.)

Note the following exceptions concerning the function identifiers: sabs and dot, dotdwn, dotup, resp. in the case of result type real.

I	R	C	S	CS	result type / identifier
I,I					random
	S				abs
	VR,VR	VC,VC	VS,VS	VCS,VCS	dot
	VR,VR	VC,VC			dotdwn
	VR,VR	VC,VC			dotup
			S	CS	exp
			S	CS	log
			S		log10
			S	CS	sin
			S	CS	cos
			S		tanh
			S		atan
			S,S		atan2
			S	CS	sqrt

Example:

The following program is a short demonstration of the possibilities in the extended language. It computes the solution of the interval system $x = Ax + B$ of linear equations by the Jacobi method. (The function krit proves the convergence property of A.)

```
c       main program linear system

        segment x(50), y(50), b(50), a(50,50)

        read (5,100) a,b.y

100     format ( )

        if (.not. krit (a)) goto 2000

1000    x=y

        y=(a*x+b) .is. x

        if (x .ne. y) goto 1000

        write (6,100) x

2000    stop

        end
```

REFERENCES

[1] Bohlender, G. (1978). Genaue Berechnung mehrfacher Summen, Produkte und Wurzeln von Gleitkommazahlen und allgemeine Arithmetik in höheren Programmiersprachen, Dissertation, Universität Karlsruhe.

[2] Bohlender, G., E. Kaucher, R. Klatte, U. W. Kulisch, W. L. Miranker, Ch. Ullrich and J. Wolff v. Gudenberg. (1980). FORTRAN for Contemporary Numerical Computation, IBM Research Report RC 8348, and Computing 26, 277-314 (1981).

[3] Kaucher, E., Klatte, R., Ullrich, Ch. (1978). Benutzerfreundliche Darstellung der Syntax von *PASCAL* durch Syntaxdiagramme, Applied Computer Science, Berichte zur praktischen Informatik 11, 43-62, Hanser-Verlag, München.

[4] Kaucher, E., Klatte, R., Ullrich, Ch. (1978). Neuere Methoden zur Beschreibung von Programmiersprachen, Jahrbuch Überblicke Mathematik, Bibliographisches Institut, Mannheim.

[5] Kaucher, E., Klatte, R., Ullrich, Ch. (1978). Höhere Programmiersprachen ALGOL, FORTRAN, PASCAL in einheitlicher und übersichtlicher Darstellung, Reihe Informatik, Band 24, Wissenschaftsverlag des Bibliographischen Instituts Mannheim.

[6] Kaucher, E., Klatte, R., Ullrich, Ch. (1980). Programmiersprachen im Griff, Band 1. FORTRAN, Bibliographisches Institut, Mannheim,

[7] Kulisch, U. (1976). Grundlagen des Numerischen Rechnens-Mathematische Begründung der Rechnerarithmetik, Reihe Informatik, Band 19, Wissenschaftsverlag des Bibliographischen Instituts Mannheim.

[8] Kulisch, U., Miranker, W. L. (1980). Computer Arithmetic in Theory and Practice, Academic Press.

[9] ISO (International Organization for Standardization), ISO/R 1539 (July 1972).

AN INTRODUCTION TO MATRIX PASCAL:
A PASCAL EXTENSION FOR SCIENTIFIC COMPUTATION

J. Wolff von Gudenberg

Institute for Applied Mathematics
University of Karlsruhe
Karlsruhe, West Germany

To be a useful tool in scientific computation a programming language should provide an operator concept for the fundamental numeric data types of scientific computation. Such types include real and complex numbers, real and complex intervals as well as vectors and matrices over all of these. To this end PASCAL has been extended by means of a universal operator concept as well as a concept for functions with arbitrary result type. New standard types, operators, functions and procedures have been added to the language. In this paper we introduce only those parts of the extended language which are needed to employ and understand the operator and function concept for structured data types. In part A we introduce the new standard data types. In part B we deal with the evaluation of expressions and present the new standard operators. An introduction to the operator declaration is given in part C. We then describe the universal operator concept which allows the user to define his own operators and functions appropriate to his particular problem. The concept of maximum accuracy is extended to the evaluation of expressions containing general scalar products. The corresponding operands are presented in part E. We finally consider the problem to achieve maximum accuracy in evaluating the standard functions for all scalar types.

The description is given by means of easily traceable syntax diagrams. Many instructive examples are included.

A. DATA TYPES

In the programming language PASCAL, all the entities used in the program have to be declared. It is possible and sometimes necessary for the user to define and name his own types.[†]

P5 : TYPE DEFINITION (TYPEDEF)

$$\theta = \{ I,\ B,\ CH,\ CD,\ R,\ \underline{C},\ \underline{S},\ \underline{CS},\ P,\ ST,\ A,\ REC,\ SET \}$$

$$\mathcal{T} \epsilon \theta \cup \{F, TF\}$$ — $\mathcal{T}$ TYPE IDENTIFIER —(=)— $\mathcal{T}$ TYPE —

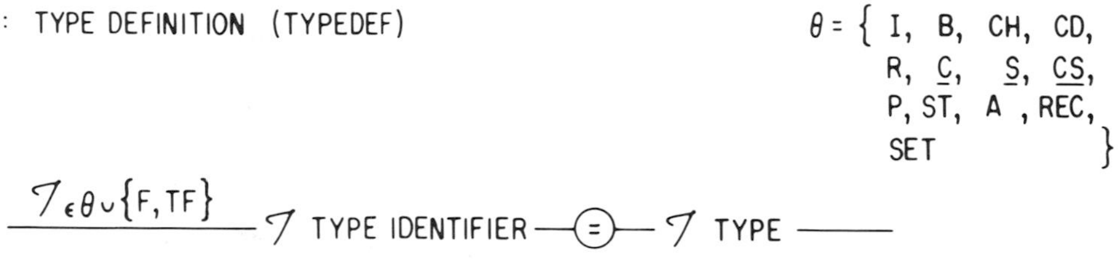

In addition to the base types, one may define subrange types of ordinal types.

[†] The diagrams employed here are taken from [2].

P7 : $\mathcal{T}$ ORDINAL TYPE $\qquad \mathcal{T} \in \Theta_{ORD} = \{I, B, CH, CD\}$

ORDTYPE

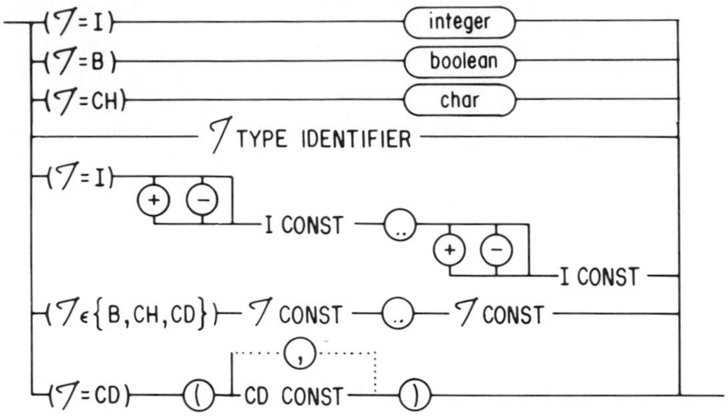

Example

<u>type</u> number = integer;

index = 1..10;

We consider two ways of defining structured types in standard PASCAL:

- arrays
- records

An array describes a fixed number of entities of the same base types. The single elements are addressed by one or more indices.

Example

<u>type</u> index = 1..10;

bow = <u>array</u>[3..15] <u>of</u> boolean;

mat = <u>array</u>[index,index] <u>of</u> real;

With <u>var</u> *a*: mat; *f*: bow;

f[4] denotes the second element of *f* and

a[3,4] the fourth element in the third row of *a*.

A record describes a fixed number of entities of different base types. The particular components are addressed by their names.

Example: type money = record dollar, cent: integer; end;

car = record model: array[1..20] of char;

year: integer;

price: money

end;

With var p: car; p.model[3] denotes the third letter of the model name, p.year is an integer value as well as p.price.dollar.

P6 : τ TYPE $\quad \tau \in \theta \cup \{F, TF\}$, $\theta = \{I, B, CH, CD, R, \underline{C}, \underline{S}, \underline{CS}, P, ST, A, REC, SET\}$ $\quad \theta_{ORD} = \{I, B, CH, CD\}$

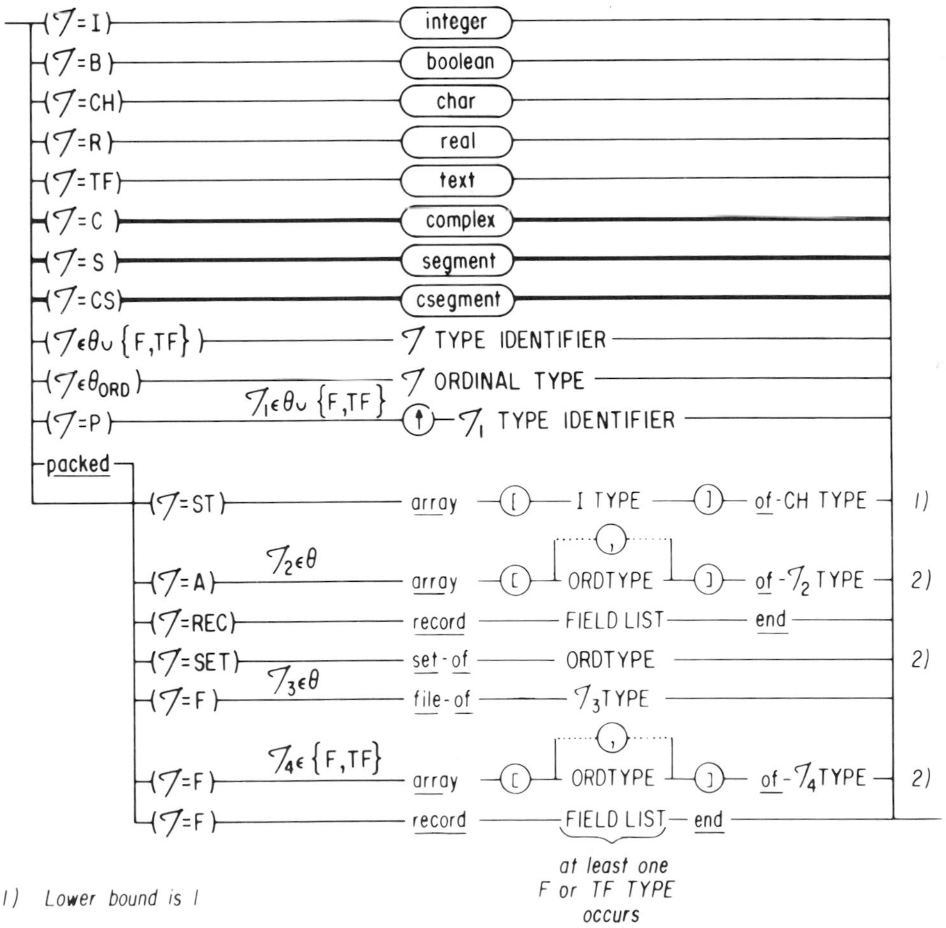

1) Lower bound is 1

2) Range of the ordinal type is implementation dependent

P8 : FIELD LIST (FL)

$\theta = \{$ I, B, CH, CD, R, $\underline{C}$, $\underline{S}$, $\underline{CS}$, P, ST, A, REC, SET $\}$

$\theta_{ORD} = \{$I, B, CH, CD$\}$

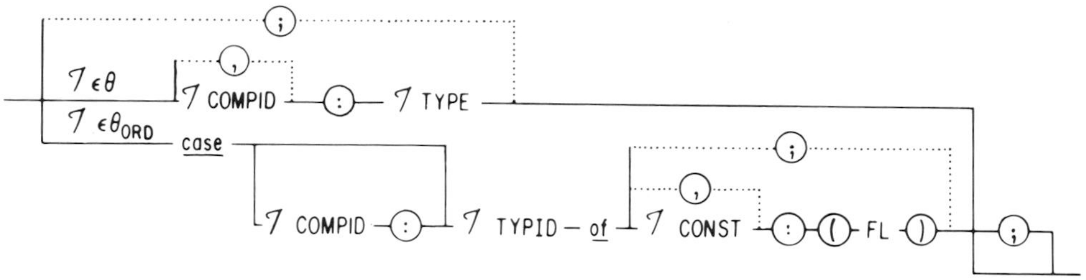

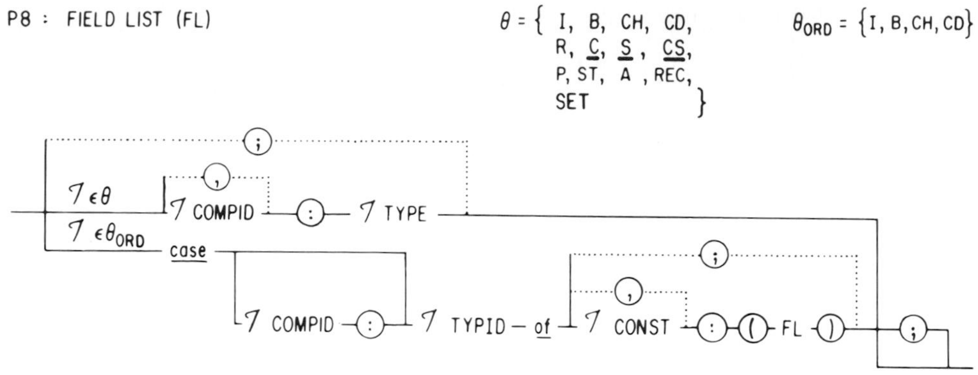

Whereas standard PASCAL provides only for arrays with fixed index bounds, in MATRIX PASCAL three different kinds of dynamic arrays, such as vector, matrix and diagonal may be used. The index type of these dynamic types is integer, the index bounds may be evaluated at running time but must be known when variables of these types are declared. Therefore, in the main program the index bounds may only contain constants whereas in procedures formal parameters and global entities may occur. Every non-dynamic type except file may occur as component type of a dynamic type.

P13: 7 DYNAMIC TYPE (7 DYNTYPE)

$7 \in \theta_{DYN} = \{$VEC, MAT, DIAG$\}$

$\theta = \{$I, B, CH, CD, R, C, S, CS, P, ST, A, REC, SET$\}$

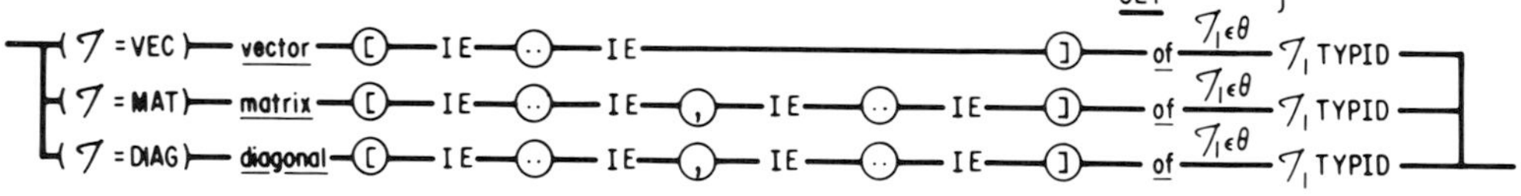

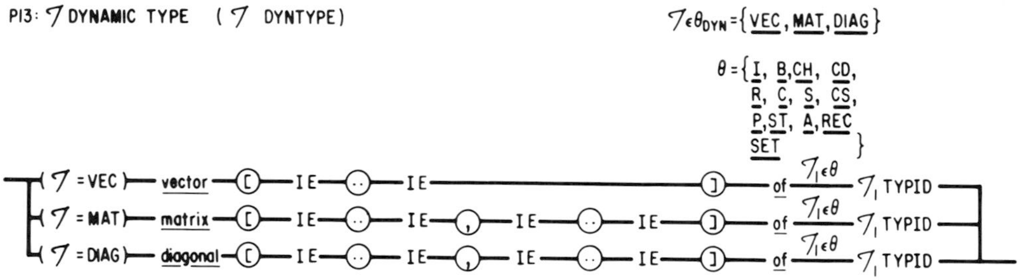

Using the type concept of MATRIX PASCAL, we can define the new numerical data types listed in the following table:

Datatypes for Scientific Computation

$\mathcal{T}$	name	definition		
CH	char			
B	boolean			*standard*
I	integer			*data*
R	real			*types*
S	segment	record	inf, sup: real end	
C	complex	record	re, im: real end	
CS	csegment	record	re, im: segment end	
VR		vector	[1..*m*] of real;	
MR		matrix	[1..*m*,1..*m*] of real;	
VC		vector	[1..*m*] of complex;	*user defined*
MC		matrix	[1..*m*,1..m] of complex;	*data*
VS		vector	[1..*m*] of segment;	*types*
MS		matrix	[1..*m*,1..*m*] of segment;	
VCS		vector	[1..*m*] of csegment;	
MCS		matrix	[1..*m*,1..*m*] of csegment;	

In general, variables are only assignment compatible when their type names are identical or defined to be equal or if the variables appear in the same variable declaration. As an exception to this rule note that integer is assignment compatible with real.

For the dynamic types the following compatibility rules hold: Rows and columns of variables of type matrix and diagonal are of type vector. Rows can be addressed directly, for instance, as $a[i]$, columns as $a[*,i]$.

Variables of the types vector resp. matrix resp. diagonal are assignment compatible if they have the same component type and are of the same size.

The types vector, matrix and diagonal are not permitted as component types of structured data types (array, record and file).

The type diagonal is predefined as a special matrix type, variables of this type represent diagonal matrices. They are to be declared, for instance, as follows:

var k: diagonal [1..m,1..m] of segment;

Only the diagonal components for matrices of type diagonal are kept in storage.

B. EXPRESSIONS

For scientific computation we need floating-point operations with a monotone, antisymmetric rounding as well as with the upwardly or downwardly directed rounding, resp. We also need operators for the operations in the higher numerical spaces. The standard operators of MATRIX PASCAL are displayed in the following diagram.

P11 𝒯 OPERATOR

=, <>, <, <=, >, >=, in | ><	comparing	operators
+, −, or | +>, +<, −>, −<, +*	adding	
*, /, mod, div, and | *>, *<, />, /<, **	multiplying	
+, −, not	unary	

The additional operators are listed to the right of the vertical bar. The operators +<, +>, −<, −>, *<, *>, /<, /> denote the arithmetic operations with rounding downwardly (<) resp. rounding upwardly (>). The operators ><, +*, ** are provided for segment arithmetic. >< denotes the operator "disjoint", +* the convex hull and ** the intersection of two segments.

The four rows in the above operator table express the 4 levels of priority, i.e., comparison operators have lowest priority; unary the highest.

Examples

1) if (3* − x in $a + b$) or (a><b) then a:= a**c

corresponds to the statement

if $3*(-x) \in a + b$ or $a \cap b = \emptyset$ then $a := a \cap c$

where a,b,c denote intervals and x a real number

2)

$\ell := 4 * < s / < (a + > b)$

computes a lower bound for $\dfrac{4s}{a+b}$

3) Evaluation of a polynomial according to the Horner scheme

$$z := \sum_{i=0}^{3} a_i x^i$$

$$z := ((a[3]*x + a[2])*x + a[1])*x + a[0].$$

The last expression has the same form whether the a_i, x and z are all real or all complex. The operators are provided for all numerical types.

C. PROCEDURES, FUNCTIONS, OPERATORS

We have seen that all the spaces of numerical computation may be introduced by the type concept of MATRIX PASCAL. Now we will consider the problem of introducing the operators for these data types.

Let us consider the segment addition as an example

$$A = [\underline{a}, \bar{a}],\ B = [\underline{b}, \bar{b}],\ A + B = [\underline{a} + \underline{b},\ \bar{a} + \bar{b}].$$

In rounded arithmetic this reads

$$A \mathbin{\Diamond\!\!\!\!+} B := [\underline{a} \mathbin{\triangledown\!\!\!\!+} \underline{b},\ \bar{a} \mathbin{\vartriangle\!\!\!\!+} \bar{b}].$$

We assume that the operations with directed roundings are known to the compiler.

1) Then the segment addition can be implemented by the following procedure

```
procedure iadd (a,b: segment; var res: segment);
begin
        res. inf:= a.inf + <b.inf;
        res.sup:= a.sup + >b.sup
```

```
end;
```

If we wish to compute the sum

$$S := A + B + C + D \qquad (*)$$

we have to write the following sequence of procedure calls

```
    iadd(a,b,z1);
    iadd(z1,c,z2);
    iadd(z2,d,s);
```

We see that the short expression is spread over 3 lines and hardly recognizable.

2) It would be better to give a function rather than a procedure. This would mean an extension of standard PASCAL where normally no functions with structured result type are permitted. If we allow this extension our problem can be solved as follows:

```
function iadd(a,b: segment): segment;
begin
      iadd.inf:= a.inf + <b.inf;
      iadd.sup:= a.sup + >b.sup;
end;

                    s:= iadd(iadd(iadd(a,b),c),d).
```

Now we require only one statement for the computation of the sum. The notation is already very close to (*). The only difference is that a function name is used instead of an operator, and that the function name is written in front of the operands instead of between them.

3) The operator declaration eliminates this notational difference:

```
operator . + (a,b: segment) iadd: segment;
begin
iadd.inf:= a.inf + <b.inf;
iadd.sup:= a.sup + >b.sup;
end;
        s:= a + b + c + d
```

MATRIX PASCAL is equipped with operators for all additional numerical data types. These operators are to be considered as standard operators of the extended language. They are predefined and precompiled. The new numerical standard operators of MATRIX PASCAL can be defined in MATRIX PASCAL, if a few (15) fundamental routines are available [3].

As an example we consider the multiplication of two complex matrices $A = (a_{ij})$ and $B = (b_{ij})$, $A*B = \left(\sum_{j=1}^{n} a_{ij} b_{jk}\right)$

```
operator*(a: matrix [n1..m1, n2..m2] of complex;
          b: matrix [n3..m3, n4..m4] of complex)
     cprod: matrix [n1..m1, n4..m4] of complex;

  var row, col: integer;

  begin                  {assume that m2 - n2 = m3 - n3}
     for row:= n1 to m1 do
        for col:= n4 to m4 do
           cprod [row,col]:= a[row] * b[*,col]
           {scalar product of row-th row of a with col-th column of b}
  end;
```

D. UNIVERSAL OPERATOR CONCEPT

From our experience with the implementation of the great number of operators for all numerical data types, we concluded that it is easier to implement a universal function and operator concept than to introduce numerous new standard operators. This function and operator concept is at the user's disposal, so he can define operators adequate to each particular problem. We list a few examples: evaluation with complex numbers in polar coordinates, string handling, all the operators of differential geometry.

The following four diagrams describe the declaration of procedures, functions and operators in MATRIX PASCAL. Note that all types may be result types of functions and operators.

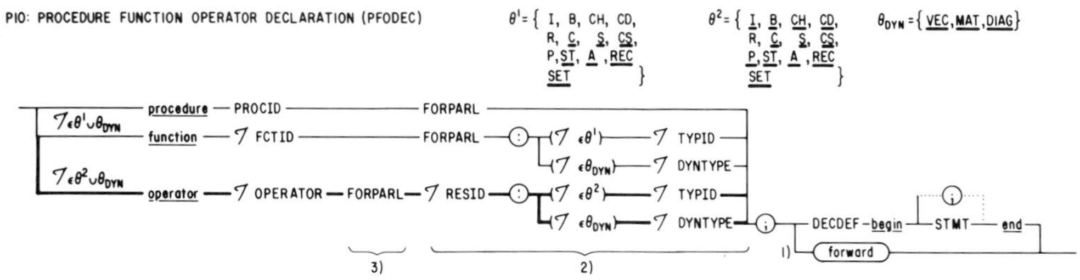

(1) Incomplete PFODEC
(2) Is missing in a completion of an incomplete PFODEC
(3) One or two parameters.

PI2: FORMAL PARAMETER LIST (FORPARL)

θ_{DYN} = { VEC, MAT, DIAG }

θ = { I, B, CH, CD, R, C, S, CS, P, ST, A, REC SET }

θ' = { I, B, CH, CD, R, C, S, CS, P, ST, A, REC SET }

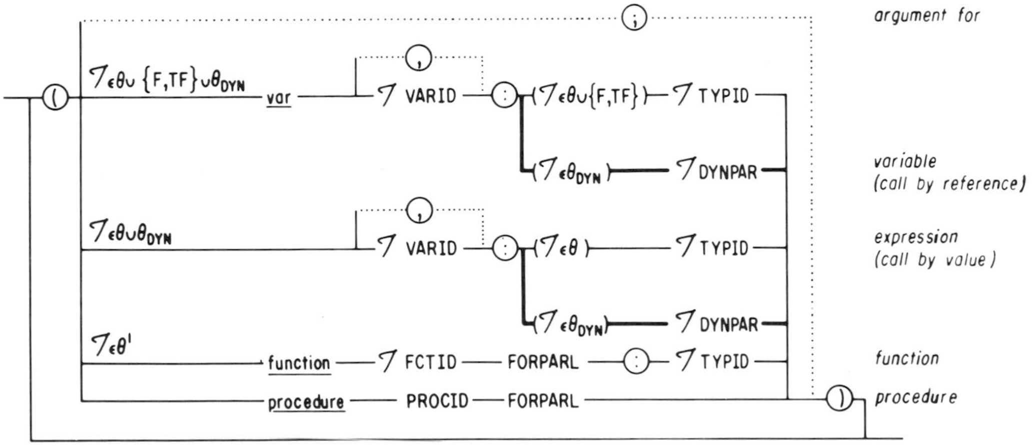

In an operator declaration at least one, at most two parameter.

PI3: $\mathcal{T}$ DYNAMIC TYPE ($\mathcal{T}$ DYNTYPE)

$\mathcal{T}\in\theta_{DYN}$ = { VEC, MAT, DIAG }

θ = { I, B, CH, CD, R, C, S, CS, P, ST, A, REC SET }

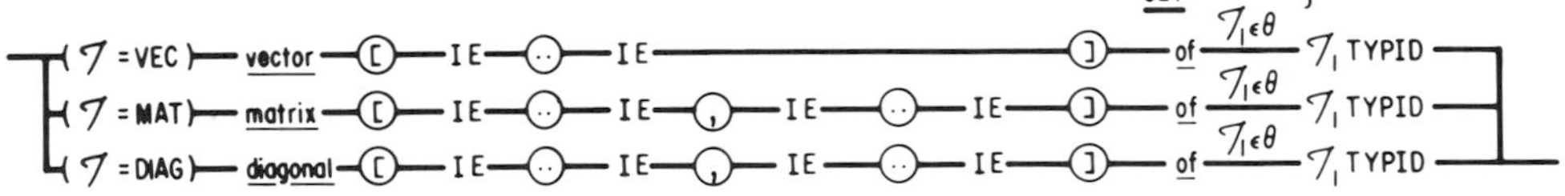

PI4: $\mathcal{T}$ DYNAMIC PARAMETER ($\mathcal{T}$ DYNPAR)

$\mathcal{T}\in\theta_{DYN}$ = { VEC, MAT, DIAG }

θ = { I, B, CH, CD, R, C, S, CS, P, ST, A, REC SET }

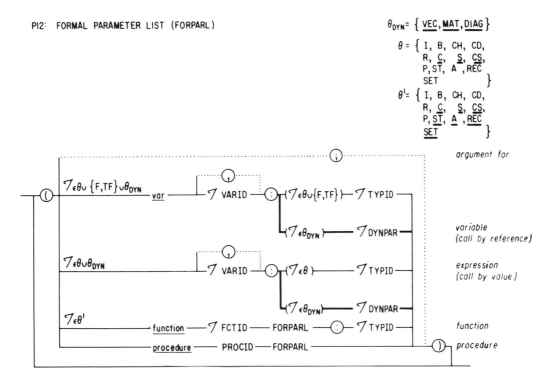

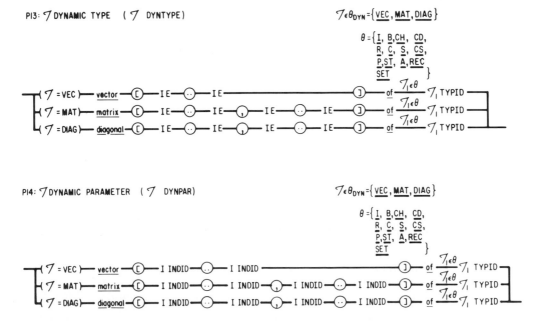

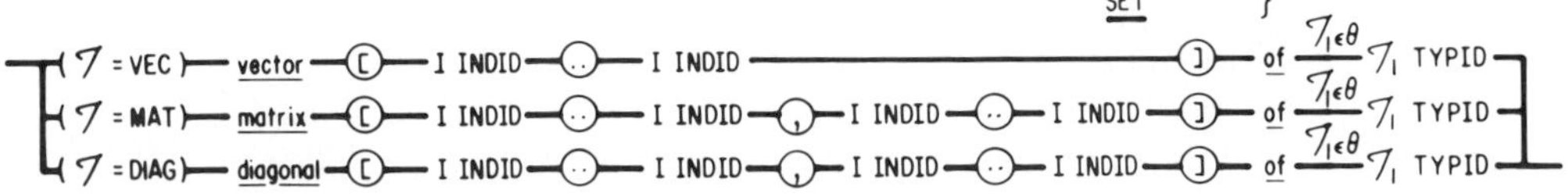

Remarks

Dynamic Types

- The index identifiers of all dynamic parameters must be different. They define local constants.
- When the procedure, function, or operator is called these constants are assigned the values of the index bounds of the actual parameter.
- The index expressions for a dynamic result are evaluated prior to the execution of the function block but after the evaluation of the parameter list.
- Only constants, global entities, formal parameters, or index identifiers of dynamic parameters may occur in an index expression for a dynamic result.

Functions with arbitrary result type

- An assignment to the function identifier must have occurred before leaving the body of the function.
- Assignment to components of the function identifier is also possible.
- Identification is accomplished via the function identifier only.
- Actual parameters must be assignment compatible to the formal parameters.
- Number and type of formal parameters may not be changed.

Operators

- The formal parameter list contains one or two parameters.
- The result identifier corresponds to the function identifier.

- A binary operator is called by writing the symbol between the two parameters.
- A unary operator is called by writing the symbol in front of the parameter.
- All existing operators symbols, listed in P11, may be overloaded with an additional meaning or redefined.
- An operator is distinguished by its symbol and the type and order of its parameters.
- In particular, it is possible to use the same symbol for different parameter types.
- Arbitrary result types are permitted.
- Actual parameters must be of the same type as formal parameters.
- In case of dynamic parameters, this means, that the number of indices and the component type must be the same.

Example

```
Appending a vector to a matrix
program   ex1;
var a     : matrix [1..10, 1..10] of real;
    aa    : matrix [1..10, 1..11] of real;
    b     : vector [1..10] of real;
operator  + (a: matrix [m1..n1, m2..n2] of real;
             b: vector [m3..n3] of real)
             res: matrix [m1..n1, m2..n2 + 1] of real;
var j     : integer;
```

```
begin
        for j:= m2 to n2 do
          res [*,j]:= a[*,j];
          {copy a column by column}
        if n1 - m1 = n3 - m3 then
          res [*,n2 + 1]:= b
          {append b if dimension matches}
        else
          res [*,n2 + 1]:= vrnull (n1 - m1);
          {append zeros}
end;
aa:= a + b;{yields a 10×11 matrix}
```

E. EXPRESSIONS WITH MAXIMUM ACCURACY

MATRIX PASCAL is equipped with the dynamic types and new standard operators for all the spaces which occur in numerical computation [3]. But this is only the first step to obtain maximum accuracy not only in operations but in expressions and algorithms as well. For most problems of numerical analysis, it suffices to evaluate expressions involving scalar products with maximum accuracy. MATRIX PASCAL is therefore furnished with a concept to evaluate specific expressions with maximum accuracy.

These expressions are written in parentheses with a rounding symbol in front of them. We use the following rounding symbols.

rounding symbol	alternative symbol	meaning
$\square$	#=	monotone, antisymmetric rounding
$\triangledown$	# <	downwardly directed rounding
$\triangle$	# >	upwardly directed rounding
$\lozenge$	# #	rounding to the closest segment which contains the value

P26: $\mathcal{T}$ OPD ($\mathcal{T}$ OPERAND) $\qquad \mathcal{T} \in \theta \cup \{F, TF\}$

θ = {I, B, CH, CD, R, C, S, CS, P, ST, A, REC, SET, VEC, MAT, DIAG}

θ_{EMA} = {R, VR, MR, C, VC, MC}

θ_{CON} = {I, B, CH, CD, R, C, S, CS,}

θ_{SEG} = {S, VS, MS, DS, CS, VCS, MCS, DCS,}

θ_{SFCT} = {I, B, CH, CD, R, C, S, CS, VR, VC, VS, VCS, MR, MC, MS, MCS, DR, DC, DS, DCS}

θ_{SET} = {SI, SB, SCH, SCD}

θ_{ORD} = {I, B, CH, CD}

$\beta\mathcal{T}$ denotes the basic type $\beta\mathcal{T} \in \theta_{EMA}$ of the segment type $\mathcal{T} \in \theta_{SEG}$

$\gamma\mathcal{T}$ denotes the basic type $\gamma\mathcal{T} \in \theta_{ORD}$ of the set type $\mathcal{T} \in \theta_{SET}$

β: $\theta_{SEG} \rightarrow \theta_{EMA}$

S → R
VS → VR
MS → MR
CS → C
VCS → VC
MCS → MC

γ: $\theta_{SET} \rightarrow \theta_{ORD}$

SI → I
SB → B
SCH → CH
SCD → CD

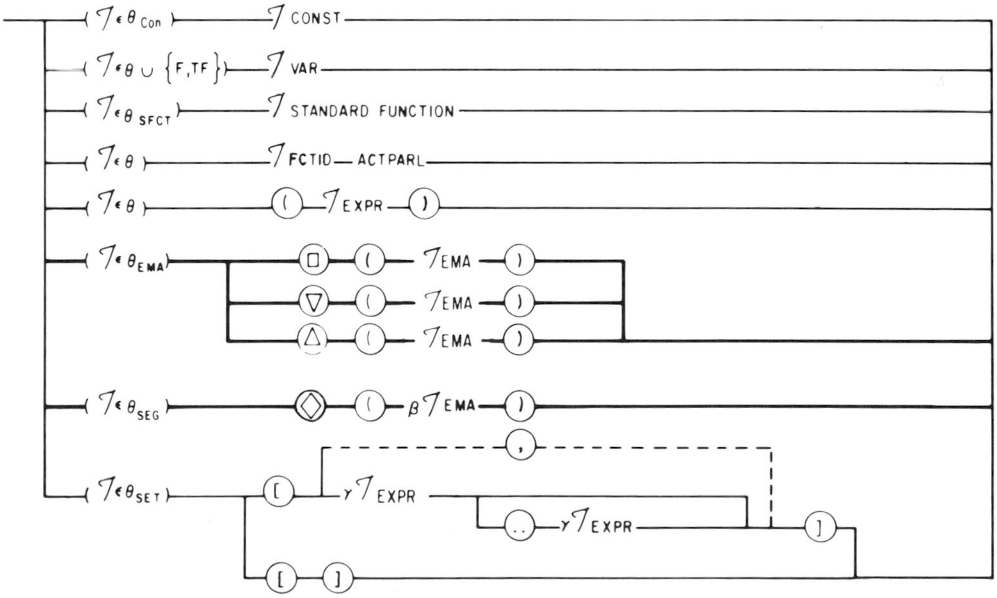

P27: R EMA (R EXPRESSION WITH MAXIMUM ACCURACY)

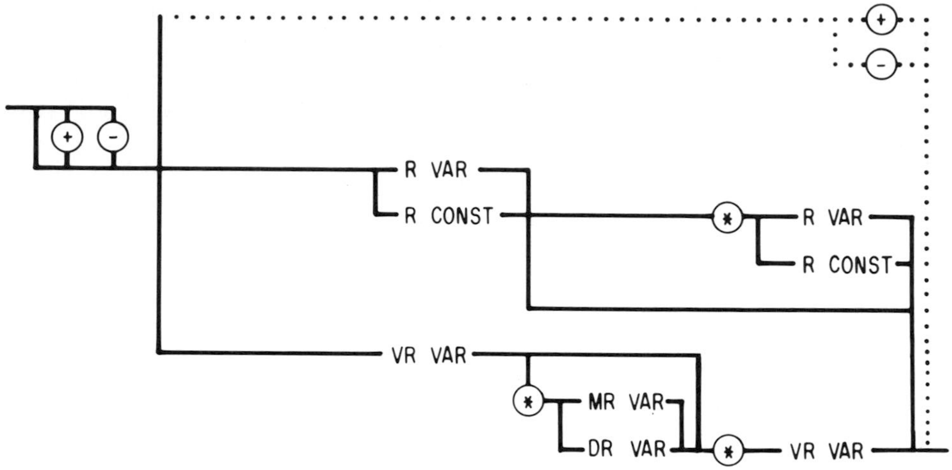

P28: C EMA (C EXPRESSION WITH MAXIMUM ACCURACY)

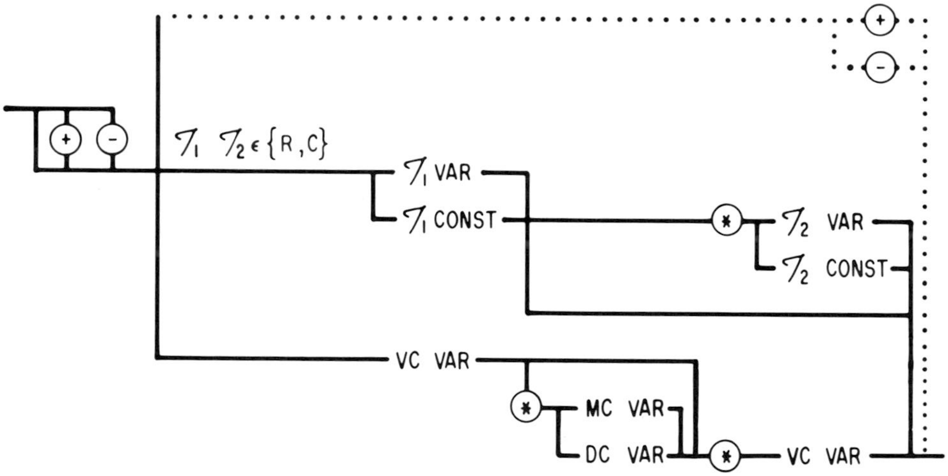

at least one C or VC VARIABLE or CONSTANT occurs

P29: VR EMA (VR EXPRESSION WITH MAXIMUM ACCURACY)

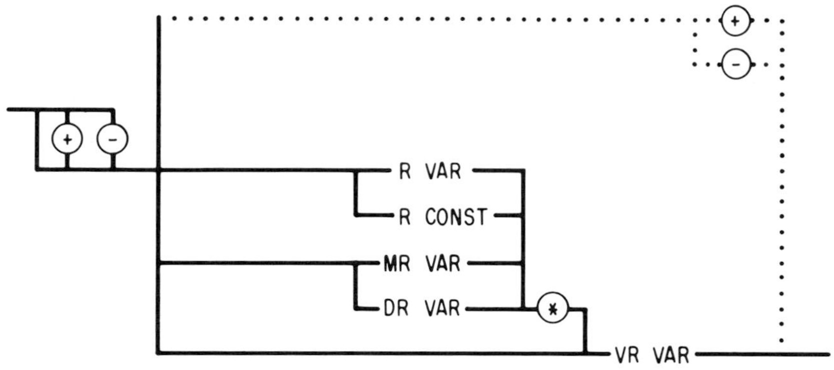

P30: MR EMA (MR EXPRESSION WITH MAXIMUM ACCURACY)

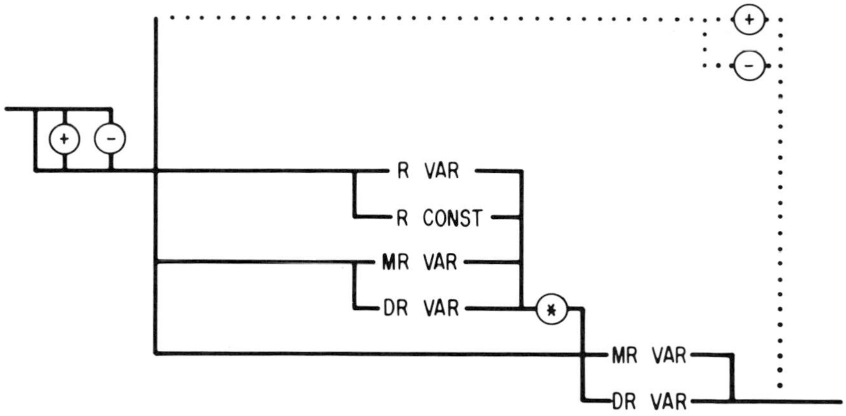

P31: VC EMA (VC EXPRESSION WITH MAXIMUM ACCURACY)

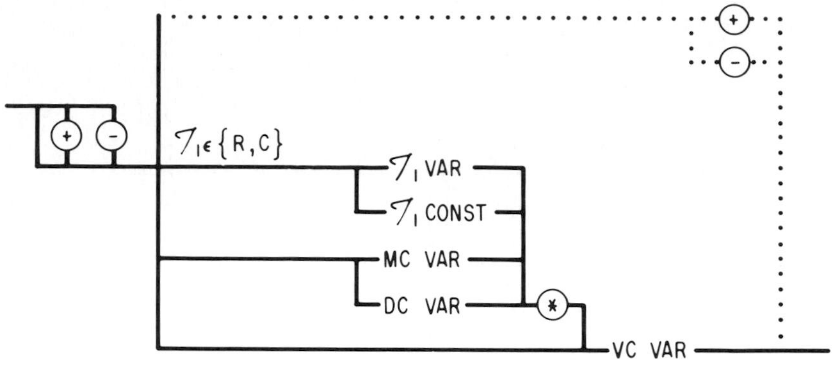

P32: MC EMA (MC EXPRESSION WITH MAXIMUM ACCURACY)

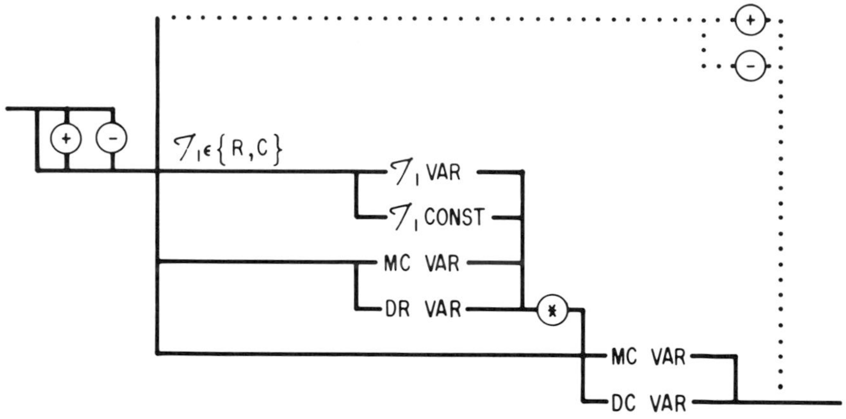

Constants occurring in EMAs must be exactly representable or an explicit rounding symbol must indicate the manner of conversion. For each of these expressions, called EMAs, (expression with maximum accuracy) it is guaranteed that no floating-point number lies between the computed and the correct result. With this new kind of operands the user of MATRIX PASCAL has a convenient and powerful tool to write algorithms which yield maximum accuracy and guaranteed results.

A few examples may illustrate the use of the EMAs:

The multiplication of two complex matrices $A = (a_{ij})$ and $B = (b_{ij})$ now reads

```
operator*   (a: matrix [n1..m1, n2..m2] of complex;
             b: matrix [n3..m3, n4..m4] of complex)
         cprod: matrix [n1..m1, n4..m4] of complex;
begin
            cprod:= □(A*B)
end;
```

The solution u of a linear system of the form

$$u = Au + b$$

with a matrix $A = (a_{ij})$ and a vector $b = (b_j)$ may be obtained by the Jacobi method,

$$x^{k+1} = Ax^k + b, \quad k = 0,1,\ldots .$$

In MATRIX PASCAL the expression for the iteration step can be written in the form $x := \square(A*x + b)$. Then it is evaluated with maximum accuracy.

The following function computes the inverse of a given dynamic square matrix A according to the Schultz-Method. $X1$ is a given approximation to the inverse.

```
function inv(var A,X1: matrix [m1..n1, m2..n2] of real):
                       matrix [m1..n1, m1..n1] of segment;
var          R,Y,Z,Y0: matrix [m1..n1, m1..n1] of segment;
                    I: diagonal [m1..n1, m1..n1] of real);
begin              I:= drid (n1 − m1);                    {identity matrix}
                   R:= ◊(2*I − X1*A);
                   Y0:= ◊(I − X1*A);
                   Z:= Y0;
            repeat Y:= Z;
                   Z:= Y0 + R*Y
             until Z in Y;
            inv := X1 + Z
end
```

F. STANDARD FUNCTIONS

In arithmetic expressions not only variables and constants occur but also standard functions and function calls. For an optimal arithmetic the standard functions as well as the standard operators must be computed with maximum accuracy.

The requirement of maximum accuracy for standard functions is

$$\underline{a} < f(x) < \overline{a} \tag{1}$$

for each standard function f and three successive floating-point numbers $\underline{a}$, a, $\overline{a}$, where $a = \boxed{f}\,(x)$ the computer evaluation of f. (1) means that no floating-point number lies between the computed and the exact result. Although condition (1) is clear and easy to state, it is hard to prove for each specific function. Therefore, we derive a sufficient condition for (1). Let $\tilde{f}(x)$ denote the computed result in a finer floating-point screen $\tilde{F}$ and $\bigcirc$: $\tilde{F} \rightarrow F$ the rounding to the nearest floating-point number. Then (1) holds if

$$\left|\frac{f(x)-\tilde{f}(x)}{f(x)}\right| < \frac{1}{2}b^{-\ell} \quad \text{and} \quad a := \bigcirc\tilde{f}(x), \tag{2}$$

where b is the base and ℓ the mantissa-length of the floating-point system F.

Proof:

We have to show $\underline{a}<f(x)<\bar{a}$ where $\underline{a}$, a, $\bar{a}$ are successive floating-point numbers. We consider the case $f(x)\leq\tilde{f}(x)$. Given two successive floating-point numbers, b, d with $b\leq f(x)<d$ and $c := \frac{b+d}{2}$, we distinguish the following cases

(i) $f(x)=b \Rightarrow \tilde{f}(x)<c \Rightarrow \bigcirc\tilde{f}(x)=b \Rightarrow$ (1) with $a=b$

(ii) $b<f(x)\leq c$ then either

(α) $\tilde{f}(x)<c \Rightarrow \bigcirc\tilde{f}(x)=b \Rightarrow$ (1) with $a=b$

or (β) $d>\tilde{f}(x)\geq c \Rightarrow \bigcirc\tilde{f}(x)=d \Rightarrow$ (1) with $a=d$

(iii) $f(x)>c$ then either

(α) $\tilde{f}(x)<c \Rightarrow \bigcirc\tilde{f}(x)=b \Rightarrow$ (1) with $a=b$

or (β) $\tilde{f}(x)\geq c \Rightarrow \bigcirc\tilde{f}(x)=d \Rightarrow$ (1) with $a=d$

The case $f(x)>\tilde{f}(x)$ is treated analogously. ■

The best way to implement complex functions is to calculate appropriate real functions and use well known formulas to compute the real and imaginary part. Because of the rounding error in these final computations (2) does not suffice for the real functions.

We therefore demand the following stronger property for the floating-point functions

$$\left|\frac{f(x)-\tilde{f}(x)}{f(x)}\right| < \frac{1}{2}b^{-(\ell+k)} \tag{3}$$

where k is an integer, which should be chosen, so that the complex versions can be obtained with maximum accuracy. It can be shown that $k = 1$ suffices for the functions sqrt, exp, ln, arctan, sin and cos.

Note that usually the strongest requirement is (3) with $k = -1$.

The condition (1) of maximum accuracy is expressed slightly differently for interval functions. Of course, we now must allow that the bounds be valid results.

$$\underline{a} \leq f(x) \leq \overline{a} \tag{4}$$

for every real function f and three successive floating-point numbers $\underline{a}$, a, $\overline{a}$ where $a = \boxed{f}\,(x)$. Obviously (4) is a weaker condition than (1), we therefore can fulfill (4) using monotone directed roundings instead of the rounding to the nearest number.

REFERENCES

[1] Bohlender, G., Grüner, K. Realization of an optimal Computer Arithmetic, this volume.

[2] Bohlender, G., Böhm, H., Grüner, K., Kaucher, E., Klatte, R., Krämer, W., Kulisch, U., Miranker, W. L., Rump, S. M., Ullrich, Ch., Wolff von Gudenberg, J. (1982). MATRIX PASCAL IBM Research Report No. 9577 and this volume.

[3] Kulisch, U. A New Arithmetic for Scientific Computation, this volume.

[4] Ullrich, Ch. A FORTRAN Extension for Scientific Computation, this volume.

[5] Wolff von Gudenberg, J. (1980). Einbettung allgemeiner Rechnerarithmetik in PASCAL mittels eines Operatorkonzeptes und Implementierung der Standardfunktionen mit optimaler Genauigkeit, Dissertation, Universität Karlsruhe.

REALIZATION OF AN OPTIMAL COMPUTER ARITHMETIC

G. Bohlender and K. Grüner

Institute for Applied Mathematics
University of Karlsruhe
Karlsruhe, West Germany

For numerical applications a mathematically well founded computer arithmetic is necessary. In addition to the usual floating-point operations this arithmetic must also take into consideration the higher structures built upon these, e.g., complex numbers, intervals, matrices and vectors. In all cases the operations for these structures are deduced by a uniform construction principle, namely, semimorphism [6], by means of which mathematical structure is preserved as much as possible. This also results in optimal accuracy.

For the implementation of the arithmetic many operations must be provided. This can be achieved only when the arithmetic as a whole is built up by a modular concept. The scalar product for any collection of floating-point numbers is introduced as an elementary operation. By means of these operations, all others in the higher spaces can be constructed in such a way that the requirement of semimorphism is also realized. The whole arithmetic is imbedded in a higher programming language and both are fully implemented on a microcomputer.

CONTENTS

1. INTRODUCTION - MATHEMATICAL FOUNDATIONS

Until recently the design of floating-point arithmetic was largely influenced by hardware requirements such as word length, economic use of storage space, fast and easy execution, etc. Therefore, the mathematical structures and the accuracy of the operations were not as

A NEW APPROACH
TO SCIENTIFIC COMPUTATION

ISBN 0-12-428660-7

carefully considered as they should have been. But the whole of computer arithmetic determined by mathematical aspects can be realized in a simple and general way using fast algorithms. To achieve this in all the structures which are considered, we define the operations by the so-called horizontal method of semimorphism [8]. By this uniform and general construction principle all spaces commonly occurring in numerical computation are treated and thus results with optimal accuracy are guaranteed.

Figure 1 gives an overview of the mathematical spaces occurring in numerical computation. To the right of each space, we list the operations occurring therein. Clearly, the space $\mathbb{R}$ of the real numbers is not sufficient to describe numerical algorithms. In addition, one needs the space of the complex numbers $\mathbb{C}$ and the spaces $V\mathbb{R}$, $V\mathbb{C}$, $M\mathbb{R}$, $M\mathbb{C}$ of the vectors and matrices composed of these numbers respectively. Many recent investigations have shown that for obtaining accurate bounds on solutions, interval arithmetic is indispensable. Thus we include the spaces $I\mathbb{R}$, $I\mathbb{C}$, $IV\mathbb{R}$, $IV\mathbb{C}$, $IM\mathbb{R}$ and $IM\mathbb{C}$. The structures in the left column of figure 1 are not exactly representable on a computer. They must be approximated using the finite subsets in the middle column of figure 1.

$\mathbb{R}$	$\supset$	R	$+,-,*,/$
$V\mathbb{R}$	$\supset$	VR	$+,-$
$M\mathbb{R}$	$\supset$	MR	$+,-,*$
$I\mathbb{R}$	$\supset$	IR	$+,-,*,/,\underline{\cup},\cap$
$IV\mathbb{R}$	$\supset$	IVR	$+,-,\underline{\cup},\cap$
$IM\mathbb{R}$	$\supset$	IMR	$+,-,*,\underline{\cup},\cap$
$\mathbb{C}$	$\supset$	CR	$+,-,*,/$
$V\mathbb{C}$	$\supset$	VCR	$+,-$
$M\mathbb{C}$	$\supset$	MCR	$+,-,*$
$I\mathbb{C}$	$\supset$	ICR	$+,-,*,/,\underline{\cup},\cap$
$IV\mathbb{C}$	$\supset$	$IVCR$	$+,-,\underline{\cup},\cap$
$IM\mathbb{C}$	$\supset$	$IMCR$	$+,-,*,\underline{\cup},\cap$

Figure 1: Overview of the spaces and operations occurring in numerical computation

It can be assumed that the operations are known in all spaces $\mathbb{K}$ of the left column (figure 1). Let a subspace of $\mathbb{K}$ representable on a computer be denoted by K. Then for each space $\mathbb{K}$, we define the corresponding operations in K according to the uniform construction principle of the horizontal method ([6], [9]):

$$\text{(RG)} \quad \bigwedge_{\circ\in\{+,-,*,/\}} \quad \bigwedge_{\square\in\{\triangledown,\triangle,\bigcirc\}} \quad \bigwedge_{x,y\in K\subseteq\mathbb{K}} x \boxdot y := \square(x\circ y).$$

Everything which follows is valid for all roundings, but we confine ourselves to the roundings $\bigcirc$, $\triangledown$ and $\triangle$, which are the most important in practice. For it is well known that to compute guaranteed results and to prove mathematical facts on a computer, we require not only the rounding $\bigcirc$ to the next machine number but also the downwardly and upwardly directed roundings, $\triangledown$ and $\triangle$, respectively.

In addition to the inner operations in the spaces of figure 1, there are also many outer operations $\circ: \mathbb{M}\times\mathbb{K}\rightarrow\mathbb{K}$. For these outer operations, we have

$$\text{(AG)} \quad \bigwedge_{\circ\in\{+,-,*,/\}} \quad \bigwedge_{\square\in\{\triangledown,\triangle,\bigcirc\}} \quad \bigwedge_{x\in M\subseteq\mathbb{M}} \quad \bigwedge_{y\in K\subseteq\mathbb{K}} x \boxdot y := \square(x\circ y).$$

Also, the user of a computer expects that the standard functions f are available. In an optimal computer arithmetic they must fulfill the following requirement:

$$\text{(FG)} \quad \bigwedge_{f\in\{\text{sqrt},\exp,\ln,\arctan,\sin,\cos\}} \quad \bigwedge_{\square\in\{\triangledown,\triangle,\bigcirc\}} \quad \bigwedge_{x\in K\subseteq\mathbb{K}} \boxed{f}\,(x) := \square(f(x)).$$

Here $\mathbb{K}$ denotes one of the spaces $\mathbb{R}$, $\mathbb{C}$, $I\mathbb{R}$, $I\mathbb{C}$.

These considerations reduce all operations $\boxdot$ in the subsets K to the corresponding exact operation $\circ$ in $\mathbb{K}$. Evidently the nonconstructive definitions (RG), (AG) and (FG) are not realizable on a computer without further considerations. The exact result $x\circ y$ is not representable in every case. But it can be shown that for all operations it is sufficient to construct an approximate value $x\,\tilde{\circ}\,y$ with

$$\text{(RG)} \quad \bigwedge_{\circ\in\{+,-,*,/\}} \quad \bigwedge_{\square\in\{\triangledown,\triangle,\bigcirc\}} \quad \bigwedge_{x,y\in K\subseteq\mathbb{K}} x \boxdot y := \square(x\circ y) = \square(x\,\tilde{\circ}\,y).$$

Analogous properties hold for (AG) and FG). For all these operations the existence of such an approximate result must be provided in all spaces by supplying fast and concrete algorithms ([1], [3], [9], [11]).

It seems that the large quantity and the diversity of operations prevent an implementation from being easily realized. The effort is greatly reduced if the whole arithmetic is constructed in a modular fashion involving three levels.

In the following we assume that computer arithmetic includes all operations in all spaces of figure 1. We demand of an optimal arithmetic that:

- there is no machine number between the computed and the exact result
- in all numerical spaces all operations are rounded exactly; this is also valid for intervals, complex numbers, vectors, matrices, etc.
- the subspaces K must have at least the properties of a ringoid or vectoid ([8]), respectively
- numerical problems that are exactly representable must be exactly representable in the machine
- it must be possible to implement new numerical algorithms which compute guaranteed results
- all features of the arithmetic must be accessible to a higher programming language
- the required accuracy must not lead to a loss of speed
- the memory must be economically utilized and the construction must be modular
- it should be easily portable.

2. ORGANIZATION OF THE ARITHMETIC

The construction may be divided in three separate levels. Because existing computers do not allow one to easily implement the arithmetic according to the formulas (RG), (AG) and (FG), even the simplest, most fundamental operations must be taken into account in the implementation of an optimal arithmetic. These simple operations are included in the Level 1:

Level 1: Support routines for mantissas of base b

a) Fixed point operations $+,-,*,/$

b) Complement of mantissa

c) Comparison of mantissas

d) Shifting left/right for 1,2,... digits

e) Normalization, clear to zero, roundings.

Level 2 includes the elementary operations which should be available on every computer:

Level 2: Elementary operations

a) dyadic operations $\oplus$, $\ominus$, $\circledast$, $\oslash$, $\triangledown\!\!\!\!\!+$, $\triangledown\!\!\!\!\!-$, $\triangledown\!\!\!\!\!*$, $\triangledown\!\!\!\!\!/$, $\triangle\!\!\!\!\!+$, $\triangle\!\!\!\!\!-$, $\triangle\!\!\!\!\!*$, $\triangle\!\!\!\!\!/$

b) monadic operation $-$

c) comparisons $=$, $\neq$, $<$, $>$, $\leq$, $\geq$

d) scalar product for any quantity of floating-point numbers

e) in/output

f) conversions: integer $\longleftrightarrow$ floating-point and between numbers with different mantissa length.

Up to now, we have considered only the operations in R.

Level 3 provides all remaining operations in the spaces in figure1:

Level 3: Structured operations

a) interval operations

b) complex operations

c) complex interval operations

d) real matrix and vector operations

e) interval scalar product, interval-matrix and interval-vector operations

f) complex scalar product, complex matrix and vector operations

g) complex interval scalar product, complex interval-matrix and interval-vector operations

h) standard functions.

All modules of a level are independent of each other and they employ the modules of the lower levels. Thus, the levels 1 and 2 must be formulated quite basically, for example, in micro code or in assembly language. The highest level can be implemented in an extended higher programming language, such as PASCAL-SC. The scalar product is taken as an elementary level 2 operations, since, to achieve optimal accuracy it must not be composed of rounded additions and multiplications. To compute the standard functions with the necessary accuracy a second floating-point system with more digits is required.

In the following the arithmetic proceeds from a normalized floating point system. It has the base b, the mantissa length ℓ and the smallest or greatest exponent emin or emax, respectively. We denote by v, e and m the sign, the exponent and the mantissa of a floating-point number. Furthermore, we choose the sign-magnitude representation. Then the flow of a floating-point operation $z := x \boxdot y = \square(\widetilde{x \circ y})$ can be described as follows:

1. Decomposition of x and y into mantissa and exponent (if they are stored in one word)
2. Computation of vz and ez
3. Evaluation of $\widetilde{mx} := mx \,\tilde{\circ}\, my$ with guard digits; normalization if necessary
4. Rounding: $mx := \square \widetilde{mx}$
5. If 1: Composition of vz, ez and mz.

The intermediate result $\widetilde{mz}$ is computed independently of the rounding to follow. In all cases it is computed so that every rounding can be applied to it. This holds for all operations of level 2, in particular, for the scalar product and the in/output.

The intermediate result of minimum length has the form:

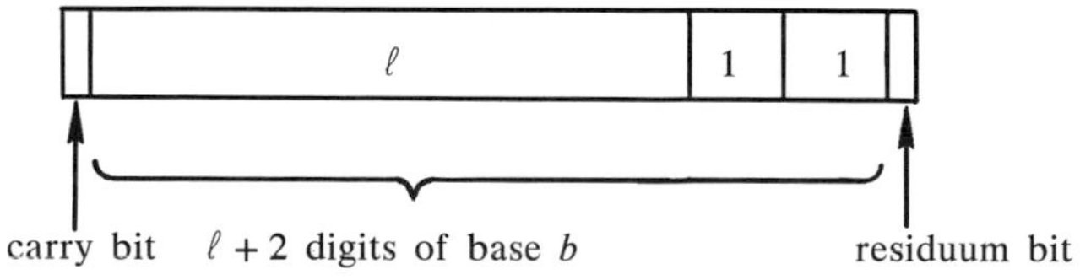

The residuum bit Rm is set if the exact result cannot be represented by $\ell + 2$ digits. Rm is essential for implementing the directed roundings $\bigtriangledown$ and $\bigtriangleup$. Furthermore, the complement is also essentially influenced by the residuum bit.

In place of the form of the intermediate result as described above, an extended format is also possible. It is characterized by $2\ell + 1$ digits of the base b and by one carry bit. In this way, some decisions in the algorithms for the operations $\tilde{\circ}, \circ \in \{+, -, *, /\}$ can be avoided. In the following we restrict ourselves to the more general format of minimum length as described above.

3. IMPLEMENTATION OF THE ELEMENTARY OPERATIONS

We now describe some essential features of the modular concept by implementing the operations in levels 1 and 2. Herein we measure the complexity of the algorithms by units of the smallest computer operation, namely the byte addition (BA).

3.1 Addition and subtraction

In the addition of floating-point numbers we assume that $ex \geq ey$. If not, the operands x and y are exchanged. Then three cases are to be distinguished:

a) $ex - ey \geq \ell + 2$:

$ez := ex$,

if $my = 0$: $\widetilde{mz} := mx$

if $my \neq 0$: $Rmy := 1$ and $\widetilde{mz} := mx + vy \bullet b^{-(\ell+2)}$, i.e., mx is corrected by one unit in the $(\ell + 2)^{nd}$ digit depending on the sign of y.

b) $ex - ey \leq 2$:

y is adapted to x, $\widetilde{mx} := mx + my$;

$\widetilde{mz}$ can be represented exactly in the given format, the residuum bit remains unoccupied.

c) $2 < ex - ey \leq \ell + 1$:

y is adapted to x; Rmy is set if any digit different from zero is shifted out of my by adaptation. Once the residuum bit is set it may not be influenced by further shift operations. $\widetilde{mz} := mx + my + vy \bullet Rmy \bullet b^{-(\ell+2)}$.

Since mantissas have to be added according to their sign, it is convenient to distinguish the following two cases:

a) $vx = vy$: Then $\tilde{mz}$ is computed by $\tilde{mz} := mx + my + Rmy \bullet b^{-(\ell+2)}$. $\tilde{mz}$ is to be corrected only if a carry bit arises from addition.

b) $vx \neq vy$: The negative mantissa is subtracted from the positive. If the result $\tilde{mz}$ is negative then $\tilde{mz}$ is to be complemented. If some digits are cancelled $\tilde{mx}$ is to be normalized.

The subtraction of floating-point numbers is reduced to addition by changing the sign of y.

3.2 Multiplication

The exponents are added. For the multiplication of the mantissas the following procedures are to be followed:

I. The multiplication of two mantissas by successive addition requires $2.25\ell^2 + 5\ell$ *BAs*, on the average. This procedure requires the smallest amount of storage.

II. Using a multiplication table with the multiples of 1 to $(b - 1)$, the operations reduce to $2\ell^2$ *BAs* in number. The table requires a memory of nearly 300 bytes if $b = 10$.

III. If $\ell \geq 8$ it turns out to be fastest to multiply by means of a table of multiples, in which case one proceeds as follows: We require $0.75\ell^2 + 5\ell + 8$ *BAs* and for the multiplication table $4\ell + 8$ bytes of storage if $b = 10$. Let $mx = 0.x_1x_2...x_\ell$ and $my = 0.y_1y_2...y_\ell$ be the two mantissas. We compute the exact mantissa product

$$mx * my = \sum_{i=1}^{\ell} x_i * my * b^{-i}$$

in the following steps:

1) construct the table of all multiples of my, i.e., compute the 2nd to $(b - 1)^{st}$ multiple by repeated addition of the mantissa

2) for $i = 1,...,\ell$: extract the x_i-multiple from the table, add it to the intermediate result and shift it by one digit.

In some cases an acceleration is possible by construction of a second table with the bth to $b(b - 1)^{st}$ multiples of my or by displaced addition of the partial products.

3.3 Division

First the exponents are subtracted: $ez := ex - ey$. To obtain a normalized result we shift mx by one digit to the right (correspondingly correcting ez), if $|mx| \geq |my|$. By inverting the procedures I and III for the mantissa multiplication we obtain the following procedures for mantissa division.

I. The simplest method is the procedure using additions and subtractions without resetting the remainder; it requires $2.5\ell^2 + 11\ell + 11BAs$ on the average.

II. The mantissa division by means of two tables of multiples requires $1.5\ell^2 + 17.5\ell + 28BAs$. Here we evaluate the fraction mx/my as follows:

1. Construct the table containing the multiples of the denominator my (the multiples with 1 to $(b - 1)$ and b to $b(b - 1)$), and set $i = 1$.

2. Look for the greatest multiple $k \leq mx$ in one of the tables.

3. Subtract k from mx. The index of k in the table determines the ith digit of the quotient.

4. If i is even, shift my by two digits.

5. End, if $i = \ell + 2$ or $mx = 0$. Otherwise, increment i and go to 2.

In order to perform all roundings correctly, even for multiplication and division, the residuum bit Rmz must also be set by these mantissa operations if there is any non-digit digit beyond the $(\ell + 2)^{nd}$ digit.

3.4 Scalar product

For the realization of an optimal arithmetic, the scalar product

$$S := \sum_{i=1}^{n} a_i \bullet b_i$$

turns out to be so important, that it must be considered a fundamental operation of arithmetic. Scalar products occur, in particular, in matrix and vector products (real, complex and involving real and complex intervals), in products and quotients of complex floating-point numbers, in the evaluation of residues and therefore in numerous algorithms of numerical mathematics ([4], [10]).

The usual computation of scalar products in the form

$$\tilde{s} := a_1 ⊡ b_1 ⊞ a_2 ⊡ b_2 ⊞ \ldots ⊞ a_n ⊡ b_n$$

results in $2n - 1$ rounding errors. Thus the result is excessively degraded by rounding errors and the error analysis of complicated algorithms is rendered nearly impossible. Even in very simple cases catastrophic loss of precision can occur: For example, let $n = 3$, $a_1 = a_2 = a_3 = 1$, $b_1 = 1$, $b_2 = 1e30$, $b_3 = -1e30$. Then the exact result is $S = 1$, whereas the computed result is $\tilde{s} = 0$. It is possible for this to occur even if the intermediate results are computed with double precision.

Therefore, an algorithm for the scalar product is needed which computes the result with only a single rounding error. Three such algorithms are briefly described and compared in the following.

I. Order and Add

In this algorithm the summands are first ordered and then added in an appropriate sequence. The following steps have to be executed

1) compute $x_i := a_i \bullet b_i$ exactly (with double precision)

2) order the products x_i according to their exponents ex_i: $ex_1 > ex_2 > \ldots > ex_n$

3) add $s_1 := x_1 + x_2 + \ldots + x_p$ including as many terms as can be handled without rounding errors in a "short accumulator" for mantissas of length $L = 2\ell$.

4) add $s_2 := x_n \tilde{+} x_{n+1} \tilde{+} \ldots \tilde{+} x_{p+1} \tilde{+} s_1$ with rounded operations $\tilde{+}$; because of step 3, no catastrophic cancellation can occur

5) round $s := \square s_2$.

The approximation s_2 for the exact scalar product S is precise enough to guarantee the property

$$\bigwedge_{\square \in \{\nabla, \Delta, \bigcirc\}} s := \square s_2 = \square S.$$

The proof of this property and a more detailed description of this algorithm can be found in [7]. The steps 3), 4) and 5) can be illustrated by shifting the products x_i to the left according to their exponent:

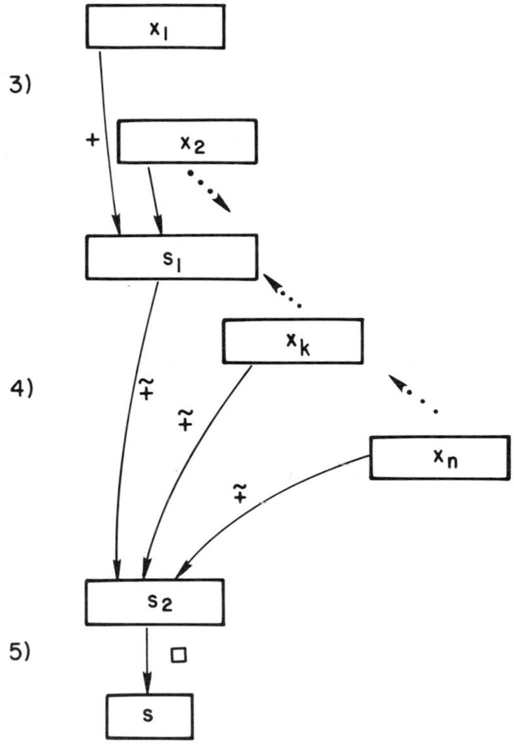

II. Addition with remainder

In this algorithm the products are added in their original order and the remainders of these additions are stored for a potential subsequent correction of the result:

1) compute $x_i^{(0)} := a_i \cdot b_i$ exactly (with double precision)

2) repeat the following steps for $k = 1,\ldots,n$, until the remainders $x_1^{(k)},\ldots,x_{n-1}^{(k)}$ can no longer influence the result $x_n^{(k)}$:

$s_1 := x_1^{(k-1)} \tilde{+} x_2^{(k-1)}$, store remainder in $x_1^{(k)}$

.....

$s_i := s_{i-1} \tilde{+} x_i^{(k-1)}$, store remainder in $x_i^{(k)}$ (for $i = 2,\ldots,n-1$)

.....

$x_n^{(k)} := s_{n-1}$

3) round $s := \square x_n^{(k)}$.

The iteration process in 2) can be illustrated as follows:

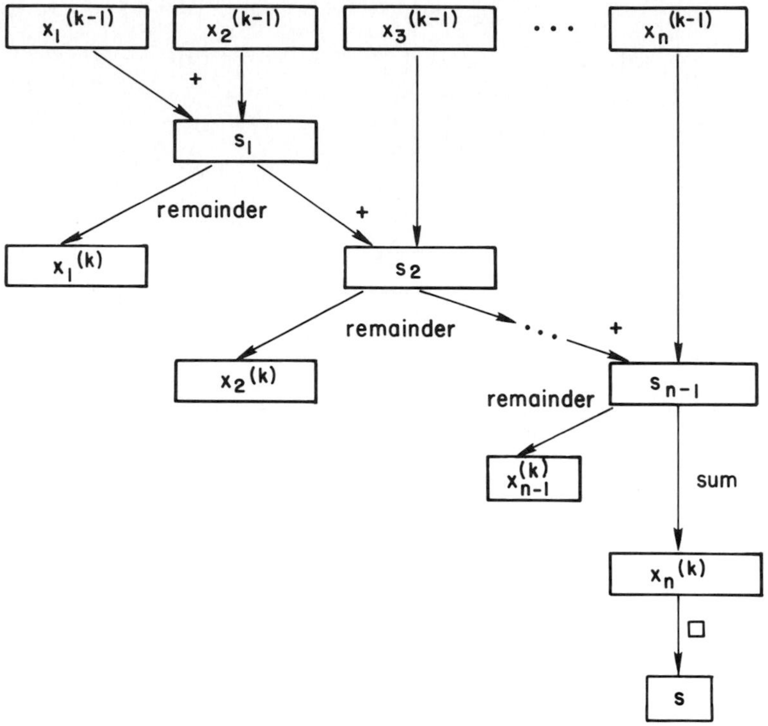

It can be shown that the property

$$\sum x_i^{(k)} = \sum x_i^{(0)}$$

holds for all $k = 1,...,n$ and that the approximation $x_n^{(k)}$ is sufficiently precise after at most $k = n$ iterations, i.e., the remainders $x_1^{(k)},...,x_{n-1}^{(k)}$ can no longer influence s, and the following property holds:

$$\bigwedge_{\square \in \{\nabla, \Delta, \bigcirc\}} s := \square x_n^{(k)} = \square S.$$

With an appropriate estimation of the sum of the remainders $\sum_{i=1}^{n-1} x_i^{(k)}$ the iteration can be terminated as soon as the required level of accuracy is attained. Typically - if no cancellation occurs - this is the case for $k = 1$. The details of this algorithm are described in [1].

III. Exact addition in long storage

In this algorithm a long storage space A is used which covers the range of all feasible exponents. The following steps are executed:

1) clear long storage A

2) compute $x_i := a_i \cdot b_i$ exactly (with double precision)

3) add x_i exactly to long storage A

4) round $s := \square A$

The floating-point products x_i are mapped into a fixed-point system of sufficient length:

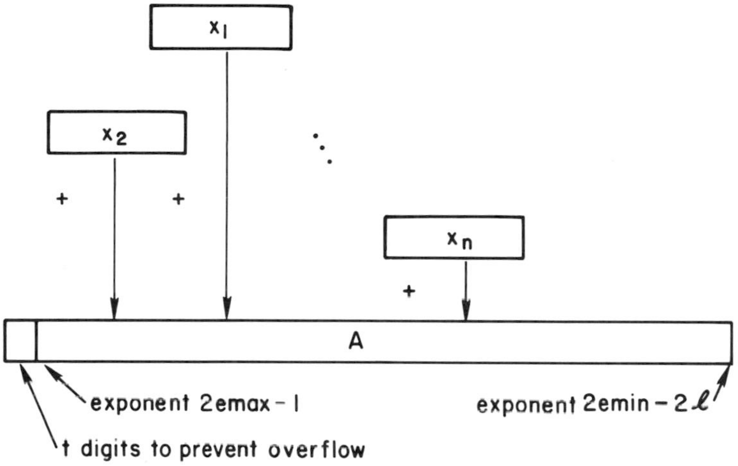

A total of $|2 \cdot \mathrm{emin} - 2\ell| + 2 \cdot \mathrm{emax} + t$ digits are needed for the long storage space. The leading t digits are used to prevent overflow; therefore, $n \leq b^t$ summands can be added exactly in A. This storage can be represented as a sequence of storage words, in a software realization e.g., as an array of integers.

If the contents of the storage space A and the next summand x_i have different signs, a carry may propagate over a significant part of the long storage, wasting execution time. This motivates the usage of two such long storage spaces A^+ and A^- for positive and negative summands, respectively. At the end of a scalar product the difference $A^+ - A^-$ has to be computed up to at least $\ell + 1$ significant digits plus a residue information bit which indicates whether any of the chopped-off digits in the exact difference is nonzero. Of course in this subtraction, some or even all of the leading digits of A^+ and A^- may cancel.

In many cases the usage of pointers turns out to be advantageous for indicating the nonzero portions of the storage spaces A^+ and A^-. Thereby the subtraction at the end of the algorithm and the clearing of the storage spaces is simplified.

There are other realizations in which only one long storage space is required. This can be accomplished by using signed partial words or redundant arithmetic.

We now briefly compare these three algorithms. Algorithms I and II need storage space for the n floating-point products x_i whereas algorithm III needs the storage spaces A^+ and A^-. Therefore in I and II the storage space is proportional to the dimension n of the scalar product, whereas in III it is proportional to the exponent range of the floating-point system.

Algorithm I is relatively slow for large dimensions n, since the execution time for the ordering is proportional to $n \log n$. Algorithms II and III are faster, provided that few iterations occur in algorithm II.

In many applications it is necessary to add additional summands to a scalar product. The new result must be computed with the same accuracy as the original scalar product. This can be achieved most easily by algorithm III.

4. OPERATIONS IN THE HIGHER SPACES

The structured operations of level 3 can be built up from operations of level 2. Here only the arithmetic operations are considered. All spaces considered are ordered and corresponding comparison relations are supplied. Furthermore, we only consider the case where both operands are of the same numerical type; otherwise one of the operands has first to be converted by a transfer function into the type of the second operand.

4.1 Interval operations

Let $[\underline{a},\overline{a}]$, $[\underline{b},\overline{b}] \in IR$ be two floating-point intervals. Then definition (RG) leads to the following formulas for the interval operations:

$$[\underline{a},\overline{a}] \mathbin{\Diamond\!\!\!\!\!+} [\underline{b},\overline{b}] = [\underline{a} \mathbin{\nabla\!\!\!\!\!+} \underline{b}, \overline{a} \mathbin{\triangle\!\!\!\!\!+} \overline{b}]$$

$$[\underline{a},\overline{a}] \mathbin{\Diamond\!\!\!\!\!-} [\underline{b},\overline{b}] = [\underline{a} \mathbin{\nabla\!\!\!\!\!-} \overline{b}, \overline{a} \mathbin{\triangle\!\!\!\!\!-} \underline{b}]$$

$$[\underline{a},\overline{a}] \mathbin{\Diamond\!\!\!\!\!*} [\underline{b},\overline{b}]$$

$$= [\ \min \{\underline{a}, \mathbin{\nabla\!\!\!\!\!*} \underline{b}, \underline{a} \mathbin{\nabla\!\!\!\!\!*} \overline{b}, \overline{a} \mathbin{\nabla\!\!\!\!\!*} \underline{b}, \overline{a} \mathbin{\nabla\!\!\!\!\!*} \overline{b}\},$$

$$\max \{\underline{a}, \mathbin{\triangle\!\!\!\!\!*} \underline{b}, \underline{a} \mathbin{\triangle\!\!\!\!\!*} \overline{b}, \overline{a} \mathbin{\triangle\!\!\!\!\!*} \underline{b}, \overline{a} \mathbin{\triangle\!\!\!\!\!*} \overline{b}\}]$$

$$[\underline{a},\overline{a}] \mathbin{\Diamond\!\!\!\!\!/} [\underline{b},\overline{b}]$$

$$= [\ \min \{\underline{a}, \mathbin{\nabla\!\!\!\!\!/} \underline{b}, \underline{a} \mathbin{\nabla\!\!\!\!\!/} \overline{b}, \overline{a} \mathbin{\nabla\!\!\!\!\!/} \underline{b}, \overline{a} \mathbin{\nabla\!\!\!\!\!/} \overline{b}\},$$

$$\max \{\underline{a}, \mathbin{\triangle\!\!\!\!\!/} \underline{b}, \underline{a} \mathbin{\triangle\!\!\!\!\!/} \overline{b}, \overline{a} \mathbin{\triangle\!\!\!\!\!/} \underline{b}, \overline{a} \mathbin{\triangle\!\!\!\!\!/} \overline{b}\}].$$

For $0 \in [\underline{b},\overline{b}]$ the quotient is not defined. In the formulas for multiplication and division the minima and maxima can be computed with fewer floating-point operations, if the information about the signs of the operands is used. The above formulas can be converted into programs for the evaluation of the interval operations in a straightforward way. The same property holds for all structured operations described below.

4.2 Complex operations

Let (a^{Re},a^{Im}), $(b^{Re},b^{Im}) \in ¢R$ be two complex floating-point numbers and let $\Box \in \{\nabla,\triangle,\bigcirc\}$ be one of the roundings considered. Then definition (RG) leads to the following formulas:

$$(a^{Re},a^{Im}) \boxplus (b^{Re},b^{Im}) = (a^{Re} \boxplus b^{Re}, a^{Im} \boxplus b^{Im})$$

$$(a^{Re},a^{Im}) \boxminus (b^{Re},b^{Im}) = (a^{Re} \boxminus b^{Re}, a^{Im} \boxminus b^{Im})$$

$$(a^{Re},a^{Im}) \mathbin{\Box\!\!\!\!\!*} (b^{Re},b^{Im}) = (\Box(a^{Re}*b^{Re}-a^{Im}*b^{Im}),$$

$$\Box\,(a^{Re}*b^{Im} + a^{Im}*b^{Re}))$$

$$(a^{Re},a^{Im}) \mathbin{\Box\!\!\!\!\!/} (b^{Re},b^{Im}) = \left(\Box\,\frac{a^{Re}*b^{Re}+a^{Im}*b^{Im}}{b^{Re}*b^{Re}+b^{Im}*b^{Im})}, \Box\,\frac{a^{Im}*b^{Re}-a^{Re}*b^{Im}}{b^{Re}*b^{Re}+b^{Im}*b^{Im})}\right).$$

For $c = d = 0$ division is not defined. Addition and subtraction are executed by componentwise application of real floating-point operations. Multiplication contains two simple scalar products that can be evaluated using one of the algorithms of the preceding section.

In the computation of quotients, expressions of the form

$$q := (x_1 y_1 + x_2 y_2)/(u_1 v_1 + u_2 v_2)$$

occur with real floating-point numbers $x_i, y_i, u_i, v_i \in R$.

These can be evaluated by first computing an approximation $\tilde{q}$ with $\ell + 5$ digit accuracy. This can be accomplished by applying the scalar product algorithm to the numerator and denominator of the fraction. In most cases it can be proved that $\Box q = \Box \tilde{q}$ for all roundings $\Box \in \{\bigtriangledown, \bigtriangleup, \bigcirc\}$. Otherwise, the correct result $\Box q$ can be determined from $\tilde{q}$ and the residue

$$v := u_1 v_1 \tilde{q} + u_2 v_2 \tilde{q} - x_1 y_1 - x_2 y_2.$$

Because cancellation can occur in this expression, the scalar product algorithm has to be applied again.

4.3 Complex interval operations

Let $a^{Re}, a^{Im}, b^{Re}, b^{Im} \in IR$ be four floating-point intervals. Then the operations for the complex floating-point intervals $a = (a^{Re}, a^{Im})$, $b = (b^{Re}, b^{Im}) \in I\not{C}R$ can be executed as follows:

$$\begin{aligned}
a \mathbin{\Diamond\!\!\!+} b &= (a^{Re} \mathbin{\Diamond\!\!\!+} b^{Re},\ a^{Im} \mathbin{\Diamond\!\!\!+} b^{Im})\\
a \mathbin{\Diamond\!\!\!-} b &= (a^{Re} \mathbin{\Diamond\!\!\!-} b^{Re},\ a^{Im} \mathbin{\Diamond\!\!\!-} b^{Im}),\\
a \mathbin{\Diamond\!\!\!*} b &= \Big(\Diamond(a^{Re}*b^{Re} - a^{Im}*b^{Im}),\ \Diamond(a^{Re}*b^{Im} + a^{Im}*b^{Re})\Big)\\
a \mathbin{\Diamond\!\!\!/} b &= \Big(\Diamond((a^{Re}*b^{Re} + a^{Im}*b^{Im})\ /(b^{Re}*b^{Re} + b^{Im}*b^{Im})),\\
&\qquad \Diamond((a^{Im}*b^{Re} - a^{Re}*b^{Im})\ /(b^{Re}*b^{Re} + b^{Im}*b^{Im}))\Big).
\end{aligned}$$

Division is not defined if $0 \in b^{Re}*b^{Re} + b^{Im}*b^{Im}$. Note that addition and subtraction can be executed by componentwise applications of the corresponding interval operations, whereas for products and quotients, the scalar products and the quotients of scalar products which occur have to be computed exactly first and rounded only once afterwards. This can be done by analogy to complex operations.

4.4 Interval scalar product

Let $a,b \in IVR$ be two floating-point interval vectors of the form $a = ([\underline{a}_i,\overline{a}_i])_{i=1,\ldots,n}$, $b = ([\underline{b}_i,\overline{b}_i])_{i=1,\ldots,n}$. Then the scalar product is defined by

$$a \diamondsuit b := [\nabla \underline{s}, \Delta \overline{s}], \text{ where}$$

$$\underline{s} := \sum_{i=1}^{n} \min \{\underline{a}_i\underline{b}_i, \underline{a}_i\overline{b}_i, \overline{a}_i\underline{b}_i, \overline{a}_i\overline{b}_i\},$$

$$\overline{s} := \sum_{i=1}^{n} \max \{\underline{a}_i\underline{b}_i, \underline{a}_i\overline{b}_i, \overline{a}_i\underline{b}_i, \overline{a}_i\overline{b}_i\}.$$

Let us first consider the computation of $\underline{s}$. It is sufficient to compute two real floating-point vectors v,w with the property that

$$v_i w_i = \min \{\underline{a}_i\underline{b}_i, \underline{a}_i\overline{b}_i, \overline{a}_i\underline{b}_i, \overline{a}_i\overline{b}_i\} \quad \text{for} \quad i = 1,\ldots,n.$$

The real scalar product algorithm is then applied to these vectors using the downwardly directed rounding. As in the algorithm for interval products, the computation of the minima can be simplified by considering the signs of the operands. In the case where $0 \in [\underline{a}_i,\overline{a}_i] \cap [\underline{b}_i,\overline{b}_i]$, two products have to be computed and compared, whereas in the other cases the computation of one product suffices. This comparison can be carried out by computing and comparing the two products with double precision or by computing the scalar product of their difference, using upwardly directed rounding. This results from the following property:

$$\underline{a}_i\overline{b}_i > \overline{a}_i\underline{b}_i \Longleftrightarrow \Delta(\underline{a}_i\overline{b}_i - \overline{a}_i\underline{b}_i) > 0.$$

The second component $\overline{s}$ of the interval scalar product can be computed using similar considerations and the property

$$\underline{a}_i\underline{b}_i < \overline{a}_i\overline{b}_i \Longleftrightarrow \nabla(\underline{a}_i\underline{b}_i - \overline{a}_i\overline{b}_i) < 0.$$

Here the scalar product with downwardly directed rounding has to be applied. The comparison then yields two real floating-point vectors x,y to which the scalar product with upwardly directed rounding must be applied.

4.5 Complex scalar product and complex interval scalar product

Let $(a_i)_{i=1,\dots,n}, (b_i)_{i=1,\dots,n} \in V¢R$ be two complex vectors. Then their scalar product is

$$\square(\sum_{i=1}^{n} a_i b_i) = \left(\square(\sum_{i=1}^{n} (a_i^{Re} * b_i^{Re} - a_i^{Im} * b_i^{Im})),\ \square(\sum_{i=1}^{n} (a_i^{Re} * b_i^{Im} + a_i^{Im} * b_i^{Re}))\right).$$

In this way a complex scalar product of dimension n can be viewed as two real scalar products of dimension $2n$.

Analogously a complex interval scalar product of dimension n can be viewed as two interval scalar products of dimension $2n$.

4.6 Matrix and vector operations

All matrix and vector operations are defined by (RG) and (AG) for inner and outer operations, respectively. They can be composed componentwise out of additions, subtractions, multiplications (for scalar multiples) and scalar products (for matrix and vector products) in the corresponding component set R, IR, $¢R$, $I¢R$. In the example of the matrix product $*: MR \times MR \rightarrow MR$ this leads to

$$\bigwedge_{(a_{ij}),(b_{jk}) \in MR} \quad \bigwedge_{\square \in \{\triangledown, \triangle, \bigcirc\}} \quad (a_{ij}) \boxed{*} (b_{jk}) = (\square(\sum_{j=1}^{n} a_{ij} * b_{jk})).$$

4.7 Standard functions

The requirement (FG) for the precision of standard functions can only be fulfilled with great effort. For instance, if a function value $f(x)$ is close to a floating-point number y, then it has to be decided, whether $f(x)<y$, $f(x) = y$ or $f(x)>y$. This decision can require that an arbitrary number of decimal places of the value $f(x)$ be computed. This is impossible using a fixed word length and fixed approximation function for f. We therefore compute an approximation $\tilde{f}$ of the real standard function f that fulfills the following requirement:

$$\text{(FG')} \quad \bigwedge_{x \in D_f} |f(x) - \tilde{f}(x)| \leq |f(x)| \cdot 0.5 \cdot b^{1-\ell} \cdot b^{-3},$$

i.e., the relative error of $\tilde{f}$ is less than b^{-3} units of the last place of the floating-point system being used. In nearly all cases this gives $\square f(x) = \square \tilde{f}(x)$; in the worst case $\square f(x)$ and $\square \tilde{f}(x)$ are neighboring floating-point numbers.

The standard functions for intervals, complex numbers and complex intervals are computed with corresponding precision. Of course, in the case of intervals, the computed result always contains the precise result.

The total error in (FG') consists of approximation error, rounding errors in the computation of the approximation, errors in transformation of the range and adaptation of the result. For the required precision, the approximation has to be evaluated in a floating-point system with an increased number of places. We use rational approximations which are evaluated in the form of continued fractions. A more detailed description of the implementation of the standard function can be found in [11].

5. REALIZATION ON A MICRO COMPUTER

The entire arithmetic structure was implemented on a micro computer and embedded in an extension of the programming language PASCAL. For the realization of the requirements in section 1 the arithmetic has to be supplemented by input and output procedures of the same precision. This affects the choice of the base of the floating-point system. Numerical problems, which are usually described in the decimal system, can only be translated into the floating-point system without rounding errors if the decimal base is selected.

5.1 Number system

For this reason the following floating-point system was chosen: floating-point numbers possess $\ell = 12$ decimal digits in the mantissa and an exponent range from -99 to $+99$. Since the decimal point occurs after the first digit, the smallest positive number is $1_{10} - 99$ and the greatest is $9.99999999999_{10}99$. For the internal representation in a byte oriented computer, packed BCD representation turned out to be best suited. A floating-point number consists of the following 8 bytes:

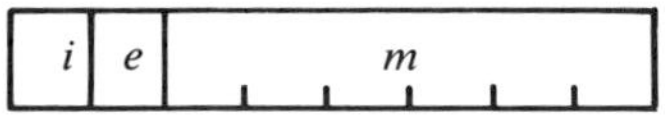

The first byte i is an infobyte. The first bit in this byte indicates the sign of the number

($1 \longleftrightarrow -, 0 \longleftrightarrow +$), the sixth bit is set if the number is zero, etc. The second byte e contains the exponent in binary representation, and the remaining bytes contain the mantissa m. The most significant two digits of the mantissa are packed into byte number eight, the least significant two digits of the mantissa are packed into byte number three. The algorithms which implement the arithmetic operations usually work with normalized mantissas, i.e., with the decimal point in front of the first digit. The internal exponent range is therefore shifted one place and is actually $-98..100$.

In addition to this format a second number format with 20 decimal digits is used for standard functions. Such a long floating-point number is represented in 12 bytes and consists of infobyte i, exponent e and mantissa m just as a normal floating-point number. Infobyte and exponent have the same format as in the case of single length floating-point numbers; the 20 decimal digits are stored in 10 bytes.

5.2 Implementation of the modules of the arithmetic

Levels 1 and 2 of the arithmetic are programmed in assembly language to achieve greatest possible execution speed. The operations of level 3 are programmed using the operator definition of an extension of PASCAL, and they have access to the lower level routines via external procedure calls. These structured operations are contained in packages which correspond roughly to the described modules of level 3. There are additional packages with additional standard functions (10^x, 2^x, $\log_{10}$, $\log_2$, tan ,sinh, cosh , tanh), packages with numerical algorithms, a package for the computation of the sum $\sum_{i=1}^{n} x_i$ of n floating-point numbers, etc. The implementation on a Z80 microprocessor needs the following storage space for the modules of levels 1 and 2 and for the standard functions:

Table 1: Storage space

Input/output	1.3 kbyte
Scalarproduct (incl. long storage)	1.2 kbyte
Remaining operations, incl. basic routines	2.6 kbyte
Standard functions, incl. long arithmetic	3.8 kbyte

For the computation of scalar products, algorithm III with two long storage spaces was chosen. In the implementation of PASCAL, the compiler automatically notes which arithmetic operations occur in a user's program. Only the required portion of the arithmetic is loaded.

For the Z80 microprocessor with 2.5 MHz frequency the execution times are listed in table 2. For $n>1$, scalar products with the internal algorithm are faster than the less precise

Table 2: Execution times

operation	datatype	times in msec		
		min	max	typical
+,−	real (12 places)	1.5	3.0	2.2
	longreal (20 places)	3.2	6.1	4.7
	interval	4.1	7.2	5.4
	complex	4.1	7.2	5.4
*	real	1.5	7.6	6.0
	longreal	3.2	15.5	12.5
	interval	17.8	40.5	23
	complex	22	49.5	45
/	real	1.5	13.9	10
	longreal	3.2	23.3	20
	interval	17.8	38.5	31
	complex	30	109	100
exact scalar-product	realvector with dimension n	8+0.05n	14+6.5n	8+5.5n
scalarproduct simulated with for-loop	realvector with dimension n	1+4.0n	1+10.5n	1+9.6n
sqrt	real	5.8	112	106
exp	real	20	142	135
ln	real	114	180	172
arctan	real	108	163	155
sin	real	2.5	172	165
cos	real	2.5	172	165

realization by means of a for-loop. The main reasons for this are:

1) the scalar product algorithm can locate the next operands simply by incrementing the old address, thus avoiding complicated index calculations,

2) some data transports from and to the stack, some range checks and some passes through the central interpreter loop are avoided,

3) in each step $s := s + a*b$ an addition and a multiplication occur; in the exact scalar product these operations are much simpler because they don't contain rounding, normalization, etc.

REFERENCES

[1] Bohlender, G. (1978). Genaue Berechnung mehrfacher Summen, Produkte und Wurzeln von Gleitkommazahlen und Arithmetik in Höheren Programmiersprachen. Dissertation, Universität Karlsruhe.

[2] Bohlender, G. (1980). Embedding Universal Computer Arithmetic in Higher Programming Languages. *Computing* 24, 149-160.

[3] Grüner, K. (1979). Allgemeine Rechnerarithmetik und deren Implementierung. Dissertation, Universität Karlsruhe.

[4] Kaucher, E. Solving functional equations with guaranteed, close bounds. This volume.

[5] Kaucher, E., Klatte, R., Ullrich, Ch. (1981). *Programmiersprachen im Griff.* Band 2. PASCAL; *Anhang* PASCAL-SC. Bibliographisches Institut Mannheim.

[6] Kulisch, U. (1976). *Grundlagen des numerischen Rechnens.* Bibliogr. Inst. Mannheim.

[7] Kulisch, U., Bohlender, G. (1976). Formalization and Implementation of Floating-point Matrix Operations. *Computing* 16, 239-261.

[8] Kulisch, U. (1981). Numerisches Rechnen wie es ist und wie es sein könnte. Jahrbuch Überblicke Mathematik. Bibliographisches Institut.

[9] Kulisch, U., Miranker, W. L. (1981). *Computer Arithmetic in Theory and Practice.* Academic Press.

[10] Rump, S. M. Solving algebraic problems with high accuracy. This volume.

[11] Wolff von Gudenberg, J. (1980). Einbettung allgemeiner Rechnerarithmetik in PASCAL mittels eines Operatorkonzeptes und Implementierung der Standardfunktionen mit optimaler Genauigkeit. Dissertation, Universität Karlsruhe.

FEATURES OF A HARDWARE IMPLEMENTATION OF AN OPTIMAL ARITHMETIC

U. Kulisch
G. Bohlender

Institute for Applied Mathematics
University of Karlsruhe
Karlsruhe, West Germany

We give a brief review of the definition of the arithmetic operations on a computer by semimorphisms. Then we display the 15 fundamental operations that are most useful and convenient for an implementation of semimorphic operations in computer representable subsets of the most commonly used linear spaces and their interval sets. Techniques for the implementation of twelve of these operations: addition, subtraction, multiplication and division with three roundings are well known and are common knowledge nowadays. The paper focuses, therefore, on the implementation of scalar products with maximum accuracy and diverse roundings. We sketch several possibilities for a hardware realization of optimal scalar products. We give an algorithmic and flow chart description of one such hardware unit and discuss the natural parallelisms in scalar products, Finally, we comment on the pipelining of the 15 fundamental arithmetic operations.

CONTENTS

1. INTRODUCTION

Integer arithmetic may be performed correctly on computers, provided that no overflow occurs. Real arithmetic, in contrast, has to be approximated. In this paper we discuss the

ISBN 0-12-428660-7

optimal way of performing this approximation for the real numbers as well as for certain product spaces over the real numbers.

In general a real number is represented as an infinite b-adic expansion, for instance,

$$\pi = 3.14159.... .$$

On computers these representations can only be approximated by a finite number ℓ of digits. So-called fixed-point or floating-point representations are customary. When one of the real operations $+, -, \times, /$ is applied on two such numbers the result, in general, is not a computer representable number (i.e., fixed-point or floating-point number) again. As an example, the multiplication of two floating-point numbers doubles the number of digits in the mantissa. If the result of an operation on two computer representable numbers is to be a computer representable number, the real operations have to be approximated. These approximations are customarily called rounded operations. On many existing computers the implementation of these computer approximations of the real operations is more or less based on intuition and not on a precise mathematical definition.

In higher mathematical structures like matrices or complex numbers the computer operations are defined by the following consideration. In analysis the operations for complex numbers and matrices, for instance, are defined in terms of operations for the real numbers through well known expressions. On most existing computers the computer arithmetic in the product spaces is defined by evaluating these expression using the given floating-point operations. It is well known that for operations in the product spaces this method results in unnecessarily large roundoff errors, an unnecessarily ramified error analysis and excessively poor roundoff error estimates. Typically even the basic floating-point operations which represent the real operations are imprecisely defined and thus they themselves do not even adhere to the customary error bounds. This defect is compounded many times by the usual approach of evaluating expressions. A remedy for this kind of imprecision is a new definition of computer arithmetic which is based on a recently developed theory ([5],[6]). There the arithmetic on computers is defined via semimorphism in all relevant cases: real numbers, real vectors, real matrices, complex numbers, complex vectors and complex matrices as well as the intervals over the real and complex numbers, vectors and matrices. We briefly discuss the requirements of a semimorphism. Let M denote any of the sets mentioned above and let $N \subseteq M$ be its computer representable subset, then the operations in N are defined as follows.[†]

(RG) $\bigwedge_{a,b \in N} a \boxast b := \square(a*b)$ for all operations $*$.

Here $\square$: $M \to N$ denotes a monotone and antisymmetric mapping, the rounding from M into N. This mapping has the following properties

(R1) $\bigwedge_{a \in N} \square a = a$ (rounding)

(R2) $\bigwedge_{a,b \in M} (a \leq b \Rightarrow \square a \leq \square b)$ (monotonicity)

(R3) $\bigwedge_{a \in M} \square(-a) = -\square a$ (antisymmetry)

In the case that M is a set of intervals, the rounding $\square$ has the additional property

(R4) $\bigwedge_{a \in M} a \leq \square a$ (upwardly directed)

In this case $\leq$ denotes the inclusion $\subseteq$. The properties (RG), (R1), (R2), (R3) and in case of interval sets (R4) as well, define a semimorphism. All operations defined by semimorphism deliver maximum accuracy in the sense that there is no element of the subset between the correct result of an operation and its approximation in the subset. Furthermore, in the case of the mapping of the real numbers into the floating-point numbers the two directed roundings $\triangledown$ and $\triangle$ are important. These are defined by (R1), (R2) and

---

[†] For an explanation of the symbols used see the appendix.

(R4) $\bigwedge_{a \in M} \nabla a \leq a$ and $\bigwedge_{a \in M} a \leq \Delta a$.

Along with these directed roundings we define the corresponding operators by

(RG) $\bigwedge_{a,b \in N} a \mathbin{\overset{*}{\triangledown}} b := \nabla(a*b)$ and $\bigwedge_{a,b \in N} a \mathbin{\overset{*}{\vartriangle}} b := \Delta(a*b)$

for all $* \in \{+,-,\times,/\}$.

The roundings ∇ and Δ as well as the operators $\overset{*}{\triangledown}$ and $\overset{*}{\vartriangle}$ are needed for reliable error estimation in numerical computations. The monotone, downwardly directed rounding ∇ maps the entire interval between two neighboring numbers of the subset N onto the computer representable lower bound of the interval. The monotone upwardly directed rounding Δ has the opposite property.

Questions of uniqueness of the directed roundings are discussed in [5] and [6]. It is also shown in [5] and [6] that operations defined by semimorphism have many desirable and useful properties.

Computers have been developed at the authors' institute which provide arithmetic operations defined by semimorphism in the spaces of real and complex numbers, vectors and matrices as well as real and complex intervals, interval vectors and interval matrices. In general, the operations defined by semimorphism turn out to be faster than the routines for these operations implemented in the customary manner.

We now turn to the question of implementing the operations defined by semimorphism by means of fast algorithms in all cases under discussion. At first sight it seems doubtful that formula (RG), in particular, can be implemented on computers at all. In order to determine the approximation $x \boxast y$, the correct result $x*y$ seems to be necessary. If $x*y$ is representable on the computer we do not have to approximate it. In general, $x*y$ will not be representable and à fortiori not executable on the computer. In this case it seems that $x*y$ can not be used to define $x \boxast y$.

Therefore, one has first to express the operations defined by semimorphism in terms of executable formulas. This can be done by means of isomorphism. For a detailed discussion see [5], [6].

In [5] and [6] it is shown that all semimorphic operations discussed in this paper can be realized on computers by a modular technique in terms of a higher programming language if

1. an operator concept or an operator notation is available for all operations in that higher level language and

2. the following 15 fundamental operations for floating-point numbers are available

$\boxplus$	$\boxminus$	$\boxtimes$	$\boxslash$	$\boxdot$
$\nabla\!\!\!\!+$	$\nabla\!\!\!\!-$	$\nabla\!\!\!\!\times$	$\nabla\!\!\!\!/$	$\nabla\!\!\!\!\cdot$
$\triangle\!\!\!\!+$	$\triangle\!\!\!\!-$	$\triangle\!\!\!\!\times$	$\triangle\!\!\!\!/$	$\triangle\!\!\!\!\cdot$

Here $\boxast$, $\in\{+,-,\times,/\}$, denotes the semimorphic operations defined by (RG) using a particular monotone and antisymmetric rounding (R1,2,3) such as rounding to the nearest number of the screen. $\nabla\!\!\!\!*$ and $\triangle\!\!\!\!*$, $*\in\{+,-,\times,/\}$, denote the operations defined by (RG) and the monotone downwardly respectively upwardly directed rounding. $\boxdot$, $\nabla\!\!\!\!\cdot$ and $\triangle\!\!\!\!\cdot$ denote three scalar products with maximum accuracy. In particular, if S is the set of floating-point numbers of a certain computer then for vectors $a=(a_i)$ and $b=(b_i)$, $a_i, b_i \in S$, $i=1(1)n$:[†]

$$\text{(RG)} \quad \bigwedge_{a,b\in S^n} a \odot b := \bigcirc\Big(\sum_{i=1}^{n} a_i \times b_i\Big),\ \bigcirc\in\{\square, \nabla, \triangle\} \qquad \text{(I)}$$

for all relevant numbers n. The multiplication and summation sign on the right hand side of (RG) mean the correct multiplication and summation in the sense of real numbers. Algorithms for the implementation of these operations can be found in [2], [5] and [6]. Traditional numerical analysis makes use of only four ($\boxplus$, $\boxminus$, $\boxtimes$ and $\boxslash$) of these 15 fundamental operations. Traditional interval arithmetic uses the eight operations $\nabla\!\!\!\!+$, $\nabla\!\!\!\!-$, $\nabla\!\!\!\!\times$, $\nabla\!\!\!\!/$ and $\triangle\!\!\!\!+$, $\triangle\!\!\!\!-$, $\triangle\!\!\!\!\times$, $\triangle\!\!\!\!/$. The latter are computer equivalents of the operations for real intervals.

[†] For an explanation of the symbols see the appendix

The newly proposed IEEE Arithmetic offers these twelve operations: $\boxed{*}$, $\triangledown\!\!\!\!\!*$, $\vartriangle\!\!\!\!\!*$, $*\in\{+,-,\times,/\}$. These twelve were systematically implemented in software on a Zuse Z23 in 1967 and in hardware on an ELEKTROLOGIKA X8 at the Institute for Applied Mathematics at the University of Karlsruhe in 1968. Both implementations were supported by a high level language "TRIPLEX ALGOL 60" published in Num. Math. in 1967 [1].

Nowadays, the implementation of the twelve operations $\boxed{*}$, $\triangledown\!\!\!\!\!*$, $\vartriangle\!\!\!\!\!*$, $*\in\{+,-,\times,/\}$, is routine and does not require any further discussion, see [5], [6]. The historical development of computation has shown that many difficulties that traditionally occur in numerical analysis as well as in interval arithmetic cannot principally be avoided unless the three scalar products $\boxdot$, $\triangledown\!\!\!\!\cdot$ and $\vartriangle\!\!\!\!\cdot$ are available on the computer for all relevant values of the dimension n. Their implementation is the purpose of this paper. These scalar products expose a new capability for Numerical Analysis with respect to an automatic error control, that is, a final verification and correction of computed results. They play a key role for an automatic error analysis and the computation of small bounds for the solution of the problem by the computer itself. Roughly speaking it may be said that interval arithmetic brings guarantees into computation while the three scalar products deliver high accuracy. Both features are desirable. They should not be confused.

2. IMPLEMENTATION OF SCALAR PRODUCTS

Optimal scalar products can be provided on a computer, as will be discussed below, by a black box technique, where the vector components a_i and b_i, $i = 1(1)n$, are the input and the scalar products the output.

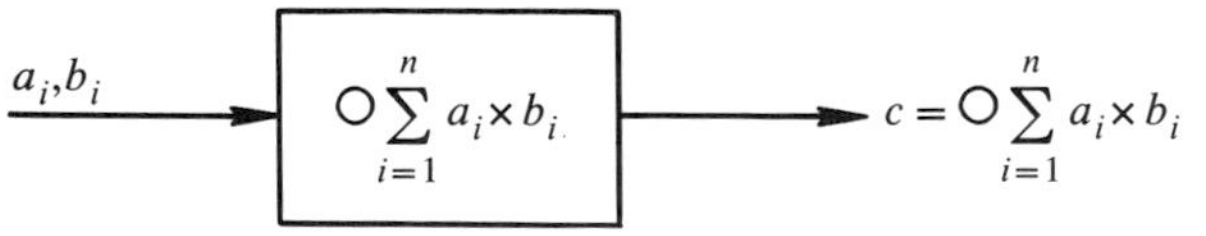

$$\bigcirc\in\{\Box,\nabla,\Delta\}.$$

The black box requires some local storage and works independently of the main store of the computer. The size of the local store depends only on the data formats in use (base of the number system, length of the mantissa and range of the exponents). In particular, it is independent of the dimension n of the two vectors $a = (a_i)$ and $b = (b_i)$ to be multiplied. *When implemented in hardware, such a scalar product unit leads to a considerable gain in speed whenever scalar products occur in a computation.*

In order to describe such a scalar product unit we use floating-point numbers in the so-called sign magnitude representation with exponents. For other representations (e.g. complement representation of the mantissa and/or characteristic in the exponents) similar considerations and arrangements are possible.

A normalized floating-point number x (in sign magnitude representation) is a real number x in the following form

$$x = *m \bullet b^e.$$

Here $* \in \{+,-\}$ is the sign of the number (sign (x)), m the mantissa (mant (x)), b the base of the number system in use and e the exponent (exp (x)). b is a natural number with $b>1$. The exponent is an integer between the two integer bounds $e1, e2$ and in general, $e1 \leq 0 \leq e2$. The mantissa m is of the form

$$m = \sum_{i=1}^{n} x[i] \bullet b^{-i}.$$

The $x[i]$ are the digits of the mantissa. They have the properties $x[i] \in \{0,1,...,b-1\}$ for all $i = 1(1)n$ and $x[1] \neq 0$. The set of all such normalized floating-point numbers does not contain zero. In order to have a unique representation of zero available we additionally assume that sign $(0) = +$, mant $(0) = 0.00...0$ (n zeros after the dot) and exp $(0) = e1$. Such a floating-point system depends on four constants b, ℓ, $e1$ and $e2$. We denote it by $S = S(b,\ell,e1,e2)$.

In order to realize (RG) in a computer the products $a_i \times b_i$ are executed correctly. This leads to a mantissa of 2ℓ digits and an exponent in the range $2e1 - 1 \leq e \leq 2e2$. Then (RG) can be generated if we implement the sum

$$\bigcirc\left(\sum_{i=1}^{n} c_i\right), \quad \bigcirc \in \{\Box, \nabla, \Delta\} \qquad \text{(II)}$$

on the computer, where the c_i, $i = 1(1)n$, denote floating-point numbers of 2ℓ digit mantissas, i.e., $c_i \epsilon S(b,2\ell,2e1 - 1,2e2)$ for all $i = 1(1)n$. We now discuss the principles of several ways to realize (I) and (II).

2.1 Long Accumulator

If one of the summands in (II) has exponent 0, its mantissa can be represented in a register of length 2ℓ. If another summand has exponent 1, it can be represented with exponent 0, if the register provides further digits on the left and the mantissa is shifted one place to the left. An exponent -1 in one of the summands of (II) causes a corresponding shift to the right. The largest exponents in magnitude, that may occur in (II), are $2e2$ and $2|e1|$. This shows that all summands can be represented (in some kind of fixed-point representation) without loss of information in a register of length $2e2 + 2\ell + 2|e1|$. If the register is built as an accumulator, all summands can even be added without loss of information.

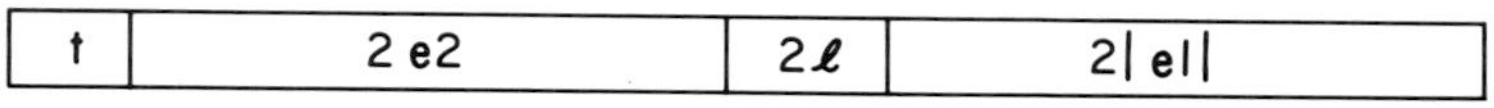

In order to catch possible overflows it is convenient to provide a few (t) more digits of base b on the left. In such an accumulator every sum (II) can be added without loss of information. As many as b^t overflows may occur and be accounted for without loss of information.

Another case of a hardware realization of (II) is given by

2.2 A Serial Adder with a Long Shift Register

We use a serial adder, which is able to add two digits of base b with proper carry handling, and a shift register of length $t + 2e2 + 2\ell + 2|e1|$.

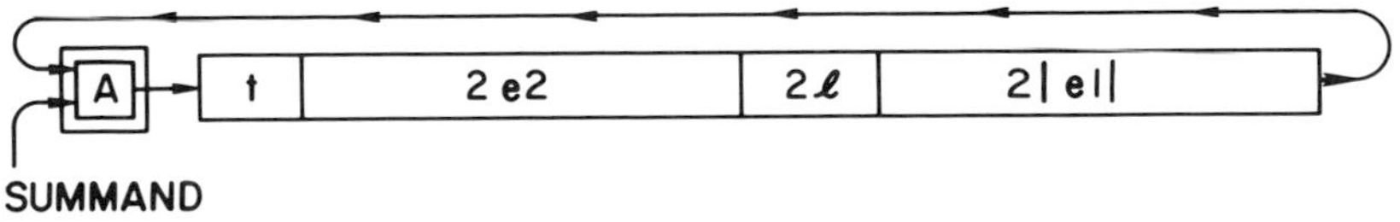

During each addition of a summand the contents of the shift register is used as one input of the adder beginning with it's right end. The next summand of (II) is the second input of the adder. It is added at a proper position as determined by it's exponent.

The case of the long accumulator, that we discussed under 2.1 is a proper generalization of the parallel adder for our purposes. The second case is a corresponding generalization of the serial adder. The implementations 2.1 and 2.2 may be viewed as extreme cases. We now sketch several other implementations which may be viewed as lying between the extreme cases.

2.3 Local Memory with an Accumulator of Length 3ℓ

In the cases 2.1 and 2.2 above, addition is executed over many digits in which the second summand is zero. This, of course, is not necessary. It suffices just to add over those digits in which the second summand is not zero. The remaining digits can be kept in a local memory. Therefore, we use a local memory of a length as in 2.1 and 2.2. The memory is organized in words of length ℓ.

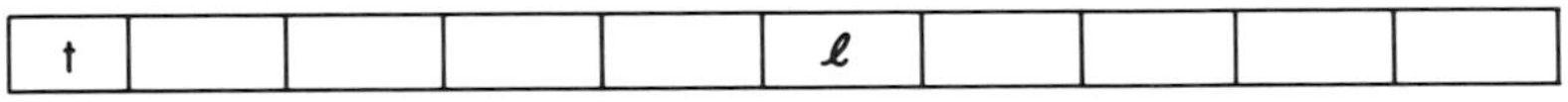

Since the summands are of length 2ℓ they fit into a certain part of this local memory of length 3ℓ. This part of the memory is determined by the exponent of the summand. We take this part of memory into an accumulator of length 3ℓ. The summand is taken into a shift register of length 3ℓ, shifted to the right position as determined by the exponent and added to the contents of the accumulator. Instead of the shift register a cross point switch may be used which serves the same purpose in parallel.

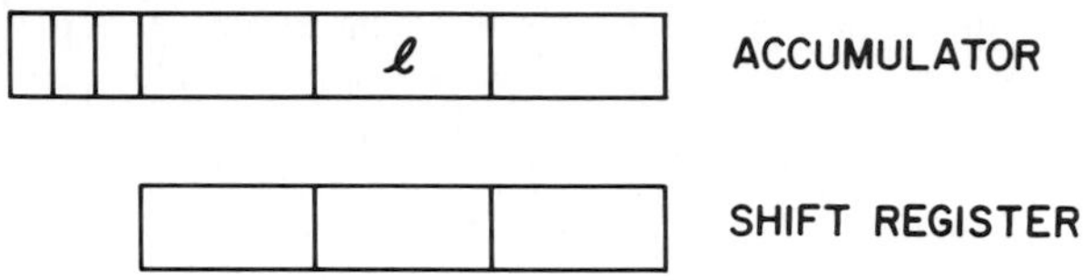

The addition into the accumulator may produce a carry. It is useful to enlarge the accumulator on its left end by a few more digits and to fill these digits with the corresponding digits of the local memory. If not all of these additional digits equal $b - 1$, they will catch a possible carry of the addition. Of course, it is possible that all these additional digits are $b - 1$. For this case a loop has to be provided that takes care of the carry and adds it to the next digits of the local memory. This loop eventually has to be traversed several times.

2.4 Local Memory with a Serial Adder and Shift Register

We use a local memory of size and organization as in 2.3. The addition, however, is executed serially. The new summand is added, corresponding to its exponent, to the proper position of the contents of the shift register.

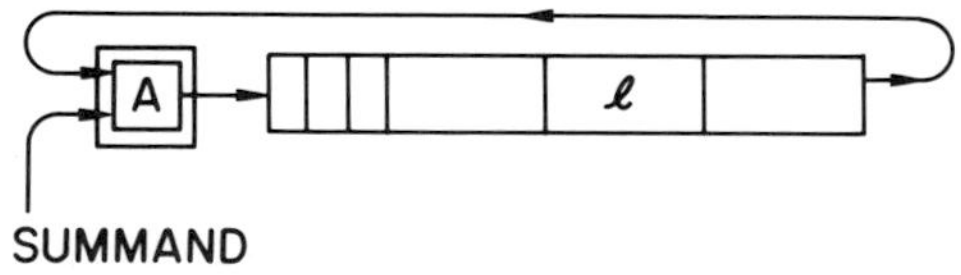

Again measures have to be provided that take care of a carry which overruns these few more digits on the left end of the shift register.

2.5 Remarks on the Implementation of Scalar Products

a) The length of the accumulator or shift register in 2.3 or 2.4, as the case may be, can be shortened further. If, for instance, the multiplications in a scalar product are executed via a table of the multiples of one factor, the multiples of this factor could immediately be added into the local memory as determined by the digits of the second factor. This enlarges the number of additions into the local memory from n to $\ell \bullet n$. The length of the summands, however, is reduced to $\ell + 1$ digits of base b. These additions can, in close analogy with the cases of 2.3 and 2.4, be executed by an accumulator or shift register of length 2ℓ and a few additional digits for the handling of possible carries.

However, this procedure exhibits some disadvantages. The addition into the local memory is relatively complicated compared to the additions that occur in a multiplication (all summands have the same sign, are of a short length and are only of a small exponent difference. No overflow can occur.)

The accumulator and shift register used under 2.3 and 2.4 can, of course, be reduced to a length of ℓ digits. Then the addition of a product c_i has to be split into three or four respectively consecutive additions.

The extreme case of this shortening of the accumulator is a software computation of the scalar product with maximum accuracy. Here, for instance, the process 2.3 is simulated by a very short accumulator perhaps of length 8 or 16 bits. The long addition is executed by several addition loops.

b) After the summation of the n products c_i, the result has still to be rounded to ℓ digits. For an execution of the rounding to the nearest number of the screen only the first $\ell + 1$ digits of the sum are needed. For a fast identification of the ℓth or $(\ell + 1)^{st}$ digit of the sum the use of a pointer is convenient. During the summation process the pointer always indicates the most significant non zero digit of the sum.

For a correct execution of the roundings ∇ and Δ it is essential to know whether after the ℓth digit of the sum there are still other non zero digits or not. This information can easily be obtained if during the summation process a second pointer is used, which always indicates the least significant non zero digit of the sum. This second pointer is very useful for high speed execution of the roundings ∇ and Δ.

c) In the cases of 2.3 and 2.4 the exponent of the products, in principle, can be computed with the same registers that perform the additions. However, as the exponents are relatively short, it may be simpler to provide a special exponent adder for the addition of the exponents. The addition of the exponents can then be executed in parallel with the multiplication of the mantissas and addition of the products. Such an exponent adder can, for example, also function as the loop counter required by the algorithm.

d) The addition of $c_1 = b^{60}$ and $c_2 = -b^{60}$ shows that carries or borrowing over many digits may occur. It may be useful, therefore, to use two local memories (accumulators or shift registers) instead of one and generate the sum of the positive and negative summands separately. At the final end of the summation process the difference of the two local memories has to be computed. (This long difference is of the computational complexity of two mantissa products so that it is negligible, in general.)

This long subtraction can be simplified, since the correct difference of the two long local memories is not necessary in order to obtain the final result. The first $\ell + 1$ significant digits of the difference together with one further bit, are sufficient for this purpose. This additional bit (called residue bit) carries the information whether there are non zero digits after the $(\ell + 1)^{st}$ digit or not. The $(\ell + 1)^{st}$ digit, for instance, is sufficient for an execution of the rounding to the nearest number of the screen. The residue bit serves for a correct execution of the roundings ∇ and Δ.

The final subtraction of the two long local memories can be simplified, if two pairs of pointers are used during the summation process, the first pointer always indicating the first, the second pointer indicating the last non zero digit of the long local memories. The first pointers, indicating the first non zero digits, allow a fast computation of the sign of the difference. The second pair of pointers, indicating the last non zero digits of the two long memories, are very useful for a fast determination of the value of the residue bit.

e) Instead of two long local memories only one long local memory together with a set of carry counters may be used. All positive or negative summands

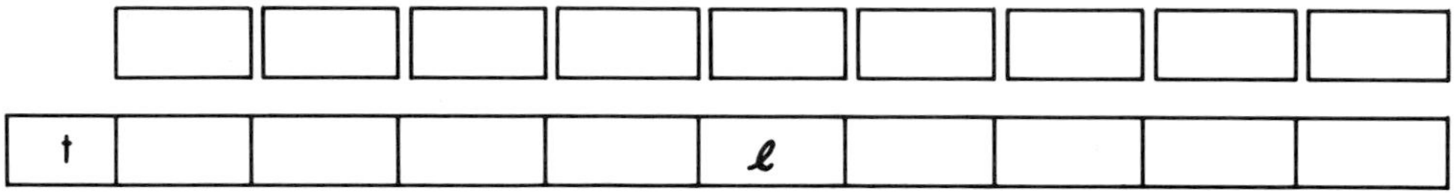

are added to the long local memory. If a carry occurs beyond a certain word of the long local memory, it is added to the next higher carry counter. Similarly, if borrowing is necessary from the next higher word it is taken as a negative carry from the next higher carry counter. This avoids a time-consuming carry processing during the summation. At the end of the summation the carries are added to the long local memory.

f) It is an essential advantage of the methods for a computation of scalar products, which we are discussing here, that the whole summation process can be executed with a bounded local memory as part of the scalar product unit. The size of this local memory is independent of the dimension n of the two vectors to be multiplied. No further use of the main store of the computer is necessary. In general, the connections to this local memory will be shorter and therefore the access time may be faster than to the main store of the computer.

g) We recommend that the additions necessary for the multiplications be already executed in an accumulator of length 3ℓ. Then the computed products can be shifted to the proper position for the next addition corresponding to the value of the exponent by a fast shift register of the same length 3ℓ or by a cross point switch.

h) We discuss the length of the long local memory (accumulator or shift register) for a few typical cases:

Let $b = 16$, $e1 = -64$, $e2 = 64$, $\ell = 14$ and $t = 8$. Then $t + 2e2 + 2\ell + 2|e1| = 292$ hexadecimal digits. These are 146 Bytes or 1168 bits.

If $b = 10$, $e1 = -99$, $e2 = 99$, $\ell = 13$, and $t = 8$, then $t + 2e2 + 2\ell + 2|e1| = 430$ decimal digits. In BCD these are 215 Bytes or 1720 bits.

If $b = 2, e1 = -1023, e2 = 1023$ $(2^{1023} \approx 10^{308})$, $\ell = 53, t = 64,$ then $t + 2e2 + 2\ell + 2|e1| = 4262$ bits.

It is suitable to add so many digits at the left and right end of the long local memory that its length becomes an integer multiple of ℓ.

i) The summation apparatuses, that we sketched above, also accommodate non normalized components a_i, b_i of the vectors a and b in the scalar products of the form $\pm 0.x[1]x[2]...x[\ell] \bullet b^{e1}$ with $x[1] = 0$.

j) Before starting the summation process the long local memory (accumulator or shift register) has to be set to zero.

Several applications, however, require another version of the summation process where the long local memory (accumulator or shift register) is not initially set to zero. Here the summands are added to the previous contents of the long local memory.

With this version of the scalar product or summation unit, respectively, for instance, expressions of the form

$$\square(AB + CD) \text{ with matrices } A,B,C,D$$

can easily be evaluated correctly. Such expressions occur, for instance, in a multiplication of block matrices. Without this additional version the matrices A,B,C,D first would have to be brought into the form

$$(A\ C) \begin{pmatrix} B \\ D \end{pmatrix}$$

by restoring, requiring an unnecessary time expenditure.

3. ALGORITHMIC AND FLOWCHART DESCRIPTION OF A HARDWARE UNIT

Now we shall discuss an algorithm, which highlights the algorithmic flow of a hardware unit for the computation of scalar products with maximum accuracy. Our model is of course only one of several possible realizations. In particular, we discuss the case of the local memory with an accumulator of length 3ℓ, which we mentioned above. We additionally assume that the unit works with two long local memories S_1 and S_2.

For two vectors $a = (a_i)$ and $b = (b_i)$ the following sum

$$c = \bigcirc\left(\sum_{i=1}^{n} a_i \times b_i\right) = \bigcirc\left(\sum_{i=1}^{n} c_i\right), \quad \text{with } c_i = a_i \times b_i$$

has to be computed. Here $a_i, b_i \in S(b,\ell,e1,e2)$, $c_i \in S(b,2\ell,2e1-1,2e2)$ for all $i = 1(1)n$ and $\bigcirc \in \{\square, \nabla, \triangle\}$.

All relevant cases $b \geq 2$ are admitted and we assume that in case $b > 2$ the multiplications are executed via a table of multiples of one of the operands.

The mantissas of the floating-point numbers a_i, b_i and c_i are denoted by ma_i, mb_i and mc_i, the exponents by ea_i, eb_i and ec_i.

The following figure highlights the organization of a hardware unit for the computation of scalar products with maximum accuracy. The flow diagram, which follows, displays the algorithmic flow of such a hardware unit. In the flow diagram, we use the usual conventions: rectangles denote statements; circles denote labels; and figures with six edges denote conditions. The statements belonging to a for-loop are framed after the opening of the loop.

The steps 2 to 5 of the flow diagram describe the multiplication of the mantissas $ma_i \bullet mb_i$. The steps 6 to 8, the addition of mc_i to a part of the long local memories S_1 or S_2 as determined by the exponent ec_i.

In step 9 the difference of the two long memories is computed.

The hardware unit could easily deliver all rounded results $\square c$, ∇c and $\triangle c$ in parallel.

Finally, we remark that a hardware unit for the computation of scalar products can, in principle, be used to execute the single floating-point operations $\boxplus$, $\mathbin{\nabla\!\!\!\!\!+}$, $\mathbin{\triangle\!\!\!\!\!+}$, $\boxminus$, $\mathbin{\nabla\!\!\!\!\!-}$, $\mathbin{\triangle\!\!\!\!\!-}$ and $\boxtimes$, $\mathbin{\nabla\!\!\!\!\!\times}$, $\mathbin{\triangle\!\!\!\!\!\times}$ also.

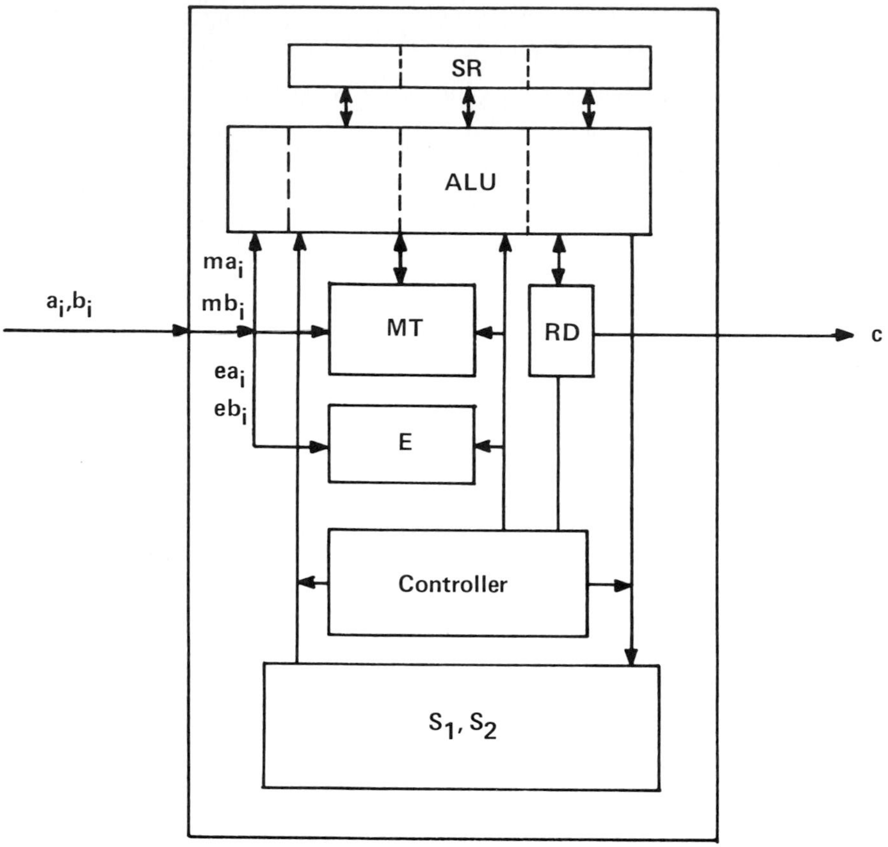

SR	speed shifter or cross point switch
ALU	arithmetic and logical unit
MT	table of multiples or multiplier
RD	rounder
E	exponent register or exp. adder
S_1,S_2	storages for sum of positive and negative summands $mc_i = ma_i \bullet mb_i$

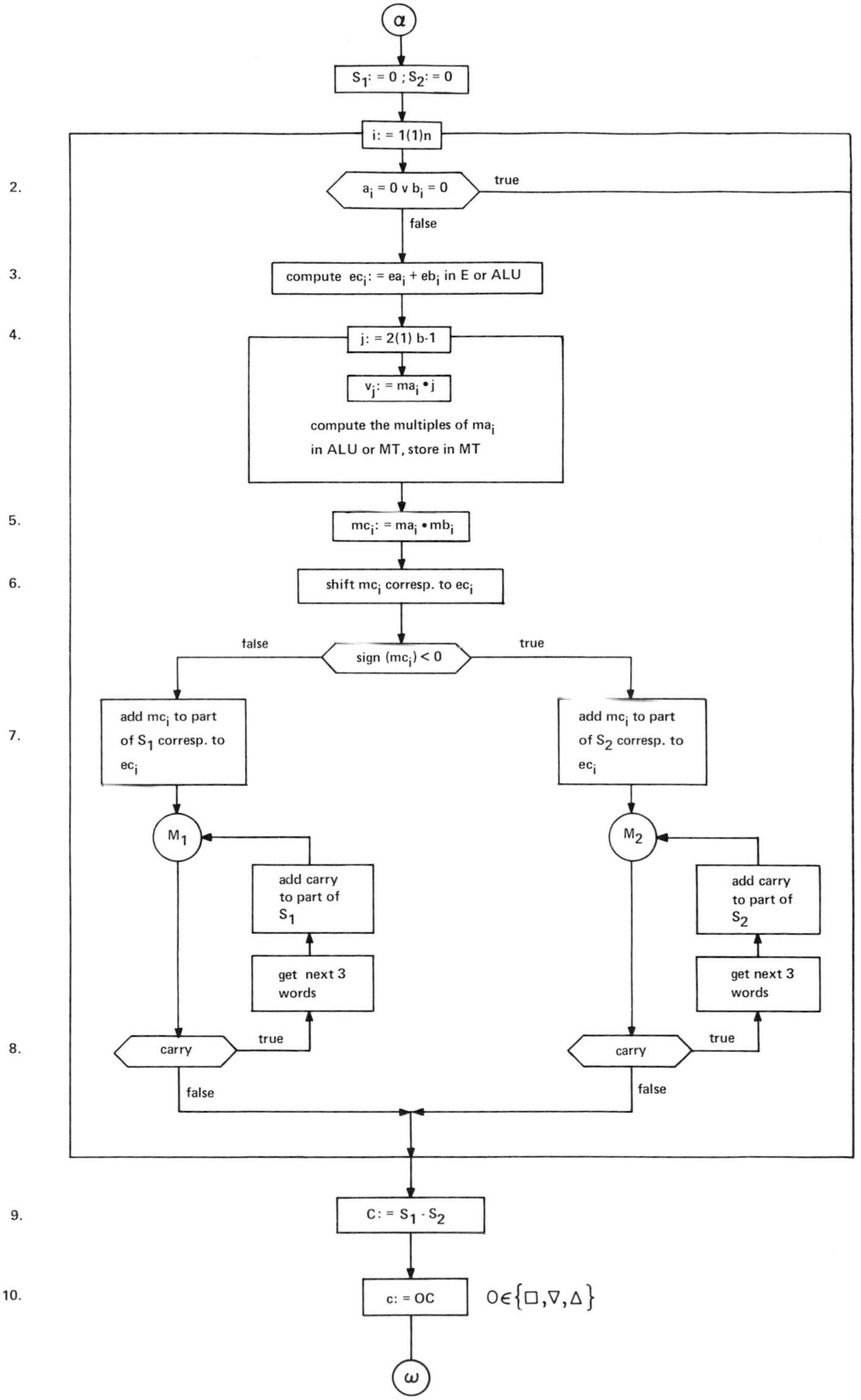
α
S1: = 0 ; S2: = 0
i: = 1(1)n
2.
ai = 0 v bi = 0
true
false
3.
compute eci: = eai + ebi in E or ALU
4.
j: = 2(1) b-1
vj: = mai • j
compute the multiples of mai
in ALU or MT, store in MT
5.
mci: = mai • mbi
6.
shift mci corresp. to eci
false
sign (mci) < 0
true
7.
add mci to part of S1 corresp. to eci
add mci to part of S2 corresp. to eci
M1
M2
add carry to part of S1
add carry to part of S2
get next 3 words
get next 3 words
8.
carry
true
false
carry
true
false
9.
C: = S1 - S2
10.
c: = OC
O∈{□,∇,Δ}
ω

The following comments are keyed to the flowdiagram by the column of numbers at its left side.

Comments:

1. The long local memories are set to zero.

2. If one of the two vector components a_i or b_i is zero, no summation is necessary.

3. Computation of the exponent of the product $a_i \bullet b_i$ in exponent adder or in ALU. The result is stored in exponent register (E).

4. Computation of the multiples of one factor (ma_i). The multiples are stored in a table of multiples (*MT*). In the case of the binary system this step is empty. If other multiplication procedures are used this step has to be altered.

5. Computation of the products of the mantissas $mc_i := ma_i \bullet mb_i$.

6. mc_i is shifted corresponding to ec_i with respect to a part of the long local memory of length 3ℓ, via a shift register or a cross point switch.

7. Positive summands are added to S_1, negative summands are added to S_2.

8. Check for carry and carry handling.

9. Computation of the difference of the two long local memories.

10. Rounding of the result to length ℓ by $\square$, $\triangledown$ or $\triangle$, or simultaneously by all three of these roundings. The rounder (*RD*) in the block diagram carries the right pointer and the rounding information bit.

We already noted that it is also necessary to provide for a scalar product computation in which the two long local memories are not set to zero, that is, where step 1 is skipped.

4. PARALLELISM IN SCALAR PRODUCTS

The computation of scalar products contains several operations which can be executed in parallel via the following scheme:

$s := 0 \quad ; ec_1 := ea_1 + eb_1; \; mc_1 := ma_1 \bullet mb_i;$

$s := s + c_1 \quad ; ec_2 := ea_2 + eb_2; \; mc_2 := ma_2 \bullet mb_2;$

– – – – – – – –

$s := s + c_{n-1} \quad ; ec_n := ea_n + eb_n; \; mc_n := ma_n \bullet mb_n;$

$s := x + c_n \quad ;$

The addition of the exponents is certainly a fast process. The main addition $x := s + c_i$ may also be made very fast, in particular, if the shift is executed via a cross point switch. The slowest of the three independent operations in one row is probably the multiplication of the mantissas. If it is much slower than the two additions, a parallel execution of the three operations offers only a marginal gain in speed. However, architectural requirements increase considerably if several operations are executed in parallel. Furthermore, the synchronization of the whole process becomes more difficult.

In complicated numerical applications an isolated scalar product hardly occurs. Such a process may be speeded up considerably, if several scalar product units are made available - which may be as simple and cheap as possible. Then, for instance, in a matrix-matrix or matrix-vector multiplication several components of the product can be computed simultaneously.

In [7] the parallel execution of of different algorithms for the computation of scalar products was studied.

5. PIPELINING OF THE ARITHMETIC OPERATIONS

Another means of accelerating arithmetic operations is pipelining. Of course, all arithmetic operations: addition, subtraction, multiplication and division as well as scalar products can be pipelined. The following figure gives a rough sketch of possible pipeline steps for these operations. Since in scalar products, multiplications are probably much slower than the additions, several multiplication units may be used to feed one addition unit. With these remarks the figures are certainly self-explanatory.

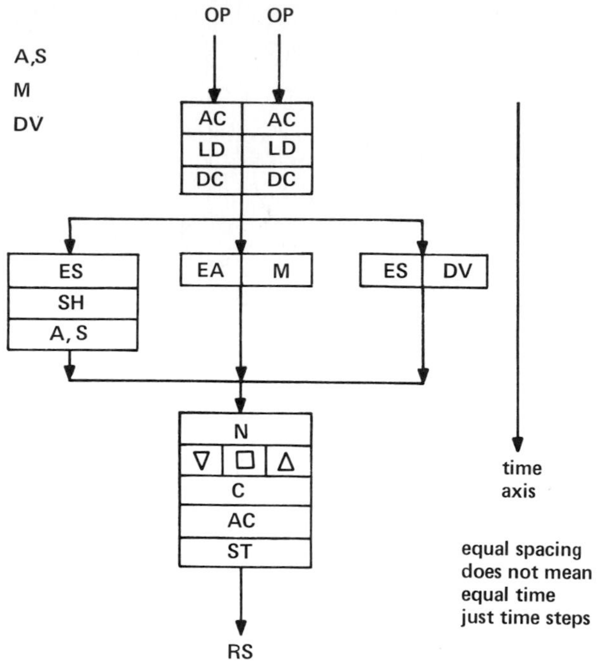

Abbreviations:

AC	address computation
ADD	addition
A,S	addition,subtraction
C	compose
CR	controller
DC	decompose
DV	division
EA	exponent addition
ES	exponent subtraction
HE	hold exponent
LD	load
LD,LST	load from local storage
M	multiplication
N	normalization
OP	operand
RS	result
SH	shift by shift register or cross point switch
ST	store
ST,LST	store into local storage
SP	scalar product

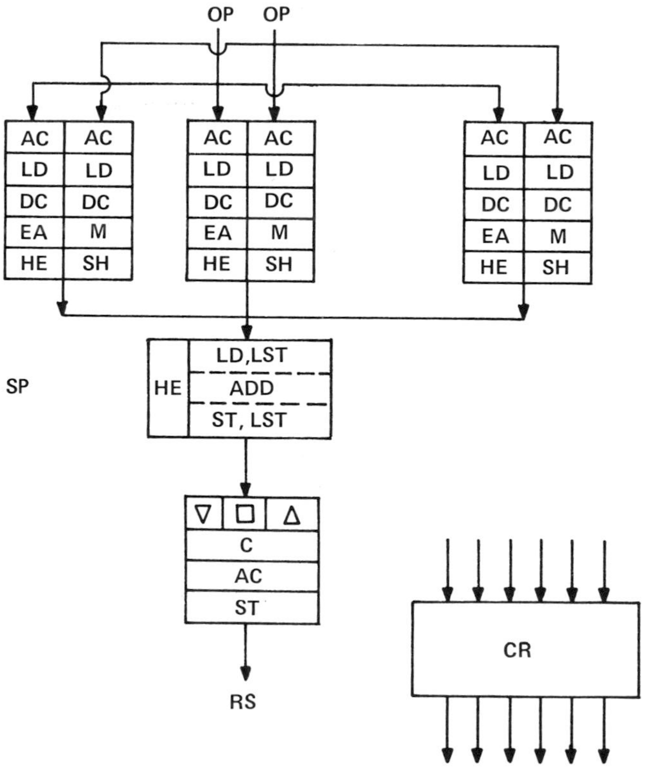

The side effect of pipelined arithmetic operations is the advantage that interval operations or complex operations become nearly as fast as single floating-point operations. Lower and upper bound of an interval operation or real and imaginary part of a complex operation occur nearly simultaneously at the exit of the pipeline.

Appendix

List of symbols and signs used:

$\mathbb{R}$	Set of real numbers
$\mathbb{N}$	Set of natural numbers $\mathbb{N} = \{1,2,3,4,...\}$
$a \in M$	a is an element of the set M
$\bigwedge_{a,b \in M}$	reads: for all $a,b \in M$
$:=$	defining equality sign. The expression on the side of the colon is per definition equal to the expression on the side of the equality sign
$\sum_{i=1}^{n} c_i$	$= c_1 + c_2 + c_3 + ... + c_n$
$\square$	rounding to the nearest number 
$\triangledown$	monotone downwardly directed rounding 
$\triangle$	monotone upwardly directed rounding 
$\bigcirc$	general rounding operator $\bigcirc \in \{\square, \triangledown, \triangle\}$
$S(b,\ell,e1,e2)$	floating – point system of base b, with ℓ digits in the mantissa, least exponent $e1$, and greatest exponent $e2$
$i = 1(1)n$	for i from 1 in steps of 1 to n
T^n	product set; set of all n-tuples $(a_1,a_2,...a_n)$ with $a_i \in T$ for all $i = 1(1)n$

REFERENCES

[1] Apostolatos, N., Kulisch, U., Krawczyk, R., Lortz, R., Nickel, K., Wippermann, H.-W. (1968). The Algorithmic Language TRIPLEX ALGOL-60, Num. Math. 11, 175-180.

[2] Bohlender, G. (1978). Genaue Berechnung mehrfacher Summen, Produkte und Wurzeln von Gleitkommazahlen und Arithmetik in Höheren Programmiersprache. Dissertation, Universität Karlsruhe.

[3] Coonan, J., et al. (1979). A proposed standard for floating-point arithmetic, SIGNUM Newsletter.

[4] INTEL 12 1586-001 (1980). The 8086 Family User's Manual, Numeric Supplement.

[5] Kulisch, U. (1976). *Grundlagen des Numerischen Rechnens-Mathematische Begründung der Rechnerarithmetik,* Bibliographisches Institut Mannheim.

[6] Kulisch, U., Miranker, W. L. (1979). *Computer Arithmetic in Theory and Practice.* IBM Research Report RC 7776. and Academic Press, New York, (1981).

[7] Leuprecht, H. and Oberaigner, W. (1982). Parallel Algorithms for the Rounding Exact Summation of Floating-Point Numbers, Computing 28, 89-104.

Additional References are given in [6].

DIFFERENTIATION AND GENERATION OF TAYLOR COEFFICIENTS IN PASCAL-SC

L. B. Rall

Mathematics Research Center
University of Wisconsin-Madison
Madison, Wisconsin

Evaluation of derivatives and Taylor coefficients of functions defined by computer programs has many applications in scientific computation. The process of automatic differentiation of such functions has as its goal the production of machine code for the evaluation of derivatives and Taylor coefficients. This is in contrast to symbolic differentiation, where the desired output is a more or less pretty formula, and numerical differentiation, which is inaccurate and unstable. Another distinction between automatic and symbolic differentiation is that the latter usually involves considerable computational overhead, while automatic differentiation can be carried out at compile time by a compiler which permits user-defined data types and operators. This paper shows how PASCAL-SC, a compiler of this type, can be used to generate the real derivative types GRADIENT, HESSIAN, TAYLOR, and the corresponding interval types IGRADIENT, IHESSIAN, ITAYLOR. Applications of these types to solution of systems of nonlinear equations, sensitivity analysis, constrained and unconstrained optimization, and the solution of initial-value problems for systems of ordinary differential equations are indicated.

1. AUTOMATION OF EVALUATION AND DIFFERENTIATION OF FUNCTIONS

The *value* $f(x)$ of a function of n variables, $x = (x_1,...,x_n)$ is obtained on a computer by the composition $f = f_1 \circ f_2 \circ ... \circ f_m$ of a finite number of functions, each usually very simple, such as an arithmetic operation or one of a small number of library functions. Using the fact that the rule for evaluation of each f_i is known, it is possible for a computer program called a compiler to produce machine code for the evaluation of a function f specified in some form similar to ordinary mathematical notation, and including the possibility of different cases depending on x and intermediate results. A key feature of such compilers is the ability to perform *formula translation,* that is, to produce machine code for the evaluation of an *expression* such as

Sponsored by the U. S. Army under Contract No. DAAG29-80-C-0041.

ISBN 0-12-428660-7

$$\text{F:= (X*Y + SIN(X) + 4.0)*(3*(Y**2) + 6);} \tag{1.1}$$

which denotes the mathematical function

$$f(x,y) = (xy + \sin x + 4)(3y^2 + 6). \tag{1.2}$$

The same strategy can be applied for the machine evaluation of derivatives $f'(x)$ or Taylor coefficients of f at x [6], [9], [10], [11], [12], [13]. The chain-rule of calculus applied to f gives

$$f'(x) = f'_1(x^{(m-1)})\bullet\ldots\bullet f'_{m-1}(x^{(1)})\bullet f'_m(x), \tag{1.3}$$

where the $'$ in general denotes Fréchet differentiation and the $\bullet$ matrix multiplication [11], and

$$x^{(k)} = f^{(k)}(x),\ f^{(k)} = f_{m-k+1}\circ\ldots\circ f_m,\quad k = 1,\ldots,m-1. \tag{1.4}$$

Since the rules for calculation of derivatives and Taylor coefficients for the basic arithmetic operations are perfectly well understood, explicit expressions are available for the functions f'_i [13], and thus machine code can be generated easily for the evaluation of derivatives or Taylor coefficients of f at points x for which it is differentiable or analytic, respectively. Most existing compilers, however, do not present this opportunity to the user. The reason is that most were developed originally only to work with integers and floating-point numbers (i.e., the types INTEGER and REAL), and are quite limited in the kinds of data and operations which can be handled.

PASCAL-SC [3], on the other hand, was developed with the needs of scientific computation in mind, and thus provides complex numbers, intervals, complex intervals, and vectors and matrices over these types, as well as the appropriate operations, in addition to integers, and floating-point numbers, vectors, and matrices [17]. This vast improvement over the ordinary type of compiler is achieved by allowing the introduction of user-defined data types and operators. This facility, as will be explained below, permits automatic differentiation and generation of Taylor coefficients, simply by introduction of the appropriate derivative data types and corresponding operators.

2. DERIVATIVE DATA TYPES

The first derivative of a function f of n variables $x = (x_1,...,x_n)$ is its *gradient vector*

$$\nabla f(x) = \left(\frac{\partial f(x)}{\partial x_1}, ..., \frac{\partial f(x)}{\partial x_n} \right); \tag{2.1}$$

similarly, the second derivative of f at x is represented by its *Hessian matrix*

$$\mathrm{HF}(x) = \left(\frac{\partial^2 f(x)}{\partial x_i \partial x_j} \right), \tag{2.2}$$

a symmetric $n \times n$ matrix [11]. Furthermore, if f has a convergent Taylor series expansion at $x = x_0$, then

$$f(x) = \sum_{k=0}^{n-1} \frac{1}{k!} f^{(k)}(x_0)(x - x_0)^k + R_n(x_0, x - x_0), \tag{2.3}$$

where, of course, $\frac{1}{0!} f^{(0)}(x_0) = f(x_0)$ [11]. In the scalar case, x is a single real variable, and $f(x)$ is approximated by the *Taylor polynomial*

$$P_{n-1}(x_0, x - x_0) = \sum_{k=0}^{n-1} \frac{1}{k!} f^{(k)}(x_0)(x - x_0)^k, \tag{2.4}$$

which in turn can be represented by the n-vector

$$p_n(x_0, t) = (f_1, f_2, ..., f_n) \tag{2.5}$$

of *normalized Taylor coefficients of f,*

$$f_i = \frac{1}{(i-1)!} f^{(i-1)}(x_0) t^{i-1}, \quad t = x - x_0. \tag{2.6}$$

Thus, in the real case, real vectors and matrices are required for the representation of derivatives and Taylor coefficients. Similarly, interval vectors and matrices will be necessary to represent the same quantities for interval-valued functions. These types are all provided in PASCAL-SC [17]. The declaration for vectors in the real case (RVECTOR) runs as follows:

```
CONST  DIM = <a positive integer>;
 TYPE  DIMTYPE = 1..DIM;                                    (2.7)
       RVECTOR = ARRAY[DIMTYPE] OF REAL;
```

To declare a real matrix (RMATRIX), one adds

```
TYPE RMATRIX = ARRAY[DIMTYPE]OF RVECTOR;                    (2.8)
```

to the above. These declarations will be basic to the following discussion.

2.1 Type GRADIENT

This basic derivative type treats the value $f(x)$ of f at x and its gradient vector $\nabla f(x)$ at the same point as the entity $(f(x),\nabla f(x))$, that is, the pair consisting of the real value of f at x, and its n-dimensional gradient vector of partial derivatives. For this purpose, the RECORD data structure of PASCAL-SC is ideal. To declare this type, (2.7) is followed by

```
TYPE GRADIENT = RECORD  F: REAL; DF: RVECTOR  END;          (2.9)
```

so that if V is a variable of type GRADIENT, then V.F will be its value, and V.DF its gradient vector. Thus, if the function v is denoted in the computer program as V, then V.F = $v(x)$ and V.DF = $\nabla v(x)$ at the current value x of the independent variables $x_1,\ldots,x_n$. Thus, V.DF[i] = $\partial v(x)/\partial x_i$, $i = 1,\ldots,n$. The ith independent variable x_i will be denoted by a GRADIENT variable, say XI, such that XI.F = x_i, the current value of x_i, and XI.DF is the ith unit vector $e_i = (0,\ldots,0,1,0,\ldots,0)$ which has its ith component equal to 1, and the others equal to zero. If it is desired to represent a constant c as a GRADIENT variable C, then C.F = c, while C.DF will be the zero vector $(0,\ldots,0)$. The user of type GRADIENT is free to name and order the independent variables arbitrarily.

2.2 Type HESSIAN

This type can be considered to be an extension of type GRADIENT. Here, operations are performed on the triple $(f(x),\nabla f(x),Hf(x))$ as the basic datum. Once again, the RE-

CORD structure is appropriate, and the declaration of this type consists of (2.7) followed by (2.8) and

```
TYPE HESSIAN = RECORD F: REAL;DF: RVECTOR;
                              HF: RMATRIX END;        (2.10)
```

with now V.F = $v(x)$, v.DF = $\nabla v(x)$, and V.HF = $Hv(x)$ for a variable V of type HESSIAN corresponding to the function v. Here, V.HF$[i,j]$ = $\partial^2 v(x)/\partial x_i \partial x_j$, for example.

2.3 Type TAYLOR

As indicated above, the quantity to be computed with in this case is the vector (2.5) of normalized Taylor coefficients (2.6). It is important in most applications to be aware of the *scale factor* $t = x - x_0$, which will be of type REAL in the scalar case. Thus, the RECORD structure will also be used for this derivative type, which is declared by (2.7) followed by

```
TYPE TAYLOR = RECORD  T: REAL; TC: RVECTOR  END;        (2.11)
```

with V.T $= x - x_0$ and V.TC$[i]$ $= \dfrac{1}{(i-1)!} v^{(i-1)}(x_0)(x - x_0)^{i-1}$ for $i = 1,\ldots,n$. The value of the Taylor polynomial (2.4) can be obtained very simply and accurately in PASCAL-SC by use of the SUM function:

```
P:= SUM(V.TC,0);        (2.12)
```

where P is of type REAL. In (2.12), the sum is rounded to the closest floating-point number; upward or downward rounding can be achieved by replacing the 0 by + 1 or − 1, respectively [17].

2.4 Interval Types IGRADIENT, IHESSIAN, ITAYLOR

Interval arithmetic, including operations for interval vectors and matrices and standard functions, is another convenient feature of PASCAL-SC [3], [17]. By the use of interval computation, inclusions of the real values of expressions such as (1.1) can be obtained [9],

[10]. The same applies to values of derivatives and Taylor coefficients, so the derivative types introduced above can also be defined over intervals. The standard declaration of an interval in PASCAL-SC is:

```
TYPE INTERVAL = RECORD  INF,SUP: REAL  END;                         (2.13)
```

and interval vectors and matrices are declared by

```
TYPE IVECTOR = ARRAY[DIMTYPE]OF INTERVAL;
                                                                    (2.14)
     IMATRIX = ARRAY[DIMTYPE]OF IVECTOR;
```

corresponding to (2.7)-(2.8). The INTERVAL versions of the real derivative types GRADIENT, HESSIAN, and TAYLOR are thus declared by

```
TYPE IGRADIENT = RECORD  IF:INTERVAL; IDF:IVECTOR  END;
      IHESSIAN = RECORD  IF:INTERVAL; IDF:IVECTOR;
                                                                    (2.15)
                         IHF:IMATRIX  END;
       ITAYLOR = RECORD  IT:INTERVAL; ITC:IVECTOR  END;
```

respectively.

3. Derivative Operators

In order for expressions to be evaluated correctly when derivative data types appear, the arithmetic operations and standard functions have to be defined in such a way as to incorporate the appropriate rules for differentiation or generation of Taylor coefficients. Furthermore, as in (1.1), provision for variables or constants of type INTEGER or REAL must be made. The following rule applies:

Variables or expressions of type REAL *or* INTEGER *will be treated as* constants *for the purpose of differentiation.*

In order to simplify the following discussion, generic variables of type INTEGER will be denoted by K, and of type REAL by R,RA,RB. The derivative types introduced in §2 will be indicated by G,H,T,IG,IH,IT, respectively, and the general derivative type by D. Thus, if

D = T, type TAYLOR is meant. Generic variables of type D will be denoted by D,DA,DB for $D \in \{G,H,T,IG,IH,IT\}$. The necessary arithmetic operators will now be listed.

3.1 Addition Operators

$$+\,D,\;\; K + D,\;\; D + K,\;\; R + D,\;\; D + R,\;\; DA + DB. \tag{3.1}$$

3.2 Subtraction Operators

$$-\,D,\;\; K - D,\;\; D - K,\;\; R - D,\;\; D - R,\;\; DA - DB. \tag{3.2}$$

3.3 Multiplication Operators

$$K*D,\;\; D*K,\;\; R*D,\;\; D*R,\;\; DA*DB. \tag{3.3}$$

3.4 Division Operators

$$K/D,\;\; D/K,\;\; R/D,\;\; D/R,\;\; DA/DB. \tag{3.4}$$

There are thus 22 operators in the above categories to be provided for each real derivative data type. For example, taking D = H, the source code for the operator symbolized by H + R (addition of a REAL to a HESSIAN) is:

```
OPERATOR + (H: HESSIAN;R: REAL) RES: HESSIAN;
        VAR U: HESSIAN;
        BEGIN U.F:= H.F + R;U.DF:= H.DF;U.HF:= H.HF;          (3.5)
                RES:= U
        END;
```

since the addition of a constant alters only the value of a variable, and not its derivatives. Rules for performing the above operations for each derivative type are given in [13], for example. Source code for each of the real data types G,H,T will be given for the operation DA*DB in the next section. A complete set of source code for arithmetic and power operations and standard functions can be found in [14] for type GRADIENT. There are 14 operators for the INTERVAL types, since operations between intervals and reals are excluded [17].

3.5 Power Operators

The provision of the power operator ** to evaluate x^y is not standard in PASCAL-SC. In order to implement this operator for derivative data types, it is necessary to define it for some combinations of the basic REAL and INTERVAL types. Thus, one needs

R**K, RA**RB, (3.6)

for type REAL, and, if I,IA,IB denote generic INTERVAL variables,

I**K, K**I, IA**IB (3.7)

for type INTERVAL. In terms of these basic power operations, the derivative operators symbolized by

D**K, K**D, D**R, R**D, DA**DB (3.8)

can be defined. Thus, five operators for each of the real derivative types are needed, in addition to the two operators (3.6) for type REAL, and the three operators (3.7) for the type INTERVAL, as well as D**K, K**D, DA**DB.

The operator R**K can be implemented by repeated squaring [13], [14], or, better yet, by the accurate algorithm described by Böhm [2]. RA**RB can be calculated using R**K for the integer part of RB, and the ordinary exponential function for the logarithm of the base multiplied by the fractional part of RB. For interval variables X,Y, X^Y is defined by

$$X^Y = [\ \min \{x^y \mid x \in X, y \in Y\},\ \ \max \{x^y \mid x \in X, y \in Y\}], \tag{3.9}$$

and the operator ** is defined accordingly. Since the inclusion

$$X^2 = [\ \min\ \{x^2 \mid a \le x \le b\},\ \ \max\ \{x^2 \mid a \le x \le b\}] \subset X \bullet X \qquad (3.10)$$

can be proper for intervals $X = [a,b]$, the standard function **ISQR** for interval arithmetic [17] is used to compute I**2, and I**K by repeated squaring.

The complete package of arithmetic operations for derivative data types thus consists of 27 operators for each real type, the two real operators (3.6), and the three interval operators (3.7), together with 17 arithmetic operators for the interval types.

The priorities of the operators given in this section, for highest to lowest, are: 1°. Unary addition and subtraction, symbolized by + D and −D, respectively; 2°. Multiplication, division, and power: *,/,**; 3°. Binary addition and subtraction ±. Persons familiar with compilers in which ** has higher priority than *,/ should beware, and use parentheses, as in (1.1).

4. EXAMPLES OF MULTIPLICATION OPERATORS IN PASCAL-SC

In order to give explicit examples, the source code for definition of the operator * for the real derivative types **GRADIENT**, **HESSIAN**, and **TAYLOR** will be given in this section, both in component form and compact form using the vector and matrix operators available in PASCAL-SC. Complete source code for the 29 operators required for type **GRADIENT** is given in [14].

4.1 GA*GB for Type GRADIENT

In component form, the source code for this operator is:

```
OPERATOR *(GA,GB: GRADIENT) RES: GRADIENT;
    VAR I: DIMTYPE;U:GRADIENT;
    BEGIN U.F:= GA.F*GB.F;
        FOR I:= 1 TO DIM DO                          (4.1)
        U.DF[I]:= GA.F*GB.DF[I] + GB.F*GA.DF[I];
        RES:= U
    END;
```

however, since multiplication of vectors by real numbers and addition of vectors are operations which are available in PASCAL-SC [17], these can be used to produce the more compact source code

```
OPERATOR *(GA,GB: GRADIENT) RES: GRADIENT;
    VAR U:GRADIENT;
    BEGIN U.F:= GA.F*GB.F;
                                                     (4.2)
        U.DF:= G.A.F*GB.DF + GB.F*GA.DF;
        RES:= U
    END;
```

4.2 HA*HB for Type HESSIAN

If XI is the ith independent variable of type HESSIAN, then XI.DF will be the ith unit vector, as for type GRADIENT, and XI.HF will be the zero matrix. If a constant C is declared as type HESSIAN, then both the vector part C.DF and the matrix part C.HF of C will be zero. It is, of course, more economical to introduce constants as type INTEGER or REAL. Multiplication of HESSIAN variables HA,HB is accomplished by the operator

```
OPERATOR *(HA,HB: HESSIAN) RES: HESSIAN;
    VAR I,J: DIMTYPE;U: HESSIAN;
    BEGIN
        U.F:= HA.F*HB.F;
        FOR I:= 1 TO DIM DO
            U.DF[I]:= HA.F*HB.DF[I] + HB.F*HA.DF[I];
        FOR I:= 1 TO DIM DO
            FOR J:= I TO DIM DO                                   (4.3)
            BEGIN
                U.HF[I,J]:= HA.F*HB.HF[I,J] + HA.DF[I]*HB.DF[J] +
                            HB.DF[I]*HA.DF[J] + HB.F*HA.HF[I,J];
                IF I<>J THEN U.HF[J,I]:= U.HF[I,J]
            END;
        RES:= U
    END;
```

in component form where the well-known formulas for the first and second derivative of products have been used [7], [13]. The calculation of U.DF here is taken directly from (4.1). The first FOR loop could be eliminated by computing U.DF [I] in the second. As before, (4.3) can be simplified by using the vector and matrix operations of PASCAL-SC, augmented by the *outer product* VA**VB of vectors defined by the operator

```
OPERATOR **(VA,VB: RVECTOR) RES: RMATRIX;
    VAR I,J: DIMTYPE;U: RMATRIX;
    BEGIN
        FOR I:= 1 TO DIM DO                                       (4.4)
            FOR J:= 1 TO DIM DO U[I,J]:= VA[I]*VB[J];
        RES:= U
    END;
```

for example. With this operation, (4.3) becomes

```
OPERATOR *(HA,HB: HESSIAN) RES: HESSIAN;
    VAR U: HESSIAN;
    BEGIN
        U.F:= HA.F*HB.F;
        U.DF:= HA.F*HB.DF + HB.F*HA.DF;                                   (4.5)
        U.HF:= HA.F*HB.HF + HA.DF**HB.DF + HB.DF**HA.DF + HB.F*HA.HF;
        RES:= U
    END;
```

which is more compact but less efficient than (4.3).

4.3 TA*TB for Type TAYLOR

Here, the well-known formula for the Taylor coefficients of the product [9], [13] are used. In symbolic form, for U:= TA*TB, one has ([13], p. 41)

$$\text{U.TC[J]} = \sum_{\text{I}=1}^{\text{J}} \text{TA.TC[I]*TB.TC[J} - \text{I} + 1], \quad \text{J} = 1,\ldots,\text{DIM}. \tag{4.6}$$

A very important advantage of the use of PASCAL-SC for this calculation is that (4.6) can be computed as the scalar product of two vectors; this is done by the standard function SCALP to the closest floating-point number, or can be rounded up or down if desired. This means, for example that U.TC[1],...,U.TC[DIM] can each be computed with only one rounding error from the components of TA.TC and TB.TC, instead of having increasing rounding error as J increases. This helps to ameliorate one of the possible bugaboos of the use of Taylor series methods in scientific computation. As can be seen in [13], a number of other formulas for the recursive generation of Taylor coefficients take the form of scalar products, so that one can take advantage of the accuracy of SCALP in their computation. Source code for the multiplication operator * for TAYLOR variables is

```
OPERATOR *(TA,TB: TAYLOR) RES: TAYLOR;
    VAR I,J,K: DIMTYPE;X,Y: RVECTOR;U: TAYLOR;
    BEGIN
        IF TA.T<>TB.T THEN
            BEGIN
                WRITELN('ERROR: MULTIPLICATION OF TAYLOR
                VARIABLES WITH UNEQUAL SCALE FACTORS');
                SVR(0)
            END;
        U.T:= TA.T;FOR I:= 1 TO DIM DO
            BEGIN X[I]:- 0;Y[I]:= 0 END;                    (4.7)
        FOR I:= 1 TO DIM DO
            BEGIN X[I]:= TA.TC[I];
                FOR J:= 1 TO I DO
                    BEGIN
                        K:= I - J + 1;Y[J]:= TB.TC[K]
                    END;
                U.TC[I]:= SCALP(X,Y,0)
            END;
        RES:= U
    END;
```

where an attempt to multiply TAYLOR variables with unequal scale factors results in printing an error message and return of control to the operating system. The initialization of the vectors X,Y could be done by the assignments X:= VRNULL, Y:= VRNULL, from a parameterless function available in PASCAL-SC to produce zero vectors [17].

The present implementation of type TAYLOR is for the scalar case, with a single independent real variable, say x, and expansions are performed at $x = x_0$. In this case, the corresponding independent variable X of type TAYLOR is defined as follows:

$$\text{X.T} = x - x_0,\ \text{X.TC[1]} = x_0,\ \text{X.TC[2]} = x - x_0,\ \text{X.TC[I]} = 0 \text{ for I>2.} \quad (4.8)$$

Thus, the assignment

```
X.TC[1]:= X.TC[1] + X.TC[2]; or X.TC[1]:= SUM(X.TC,0);
```
(4.9)

will generate a sequence $x_i = x_{i-1} + h$, $h = x - x_0$, of points of expansion of the corresponding TAYLOR variables dependent on X.

Multiplication of the interval derivative types IGRADIENT, IHESSIAN, and ITAYLOR follows the pattern given above for the corresponding real types.

5. STANDARD DERIVATIVE FUNCTIONS

The rules for differentiation of the standard functions ABS, SQR, SQRT, EXP, LN, SIN, COS, ARCTAN are well understood, as is the generation of Taylor coefficients for these functions. Thus, there is no conceptual (or practical) difficulty in providing the standard functions DABS, DSQRT, DEXP, DLN, DSIN, DCOS, DARCTAN for the derivative types D∈{G,H.T,IG,IH,IT}, and the function DSQR for the interval derivative types D∈{IG,IH,IT}.

Naming functions in this way means that in an expression for the function (1.2), Taylor coefficients are obtained automatically if one declares F,X,Y to be of type TAYLOR, and writes the assignment

```
F:= (X*Y + TSIN(X) + 4.0)*(3*(Y**2) + 6);
```
(5.1)

here, the minor nuisance of writing TSIN instead of SIN is offset by the fact that the former makes it clear that the expression in (5.1) is of type TAYLOR, and that arguments of type TAYLOR are expected.

A complete set of source code for the GRADIENT (D = G) versions of the standard functions listed above is given in [14]. For example, for the gradient cosine,

```
FUNCTION GCOS(G: GRADIENT): GRADIENT;
    VAR M: REAL;U: GRADIENT;
    BEGIN                                                         (5.2)
      M:=-SIN(G.F.);U.F.:=COS(G.F.);U.DF:=M*G.DF;GCOS:=U
    END;
```

in vector form. With the arithmetic operations and the standard functions given, it is very easy for the user to introduce others. If the gradient secant is needed, one can simply use the code

```
FUNCTION GSEC(G: GRADIENT): GRADIENT;
                                                                  (5.3)
      BEGIN GSEC:=1/GCOS(G) END;
```

for this purpose, and so on.

6. APPLICATIONS OF DERIVATIVE TYPES IN SCIENTIFIC COMPUTATION

Since the operations of differentiation and series expansion are ubiquitous in mathematical modeling, possible applications of derivative data types in scientific computation would fill a vast catalog, which would also probably never be completed. Here, only a few applications which have been implemented in software in one form or another will be mentioned. In each case, the PASCAL-SC formulation of these types and their corresponding functions and operations leads to a drastic simplification of previous efforts.

6.1 Applications of type GRADIENT and IGRADIENT

A simple application of type GRADIENT is to *sensitivity analysis*. If F is of type GRADIENT, then F.DF[i] = $\partial f/\partial x_i$ is the rate of change of the function f symbolized by F with respect to x_i. Furthermore, the gradient vector F.DF = $\nabla f(x)$ gives the direction of the fastest increase of f at the current values of the independent variables, a fact which is useful in optimization and other applications.

Given several GRADIENT variables F,G,H, their *Jacobian matrix* J will have rows J[1] = F.DF, J[2] = G.DF, J[3] = H.DF. Knowing the values F.F, G.F, and H.F of these variables as well as their Jacobian matrix makes it very easy to code the solution of systems of equations by Newton's method in terms of GRADIENT variables [7], [11], [14]. Furthermore, the type IGRADIENT can be used to obtain Lipschitz constants for functions of several variables [14].

In interval analysis [9], [10], it is well-known that the mean-value form

$$F(x) = f(x) + F'(X)(X - x) \tag{6.1}$$

can give an accurate interval inclusion of a real function f on a small interval X containing x. If f is a function of several variables, an interval inclusion $F'(X)$ of the Jacobian matrix $f'(x)$ is needed. This can be obtained automatically if f is coded in terms of its components f_i as variables FI of type IGRADIENT.

By use of PASCAL-SC and the types GRADIENT and IGRADIENT, the program NEWTON [7] for the solution of nonlinear systems of equations can be reduced from over 3,400 lines in FORTRAN to a few dozen.

6.2 Applications of type HESSIAN and IHESSIAN

As one might expect, these types appear to be extremely useful in optimization. In unconstrained optimization, the gradient vector $\nabla f(x)$ of a function $f(x) = f(x_1,...,x_n)$ will vanish at a maximum or minimum value of f, so one wants to solve the system of equations $\nabla f(x) = 0$, the Jacobian matrix of which is $Hf(x)$. If f is coded as a HESSIAN variable F, that is, by an assignment of the form

F:= F(X1,...,XN); (6.2)

then the components of the gradient vector are given automatically by F.DF, and the Hessian matrix of f by F.HF, the latter being useful in solution procedures based on Newton's method [14]. Suppose that the optimization problem is *constrained* by conditions $g(x_1,...,x_n) = 0$,

$h(x_1,...,x_n) = 0$, where g,h are coded in symbolic form by the assignments

$$\text{G:= G(X1,..,XN); H:= H(X1,...XN);} \tag{6.3}$$

as variables of type HESSIAN. Then, increasing the dimension of the problem to $n + 2$ and introducing two new independent variables $L1$, $L2$ (the *Lagrange multipliers)* as of type HESSIAN, one makes the assignment

$$\text{W:= F + L1*G + L2*H;} \tag{6.4}$$

and the solution of the unconstrained problem for W is then a solution of the constrained problem for F [14]. Once again, the Hessian matrix W.HF of W is the Jacobian matrix of the system W.DF = 0, which gives a necessary condition for an extreme value of W, and can be used in solution procedures based on Newton's method.

In the program NEWTON [7], interval Hessian matrices were used to obtain Lipschitz constants for Jacobian matrices and ultimately rigorous error bounds for approximate solutions of nonlinear systems of equations. These matrices can be obtained automatically if the type IHESSIAN is used in the computation [14].

6.3 Applications of type TAYLOR and ITAYLOR

Taylor series and the approximation of functions by Taylor polynomials appear in many places in scientific computation. One of the most successful areas of application of the idea of recursive generation of Taylor coefficients is the numerical solution of the initial-value problem for ordinary differential equations and systems of such equations [1], [4], [9], [10]. Type TAYLOR is ideally suited for production of software for this purpose. For example, suppose a Taylor polynomial approximation is sought for the solution of the equation

$$y' = (xy + \sin x + 4)(3y^2 + 6), \quad y(x_0) = y_0, \tag{6.5}$$

say on the interval $x_0 \le x \le x_T$. Part of the PASCAL-SC code for this purpose, with X,Y,

YPRIME of type TAYLOR, would run as follows for a REAL step-length of H:

```
BEGIN X.T:=H;Y.T:=H;X.TC:=VRNULL;X.TC[1]:=X0;
    WHILE X.TC[1]<=XT DO
        BEGIN Y.TC:=VRNULL;Y.TC[1]:=Y0;
            FOR I:=1 TO DIM DO
                BEGIN
                    YPRIME:=(X*Y+TSIN(X)+4)*(3*(Y**2)+6);
                    J:=I+1;                                        (6.6)
                    Y.TC[I]:=YPRIME.TC[J]*H/J
                END;
            X.TC[1]:=X.TC[1]+H;Y0:=SUM(Y.TC,0);
            WRITELN(X.TC[1],Y0)
        END;
END;
```

The same kind of coding can be expanded for systems of equations.

In the program DIFEQ [4], interval Taylor coefficients and polynomials are used for rigorous error and step-size control in the computation analogous to (6.6) for the recursive generation of Taylor coefficients of y from those for y'. In the program INTE [5], interval Taylor coefficients are used to obtain guaranteed bounds for the error terms in various rules of numerical integration. In both of these cases, the use of the type ITAYLOR would result in the reduction of programs with thousands of lines in FORTRAN and assembly language to PASCAL-SC programs of a hundred lines or so.

A challenge in the extension of type TAYLOR to the vector case is that derivatives are then multilinear operators [11], so that both storage requirements and operation times increase drastically. However, some of the present and projected "supercomputers" may well be suitable for Taylor series methods to be applied to partial differential equations along the lines indicated here, given compilers with the capabilities of PASCAL-SC which can also exploit any parallelism available in the hardware.

References

[1] Barton, D., Willers, I. M. and Zahar, R. V. M. (1971). Taylor series methods for ordinary differential equations - an evaluation, [16], pp. 369-390.

[2] Böhm, H. (1982). Evaluation of arithmetic expressions with maximum accuracy, Proceedings of the 10th IMACS World Congress on System Simulation and Scientific Computation, Vol. 1, pp. 391-393, Montreal.

[3] Bohlender, G., Grüner, K, Kaucher, E., Klatte, R., Krämer, W., Kulisch, U. W., Rump, S. M., Ullrich, Ch., Wolff von Gudenberg, J. and Miranker, W. L. (1981). PASCAL-SC: A PASCAL for contemporary scientific computation, Research Report RC 9009, IBM Thomas J. Watson Research Center, Yorktown Heights, NY

[4] Braun, J. A. and Moore, R. E. (1968). A program for the solution of differential equations using interval arithmetic (DIFEQ) for the CDC 3600 and 1604, MRC Tech. Summary Report No. 901, University of Wisconsin-Madison.

[5] Gray, Julia H., and Rall, L. B. (1975). INTE: A UNIVAC 1108/1110 program for numerical integration with rigorous error estimation, MRC Tech. Summary Report No. 1428, University of Wisconsin-Madison.

[6] Kedem, G. (1980). Automatic differentiation of computer programs, ACM Trans. Math. Software **6**, 2 150-156.

[7] Kuba D. and Rall, L. B. (1972). A UNIVAC 1108 program for obtaining rigorous error estimates for approximate solutions of systems of equations, MRC Tech. Summary Report No. 1168, University of Wisconsin-Madison.

[8] Kulisch U. and Miranker, W. L. (1981). Computer Arithmetic in Theory and Practice, Academic Press, NY.

[9] Moore, R. E. (1966). Interval Analysis, Prentice-Hall, Englewood Cliffs, NJ.

[10] Moore, R. E. (1979). Methods and Applications of Interval Analysis, SIAM Studies in Applied Mathematics 2, Society for Industrial and Applied Mathematics, Philadelphia, PA.

[11] Rall, L. B. (1979). Computational Solution of Nonlinear Operator Equations, Wiley, NY, 1969; reprinted by Krieger, Huntington, NY.

[12] Rall, L. B. (1980). Applications of software for automatic differentiation in numerical computation, Computing, Suppl. 2 141-156.

[13] Rall, L. B. (1981). Automatic Differentiation: Techniques and Applications, Lecture Notes in Computer Science No. 120, Springer-Verlag, Berlin-Heidelberg-New York.

[14] Rall, L. B. (1982). Differentiation in PASCAL-SC: Type GRADIENT, MRC Tech. Summary Report No. 2400, University of Wisconsin-Madison.

[15] Rall, L. B. (1981). Mean value and Taylor forms in interval analysis, SIAM J. Math. Anal. **14**, No. **2** (1983) (to appear); preprint: MRC Tech. Summary Report No. 2286, University of Wisconsin-Madison.

[16] Rice, J. R. (Editor) (1971). Mathematical Software, Academic Press, NY.

[17] Wolff von Gudenberg, J. (1981). Gesamte Arithmetik des PASCAL-SC Rechners: Benutzerhandbuch, Institute for Applied Mathematics, University of Karlsruhe.

MATRIX PASCAL†

G. Bohlender
H. Böhm
K. Grüner
E. Kaucher
R. Klatte
W. Krämer
U. W. Kulisch
W. L. Miranker*
S. M. Rump†
Ch. Ullrich
J. Wolff v. Gudenberg

Institute for Applied Mathematics
University of Karlsruhe
Karlsruhe, West Germany

In addition to the integers, the real and complex numbers, the segments (intervals) and complex segments as well as vectors and matrices over all of these comprise the fundamental data types of scientific computation. Here we extend the programming language PASCAL so that it accepts operands and operators of all these data types as primitives in expressions. This is done by augmenting PASCAL by means of an effective and implementable operator concept. Using this concept, operators for complex numbers, segments, complex segments as well as vectors and matrices over these data types are made available as standard operators.

All of these operators are provided with the capability to deliver maximum accuracy.

The syntax for these new operators is given by means of easily traceable syntax diagrams. The algorithms which specify these additional standard operators are listed in the last section.

In addition to the customary arithmetic operations which are employed in the commonly used linear spaces, the language allows the maximally accurate computation of the value of certain arithmetic expressions. Among these are scalar products of matrices as well as bilinear forms. Evaluation of such expressions with least significant bit accuracy is fundamental for the performance of automatic error control by the computer. The new concepts enable the language to support vector and matrix computations in a natural way. Thus it is amply adequate and well suited for use on array processors.

---

*IBM Thomas J. Watson Research Center, Yorktown Heights, New York 10598
†Present address: IBM Deutschland, Entwicklung und Forschung, 7030 Boeblingen, West Germany

ISBN 0-12-428660-7

CONTENTS

A. INTRODUCTION

The more commonly used scientific programming languages usually provide standard data types such as integer, real and boolean. Several of these languages such as PASCAL or ADA provide the user with the added capability to define additional data types customized to his applications. However some of these languages fall short of providing the user with the ability to define arithmetic operators corresponding to his customized data types. In PASCAL and other languages, for instance, operations for the customized data types must be furnished by procedures and procedure calls. The object of this paper is to introduce a new arithmetic for digital computers. For all possible numerical data types, the new arithmetic is made available in a convenient operator form. A principle feature of the new arithmetic is that it is maximally accurate in all cases (complex numbers, segments, complex segments as well as vectors and matrices over these data types). By this we mean that no computer representable element lies between the actual and computer generated result of an operation. This feature which is necessary for maximal accuracy in higher level algorithms often turns out to be sufficient as well.

For example a matrix product as traditionally implemented by means of several for-statements generally does not have maximal accuracy. This is the case even if the product is composed of maximally accurate additions and multiplications among the entries because of the accumulation of rounding errors. An operator notation $a*b$ for the matrix product does not necessarily bind the process to this accumulation of rounding errors and admits the possibility of a maximally accurate matrix product.

In addition to the customary arithmetic operations which are employed in the most commonly used linear spaces, the extended language which we introduce allows the computation of the value of certain arithmetic expressions with maximum accuracy. Among the latter are, for instance, scalar products of matrices as well as bilinear forms. Evaluation of such expressions with least significant bit accuracy is fundamental for performing automatic computer error control.

Note that our arithmetic concept considerably supersedes the arithmetic of the proposed IEEE norm. The latter does not even prescribe availability of complex multiplication with maximal accuracy. In addition to the integers, the real and complex numbers, the segments (intervals) and complex segments as well as the vectors and matrices over all of these comprise the fundamental data types of scientific computation. For reasons discussed above we extend the programming language PASCAL by means of an effective and implementable operator concept. We call the extended language MATRIX PASCAL. A compiler for the extended language including the new arithmetic has been implemented on the Z80 and the MOTOROLA 68000.

The new arithmetic and the operator concept of MATRIX PASCAL lead to a number of advantages.

- Programs become much shorter. They are easier to read and to write, and are therefore more reliable.

- Programs are easier to debug.

- Compilation time is shorter since all the operators are precompiled.

- The direct availability of scalar products and scalar product expressions (see below) saves address calculations in matrix and vector computations. Therefore, running time for programs with matrix and vector computations is usually shorter.

- The newly introduced segment types and directed roundings in combination with scalar products with maximum accuracy permit an error analysis to be performed by the computer itself. These concepts enable computation with guarantees *and* high accuracy to be performed.

- The computer can be used to verify with certainty properties of the following type: a matrix is not singular, a matrix is positive definite or a function has a zero or has no zeros in a certain interval [15].

- For historical reasons, early programming languages lacked a precise definition of the arithmetic operations and the roundings which are to be employed. This has always contributed to the unnecessary use of computers as experimental tools. As a result users of computers are still obliged to accept computational results with a feeling rather than a certainty of what they mean. To achieve even such an informal assurance about the stability of particular computations the user is obliged to make reruns and other experimental tests. One of the advantages of the extended language and the new numerical standard types and operators introduced here is the elimination of this type of vagary in computation.

Our description of MATRIX PASCAL proceeds in the following steps:

In Section B we review properties of the spaces, the operations and the data types of scientific computation.

In Section C an informal description of the new standard operators and functions is given.

In Section D we give the syntax and semantics of the language extension MATRIX PASCAL. This includes the new data types segment (interval), complex and complex segment as well as the types vector, matrix and diagonal matrix over all basic data types. Vector, matrix and diagonal matrix are so-called dynamic types. We also discuss the operator concept, the standard operators and the standard functions and embed the scalar products and scalar product expressions.

In Section D the syntax is defined in terms of a recently developed diagrammatic notation.

Section E gives some hints for the implementation of the new maximally accurate operations.

B. SPACES, OPERATIONS AND DATA TYPES

In this section we review the spaces in which numerical computations are performed, summarize the definition of the arithmetic operations in these spaces, introduce the data types of the extended language and relate them to these spaces and operations. Then we describe the role of an optimal scalar product as well as the role of scalar product expressions with maximum accuracy.

B1. Spaces of Numerical Computation

The most common spaces in which numerical computations are performed are the reals $\mathbb{R}$ and the vectors and matrices $V\mathbb{R}$ and $M\mathbb{R}$ over the real numbers. In addition there are the corresponding complex spaces $\mathbb{C}$, $V\mathbb{C}$ and $M\mathbb{C}$. All of these spaces are ordered with respect to the order relation $\leq$. (In product spaces the order relation is defined componentwise.)

The corresponding segment or interval spaces $S\mathbb{R}$, $SV\mathbb{R}$, $SM\mathbb{R}$, $S\mathbb{C}$, $SV\mathbb{C}$ and $SM\mathbb{C}$ are also included in our treatment. We use the word segment in place of the more customary word interval and the letter S in place of the letter I in the names of spaces and functions. The letter I is reserved for denoting integers.

We present a table in Figure 1, in the second column of which the spaces just cited are found. Since generic elements in these spaces are not representable in computers, each is replaced by a computer representable subset as listed in the third column of Figure 1. In that column R denotes the computer representable counterpart of $\mathbb{R}$, C the set of all pairs of elements of R, VC the set of all n-tuples of such pairs, and so forth.

The powerset of any set χ is denoted by $\mathbb{P}\chi$. The powersets of several sets of column 2 are displayed in column 1.

We indicate set-subset relations between elements of neighboring columns in Figure 1 by means of the inclusion symbol $\supset$.

1		2		3	4
		$\mathbb{R}$	$\supset$	R	$+ \ - \ * \ /$
					$\times$
		$V\mathbb{R}$	$\supset$	VR	$+ \ -$
					$\times$
		$M\mathbb{R}$	$\supset$	MR	$+ \ - \ *$
$\mathbb{P}\mathbb{R}$	$\supset$	$\boldsymbol{S}\mathbb{R}$	$\supset$	$\boldsymbol{S}R$	$+ \ - \ * \ /$
					$\times$
$\mathbb{P}V\mathbb{R}$	$\supset$	$\boldsymbol{S}V\mathbb{R}$	$\supset$	$\boldsymbol{S}VR$	$+ \ -$
					$\times$
$\mathbb{P}M\mathbb{R}$	$\supset$	$\boldsymbol{S}M\mathbb{R}$	$\supset$	$\boldsymbol{S}MR$	$+ \ - \ *$
		$\mathbb{C}$	$\supset$	C	$+ \ - \ * \ /$
					$\times$
		$V\mathbb{C}$	$\supset$	VC	$+ \ -$
					$\times$
		$M\mathbb{C}$	$\supset$	MC	$+ \ - \ *$
$\mathbb{P}\mathbb{C}$	$\supset$	$\boldsymbol{S}\mathbb{C}$	$\supset$	$\boldsymbol{S}C$	$+ \ - \ * \ /$
					$\times$
$\mathbb{P}V\mathbb{C}$	$\supset$	$\boldsymbol{S}V\mathbb{C}$	$\supset$	$\boldsymbol{S}VC$	$+ \ -$
					$\times$
$\mathbb{P}M\mathbb{C}$	$\supset$	$\boldsymbol{S}M\mathbb{C}$	$\supset$	$\boldsymbol{S}MC$	$+ \ - \ *$

Figure 1: Table of the spaces occurring in numerical computations.

B2. Arithmetic Operations in Computer Representable Spaces

Having described the sets listed in the third column of Figure 1, we turn to the arithmetic operations to be defined in these sets. These operations are supposed to approximate the operations in the corresponding sets listed in the second column. In general, the latter operations are not computer executable even for computer representable operands. The operations required for each of the computer representable sets listed in column 3 are in turn indicated in column 4. Moreover, a number of outer multiplications (e.g., a matrix-vector or a scalar-matrix multiplication) are required, and these in turn are indicated in column 4 by means of a ×-sign between rows in the figure. Our extension of PASCAL proceeds in two steps. First we augment the language by means of the new basic data types: segment, complex and complex segment as well as by means of the corresponding operators. We use the general operator concept of the language extension to provide new standard operators for the other inner and outer operations occurring in the spaces of column 3. This large collection of operations is succinctly defined in a unified manner by the theory of computer arithmetic [11], [12] which we now briefly review.

We begin with the definition of the operations in the powersets listed in column 1. Let χ be one of the sets $\mathbb{R}$, $V\mathbb{R}$, $M\mathbb{R}$, $\mathbb{C}$, $V\mathbb{C}$, $M\mathbb{C}$. Then the operations in the powerset $\mathbb{P}\chi$ are defined by

$$A \circ B := \{a \circ b \mid a \in A \wedge b \in B\}$$

for each $\circ \in \{+, -, *, /\}$ and for all pairs of subsets $A, B \in \mathbb{P}\chi$. With this definition the arithmetic operations are well defined in the leftmost element in each row of Figure 1. Now let χ be any set whatever of Figure 1 in which the operations are well defined and let Ψ be the subset of χ occurring immediately to its right in the figure. For each operation $\circ$ in χ a corresponding operation $\boxdot$ in Ψ is defined as follows.

$$(RG) \qquad \square(x \circ y) \text{ for all } x, y \in \Psi.$$

Here $\square: \chi \to \Psi$ denotes a projection, which we call a rounding, from χ into Ψ. This mapping has the following property:

$$(R1) \qquad \square(x) = x \text{ for all } x \in \Psi \qquad \text{(rounding).}$$

A rounding may have the following additional properties:

$$(R2) \qquad x \leq y \text{ implies } \square(x) \leq \square(y) \text{ for all } x, y \in \chi \qquad \text{(montonicity)}$$

$$(R3) \qquad \square(-x) = -\square(x) \text{ for all } x \in \chi \qquad \text{(antisymmetry).}$$

(In product spaces the rounding is defined componentwise.)

In the case that χ is a set of intervals, $\leq$ denotes set inclusion $\subseteq$ and the rounding has the additional property

$$(R4) \qquad x \subseteq \square(x) \text{ for all } x \in \chi \qquad \text{(upwardly directed).}$$

With the obvious modification of these rules in the case of outer operations, this completes our review of the definition of computer arithmetic.

To conveniently perform all these operations in the case of segment or interval spaces, we introduce the monotone directed roundings $\triangledown$ respectively $\triangle$ from $\mathbb{R}$ into R which are defined by $(R1)$, $(R2)$ and

$$(R4) \qquad \triangledown(x) \leq x \text{ resp. } x \leq \triangle(x) \text{ for all } x \in \mathbb{R}.$$

Along with these roundings, we will make use of the associated operations defined as follows:

$$(RG) \; x \overset{\circ}{\triangledown} y := \triangledown(x \circ y) \text{ resp. } x \overset{\circ}{\triangle} y := \triangle(x \circ y)$$

for all $x, y \in \Psi$ and $\circ \in \{+, -, \cdot, /\}$.

B3. Data Types of the Extended Language

In addition to the computer representable integers I (a subset of all of the integers $\mathbb{I}$), we consider four basic data types which are denoted by R, C, S and CS. In extended PASCAL these five types are called integer, real, complex, segment and csegment. The types segment respectively csegment denote the computer representable set of real intervals respectively complex intervals. In addition to these five basic data types we consider the following structured types in MATRIX PASCAL:

$V\chi$ vectors with components in $\chi \in \{R, C, S, CS\}$

$M\chi$ matrices with components in $\chi \in \{R, C, S, CS\}$.

A comparison of the data types just introduced with the spaces in column 3 of Figure 1 reveals certain differences as displayed in the following table.

Spaces	Types
$\boldsymbol{SR}$	S
$\boldsymbol{SVR}$	VS
$\boldsymbol{SMR}$	MS
$\boldsymbol{SC}$	CS
$\boldsymbol{SVC}$	VCS
$\boldsymbol{SMC}$	MCS

Figure 2: Table of differences in space and data type names.

These differences have been introduced both for reasons of convenience and of consistency with certain conventions. In fact spaces with the names of these data types are developed in the theory of computer arithmetic. While these two sets of spaces are formally different, no confusion should result since in any case that theory shows that corresponding spaces in each of these two sets of spaces are isomorphic.

Vectors and matrices are predefined types in MATRIX PASCAL (occasionally summarized as DYN TYPE in this paper). MATRIX PASCAL variables of these types may be declared as follows:

```
var   a: vector [1..m] of real;
      b: matrix [1..m,1..n] of real;
      c: vector [1..m] of complex;
      d: matrix [1..m,1..n] of complex;
      e: vector [1..m] of segment;
      f: matrix [1..m,1..n] of segment;
      g: vector [1..m] of csegment;
      h: matrix [1..m,1..n] of csegment;
```

In MATRIX PASCAL in general the types vector and matrix are dynamic types with index type integer. The component type may be any type of MATRIX PASCAL except the types vector, matrix and file.

In addition to the types vector and matrix, the type diagonal is predefined as a special matrix type in MATRIX PASCAL, variables of this type represent diagonal matrices. They are to be declared, for instance, as follows:

var k: diagonal [1..*m*,1..*m*] of segment;

Only the diagonal components for matrices of type diagonal are kept in storage. Rows and columns of variables of type matrix and diagonal are of type vector. Rows can be addressed directly, for instance, as $a[i]$; columns as $a[*,i]$.

Variables of the types vector resp. matrix resp. diagonal are assignment compatible if they have the same component type and are of the same size.

The types vector, matrix and diagonal are not permitted as component types of structured data types (array, record and file).

For further details concerning the types vector, matrix and diagonal we refer to the syntax diagrams.

B4. The Role of the Optimal Scalar Product

If in *(RG)* x and y are real matrices: $x = (x_{ij})$, $y = (y_{ij}) \in MR$, multiplication requires the use of the following formula:

$$x \boxast y := \square\left(\sum_{k=1}^{n} x_{ik} * y_{kj}\right) = \left(\square \sum_{k=1}^{n} x_{ik} * y_{kj}\right).$$

Here Σ and $*$ denote exact addition and multiplication for real numbers. Thus to obtain the correct rounded result specified by *(RG)* it is necessary to employ a least significant bit accurate scalar product

$$\bigcirc \sum_{k=1}^{n} a_k * b_k, \quad \bigcirc \in \{\square, \triangledown, \triangle\}.$$

A corresponding algorithm for computer implementation of such a scalar product is given in section E. This kind of scalar product is sufficient for the implementation of the operations specified by *(RG)* for all of the new numerical data types of MATRIX PASCAL that occur in rows 3, 6, 7, 9, 10, 12 of Figure 1 (see [11] and [12]).

B5. Scalar Product Expressions

With the operations defined so far including scalar products with maximal accuracy, all operations in the spaces of Figure 1 can be executed with maximal accuracy. Beyond this it is important to have the ability to evaluate arithmetic expressions with maximal accuracy in numerical mathematics. Since scalar products with maximal accuracy are already available, MATRIX PASCAL allows their employment so that by means of a simple notation certain vector and matrix expressions can be evaluated with maximal accuracy. These expressions are sums. The corresponding summands (terms) may be real or complex scalars, vectors, matrices, products of these as well as bilinear forms. The notation of such expressions (EMA, i.e., Expression with Maximum Accuracy) is given in the syntax diagrams P27-P32. The type of a term of such an expression is displayed in the following table. The subscripts occurring in this table denote the dimension of the vectors and matrices.

*	R	VR_ℓ	$MR_{\ell,m}$	C	VC_ℓ	$MC_{\ell,m}$
R	R	VR_ℓ	$MR_{\ell,m}$	C	VC_ℓ	$MC_{\ell,m}$
VR_ℓ	–	R	–	–	C	–
$MR_{k,\ell}$	–	VR_k	$MR_{k,m}$	–	VC_k	$MC_{k,m}$
C	C	VC_ℓ	$MC_{\ell,m}$	C	VC_ℓ	$MC_{\ell,m}$
VC_ℓ	–	C	–	–	C	–
$MC_{k,\ell}$	–	VC_k	$MC_{k,m}$	–	VC_k	$MC_{k,m}$

Scalar product expressions are written in parentheses with a rounding symbol in front of them. Any of the rounding symbols $\square$, ∇, $\triangle$ and $\diamondsuit$ may be used. For instance

$\square(A*x-b)$.

$\square$ denotes the implicit rounding of the computer. $\square$ is required to be a monotone and antisymmetric rounding. $\triangledown$ and $\triangle$ denote the monotone directed roundings. $\diamondsuit$ indicates that the value of the expression to be evaluated is rounded by employing $\triangledown$ and $\triangle$ to the least interval (segment), which contains the value of the expression. We refer to the syntax diagrams for further details concerning scalar product expressions.

C. INFORMAL DESCRIPTION OF THE LANGUAGE EXTENSION

We now comment informally on the syntax and semantics of the language extension. A complete and formal description of the syntax will be given in Section D.

The order of execution of the operations for the newly defined data types is analogous to that of the data types of customary PASCAL, i.e., we have four priorities. Monadic operators (+ , −) have highest priority. Multiplicative operators *, / have higher priority than the additive operators + ,−. Comparisons have the lowest priority. Parentheses are used in the customary manner.

In addition to the arithmetic operations, relational expressions, e.g., $x <= y$ are extended to the new types (see P19).

For real constants and real operations, a rounding can be specified which indicates whether the constant respectively the result of the operation shall be rounded upwardly ($\triangle$) or downwardly ($\triangledown$). If no rounding or the rounding $\square$ is specified, the constant respectively the result is rounded by the implicit rounding of the computer (see Sections B and Section E) which might be the rounding to the nearest computer representable number.

For constants the rounding symbol is placed in front of the number, e.g., $\bigtriangleup$ 0.12345e20, see P4. Rounded operations are denoted as follows, $\bigtriangledown\!\!\!\!\!\!+$, $\bigtriangledown\!\!\!\!\!\!-$, $\bigtriangledown\!\!\!\!\!\!*$, $\bigtriangledown\!\!\!\!\!\!/$, $\bigtriangleup\!\!\!\!\!\!+$, $\bigtriangleup\!\!\!\!\!\!-$, $\bigtriangleup\!\!\!\!\!\!*$ and $\bigtriangleup\!\!\!\!\!\!/$. We don't provide operations with directed roundings for product sets because such operations occur rarely. Moreover, in the cases that such operations are required the user may compose them himself by making use of the operator concept.

The symbols $\bigtriangledown$ and $\bigtriangleup$ are used to denote the directed roundings in MATRIX PASCAL. In case of constants, scalar product expressions and input-output statements (see the following paragraph) the rounding symbols $\square$, $\bigtriangledown$, $\bigtriangleup$ and $\Diamond$ may be replaced by # = , #<, #> and ##, respectively. The symbols $\bigtriangledown\!\!\!\!\!\!+$, $\bigtriangledown\!\!\!\!\!\!-$, $\bigtriangledown\!\!\!\!\!\!*$, $\bigtriangledown\!\!\!\!\!\!/$, $\bigtriangleup\!\!\!\!\!\!+$, $\bigtriangleup\!\!\!\!\!\!-$, $\bigtriangleup\!\!\!\!\!\!*$, $\bigtriangleup\!\!\!\!\!\!/$ may be replaced by + <, −<, *<, /<, + >, −>, *>, />, respectively.

In input and output statements (P17), variables of the new types are permitted and the rounding control is written after format specification. For instance, for a real variable x, read(x:$\bigtriangleup$) indicates that x is to be rounded upwardly by the input procedure, and write(x:18:0:$\bigtriangledown$) indicates that x is to be rounded downwardly by the output procedure. The segment types require the specification of intersection and convex hull as additional operations as well as a test for set membership. On a conventional keyboard, they could be expressed by ** and +* and <u>in</u>, respectively. The operation intersection has the priority of multiplication while the convex hull has the priority of addition. Note that intersection causes an error if the intersection is empty. In the computation of the lower and upper bounds respectively which comprise the output of an interval operation, downwardly and upwardly directed roundings respectively are automatically performed. Formulas which completely define the operations for the new data types are found in Section E.

All scalar product expressions are evaluated with maximal accuracy in each component. Constant operands in scalar product expressions which are not exactly convertible (i.e., machine representable) could alter the proper value of the expression. Therefore, constant operands in scalar product expressions which are not written with a rounding symbol and which are not correctly convertible cause an error message.

D. FORMAL DESCRIPTION OF MATRIX PASCAL

The formal description is composed of two parts: the syntax to be given first in subsection D1 followed by the semantics to be given in subsection D2.

D1. Syntax

Here for completeness and for the convenience of the reader, we describe the syntax of the entire language as it is extended by the operator concept, the dynamic type concept, the introduction of new numerical data types and the evaluation of expressions with maximum accuracy. We describe the syntax by means of a recently introduced diagrammatic notation [8], [9], [3], [4], [16], [17], which has a congenial compact form.

The syntax diagrams to follow specify the construction of correct program strings in the extended PASCAL, i.e., in MATRIX PASCAL. The diagrams are to be read and interpreted in the following manner:

- Syntax variables are written in upper case letters although single blanks may intervene.

- Basic symbols (i.e., the so called terminals of formal language theory) may be characters, sequences of characters or word symbols (reserved words), where characters are enclosed in circles, while word symbols are underlined; for instance (+), (:=), do. Sequences of basic symbols are enclosed in ovals (e.g., identifiers for standard functions like (cos) , (sin),...).

- Solid lines are to be traversed from left to right and from top to bottom. Dotted lines are to be traversed oppositely, i.e., from right to left and from bottom to top.

- Each syntax diagram is numbered and labeled by one or more syntax variables which it describes.

- Bold lines (and bold dotted lines) indicate extensions of the standard language. (That is, by deleting the bold lines (and the bold dotted lines) from the diagrams (as well as deleting those parts of the diagrams which are reachable only by means of the bold lines) there remains the syntax diagrams of standard PASCAL, i.e., the unextended language.)

  At the end of subsection D1 we give a glossary of the syntax variables. Familiarity with this glossary will greatly facilitate the reading and understanding of the diagrams.

A diagram is a generator of paths through itself. Some of the diagrams contain symbols usually denoted by $\mathcal{T}$ which may range over a specified domain, e.g., $\mathcal{T} \in \Theta$ as in diagram P5 (collective diagram). (Indeed sometimes several such variables and associated domains occur as in P22 where we have $\mathcal{T}_1 \in \Theta \cup \{F, TF\}$, $\mathcal{T}_2 \in \Theta$.) When such a diagram is being traversed the value of any such symbol $\mathcal{T}$ may be chosen independently from its specified domain each time that symbol $\mathcal{T}$ occurs in the traversal of the diagram. In a selective diagram e.g. P4, we write a selecting condition (membership of $\mathcal{T}$) in parentheses interrupting the corresponding line. This line may only be traversed, if the condition is fulfilled.

We use the following abbreviations for the data types:

	I	integer
	B	boolean
	CH	character
	CD	enumeration (code)
	R	real
	C	complex
	S	segment
	CS	complex segment
	P	pointer
	ST	string
	A	array
	REC	record
SET	*SI*	set of integer
	SB	set of boolean
	SCH	set of character
	SCD	set of code
VEC	*VR*	vector of real
	VC	vector of complex
	VS	vector of segment
	VCS	vector of complex segment
MAT	*MR*	matrix of real
	MC	matrix of complex
	MS	matrix of segment
	MCS	matrix of complex segment
DIAG	*DR*	diagonal matrix of real
	DC	diagonal matrix of complex
	DS	diagonal matrix of segment
	DCS	diagonal matrix of complex segment
	F	file
	TF	textfile

P1 : PROGRAM

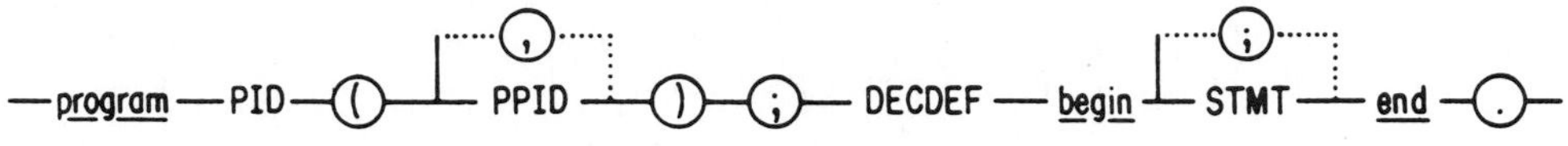

P2 : DECLARATION DEFINITION (DECDEF)

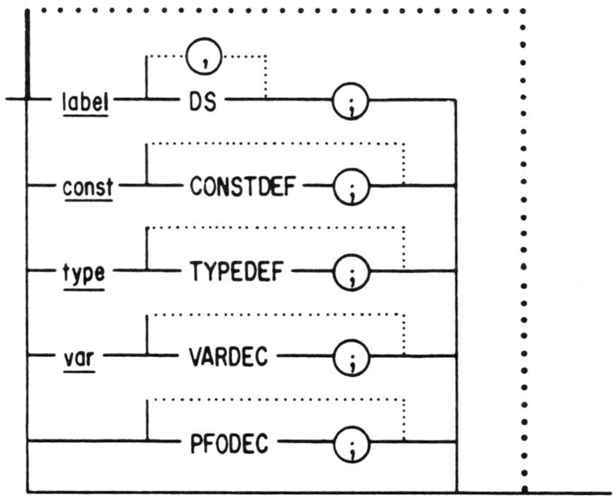

P3 : CONSTANT DEFINITION (CONSTDEF)

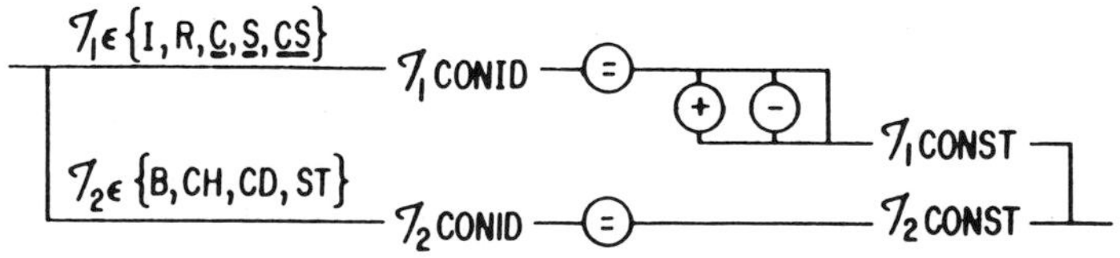

P4 : τ CONSTANT (τ CONST) $\tau \in \theta_{Con}$ $\theta_{Con} = \{I, B, CH, CD, R, \underline{C}, \underline{S}, \underline{CS}, ST\}$

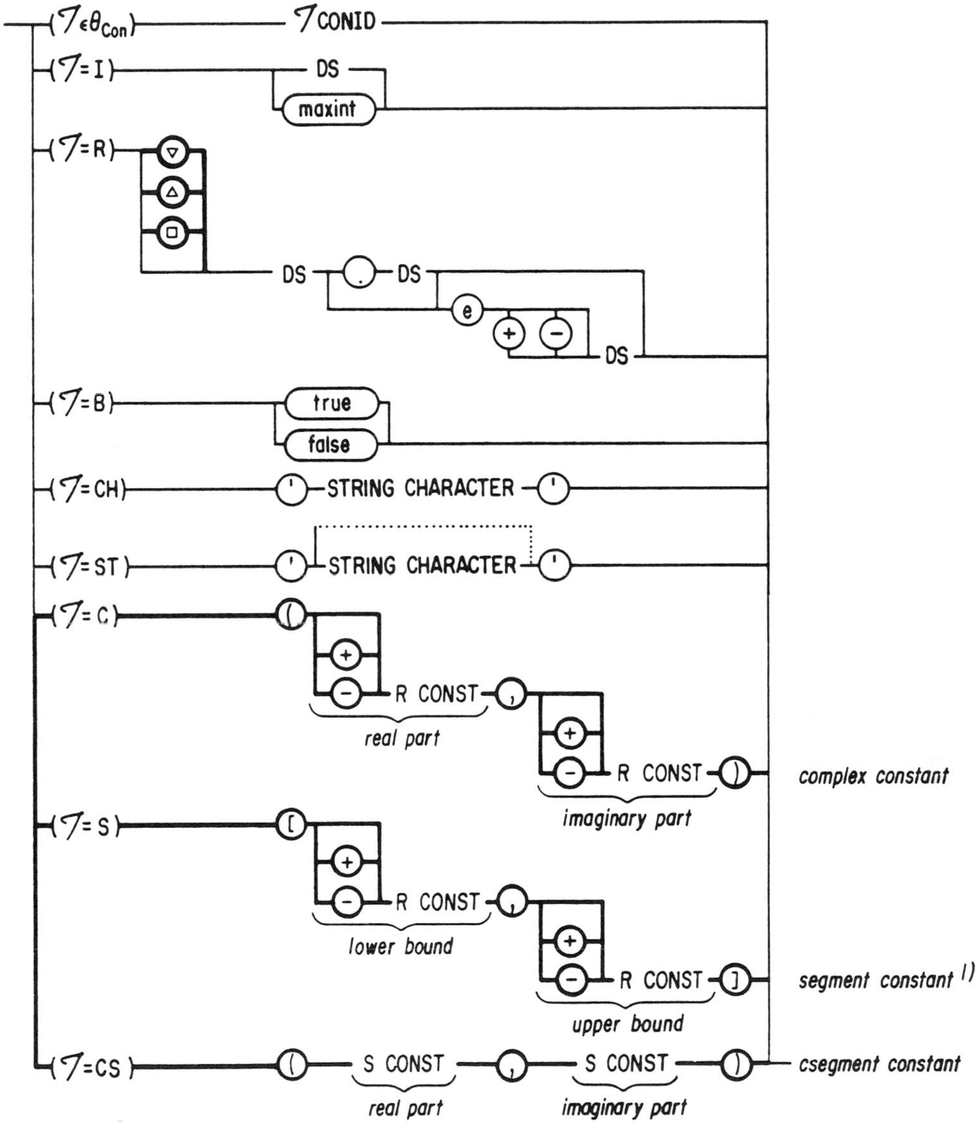

1) *If rounding control is missing in the lower resp. upper bound of a segment constant, it is automatically rounded downwardly resp. upwardly.*

P5 : TYPE DEFINITION (TYPEDEF)

$$\theta = \{ I, B, CH, CD, R, \underline{C}, \underline{S}, \underline{CS}, P, ST, A, REC, SET \}$$

$$\underline{\mathcal{T} \in \theta \cup \{F, TF\}} \quad \mathcal{T} \text{ TYPE IDENTIFIER } -(=)- \mathcal{T} \text{ TYPE}$$

P6 : $\mathcal{T}$ TYPE $\qquad \mathcal{T} \in \theta \cup \{F, TF\},\quad \theta = \{ I, B, CH, CD, R, \underline{C}, \underline{S}, \underline{CS}, P, ST, A, REC, SET \}$

$\theta_{ORD} = \{ I, B, CH, CD \}$

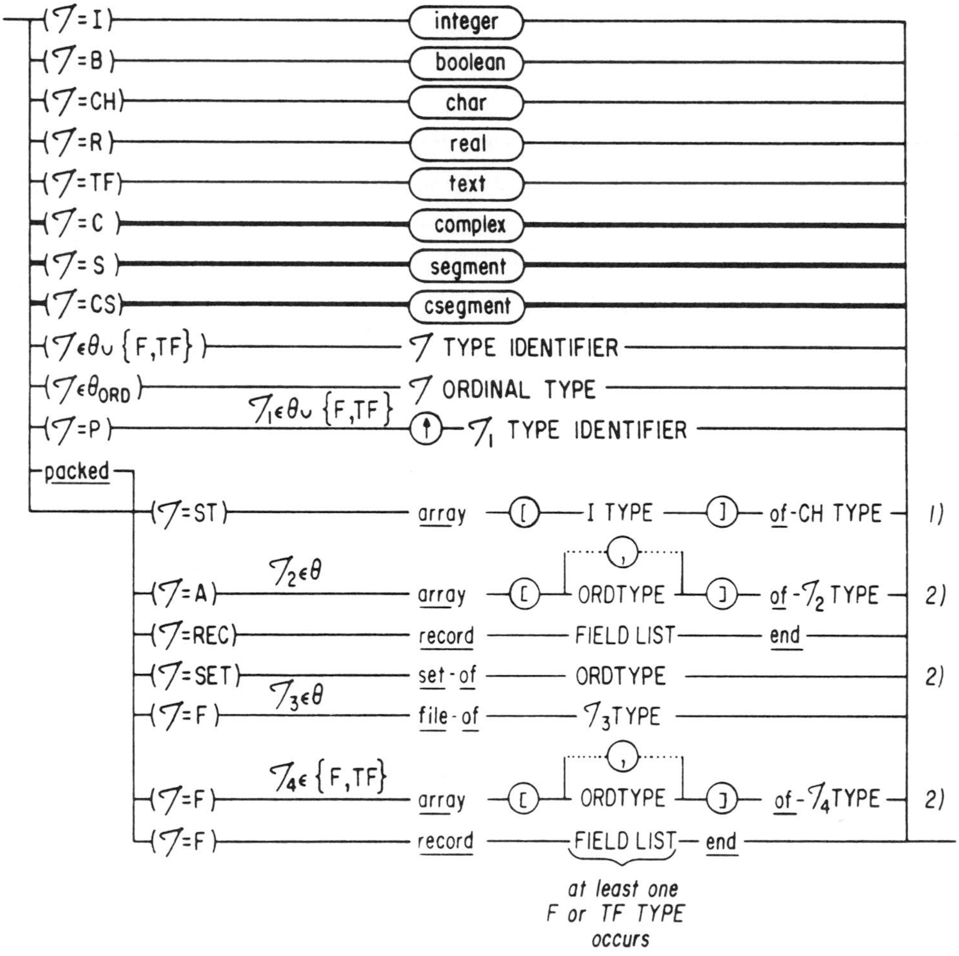

$(\mathcal{T}=I)$	integer
$(\mathcal{T}=B)$	boolean
$(\mathcal{T}=CH)$	char
$(\mathcal{T}=R)$	real
$(\mathcal{T}=TF)$	text
$(\mathcal{T}=C)$	complex
$(\mathcal{T}=S)$	segment
$(\mathcal{T}=CS)$	csegment

$(\mathcal{T} \in \theta \cup \{F, TF\})$ — $\mathcal{T}$ TYPE IDENTIFIER

$(\mathcal{T} \in \theta_{ORD})$ — $\mathcal{T}$ ORDINAL TYPE

$(\mathcal{T}=P)$ $\qquad \mathcal{T}_1 \in \theta \cup \{F, TF\}$ — $(\uparrow)$ — $\mathcal{T}_1$ TYPE IDENTIFIER

packed

$(\mathcal{T}=ST)$ — array —[— I TYPE —]— of-CH TYPE   *1)*

$\mathcal{T}_2 \in \theta$

$(\mathcal{T}=A)$ — array —[— ORDTYPE —]— of-$\mathcal{T}_2$ TYPE   *2)*

$(\mathcal{T}=REC)$ — record —— FIELD LIST —— end

$(\mathcal{T}=SET)$ — set-of —— ORDTYPE   *2)*

$\mathcal{T}_3 \in \theta$

$(\mathcal{T}=F)$ — file-of — $\mathcal{T}_3$ TYPE

$\mathcal{T}_4 \in \{F, TF\}$

$(\mathcal{T}=F)$ — array —[— ORDTYPE —]— of-$\mathcal{T}_4$ TYPE   *2)*

$(\mathcal{T}=F)$ — record —— FIELD LIST — end

at least one
F or TF TYPE
occurs

1)   Lower bound is I

2)   Range of the ordinal type is implementation dependent

P7 : $\mathcal{T}$ ORDINAL TYPE $\qquad \mathcal{T} \in \theta_{ORD} = \{I, B, CH, CD\}$

ORDTYPE

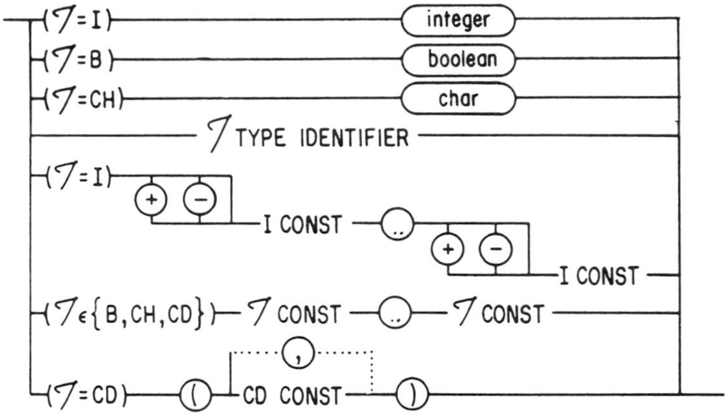

P8 : FIELD LIST (FL) $\qquad \theta = \{$ I, B, CH, CD, R, $\underline{C}$, $\underline{S}$, $\underline{CS}$, P, ST, A, REC, SET $\}$ $\qquad \theta_{ORD} = \{I, B, CH, CD\}$

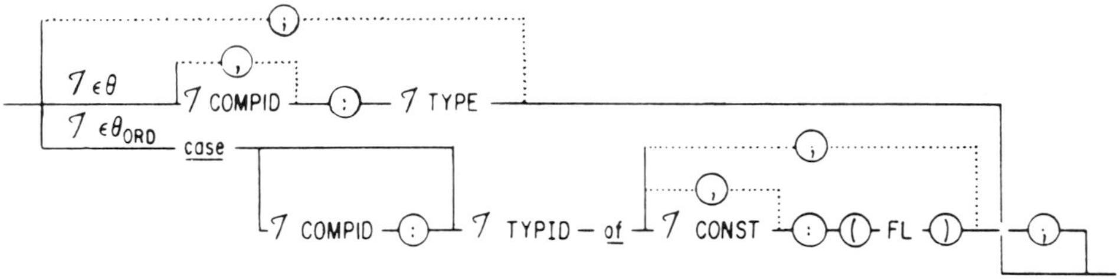

P9 : VARIABLE DECLARATION (VARDEC) $\qquad \theta = \{$ I, B, CH, CD, R, $\underline{C}$, $\underline{S}$, $\underline{CS}$, P, ST, A, REC, SET $\}$ $\qquad \theta_{DYN} = \{\underline{VEC}, \underline{MAT}, \underline{DIAG}\}$

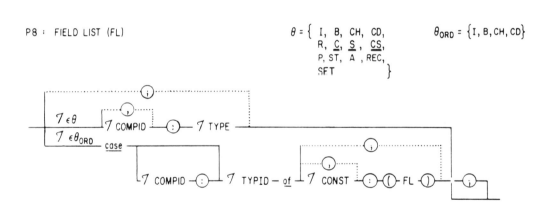

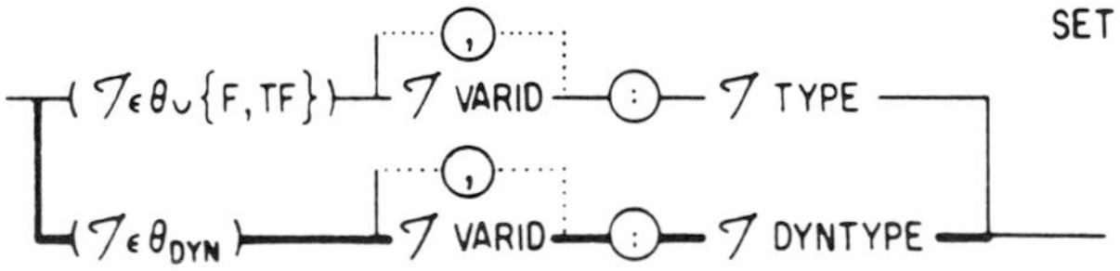

P10: PROCEDURE FUNCTION OPERATOR DECLARATION (PFODEC)

$$\theta^1 = \left\{ \begin{array}{l} \underline{I}, \ \underline{B}, \ \underline{CH}, \ \underline{CD}, \\ \underline{R}, \ \underline{C}, \ \underline{S}, \ \underline{CS}, \\ \underline{P}, \ \underline{ST}, \ \underline{A}, \ \underline{REC} \\ \underline{SET} \end{array} \right\}$$

$$\theta^2 = \left\{ \begin{array}{l} \underline{I}, \ \underline{B}, \ \underline{CH}, \ \underline{CD}, \\ \underline{R}, \ \underline{C}, \ \underline{S}, \ \underline{CS}, \\ \underline{P}, \ \underline{ST}, \ \underline{A}, \ \underline{REC} \\ \underline{SET} \end{array} \right\}$$

$$\theta_{DYN} = \left\{ \underline{VEC}, \ \underline{MAT}, \ \underline{DIAG} \right\}$$

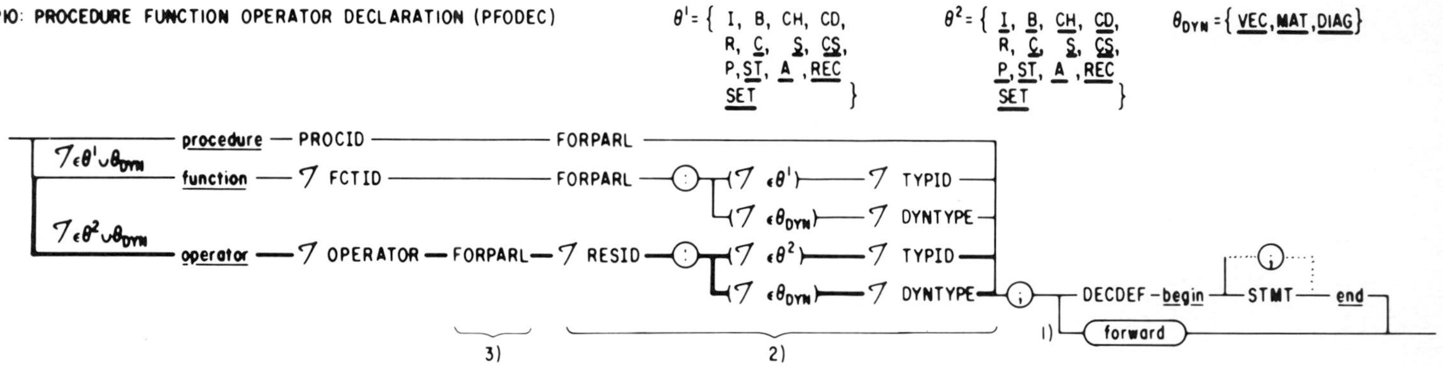

(1) Incomplete PFODEC
(2) Is missing in a completion of an incomplete PFODEC
(3) One or two parameters.

P11: $\mathcal{T}$OPERATOR ($\mathcal{T}$OP)

$\mathcal{T} \in \theta = \{$ I, B, CH, CD, R, C, S, CS, P, ST, A, REC, SET, VEC, MAT, DIAG $\}$

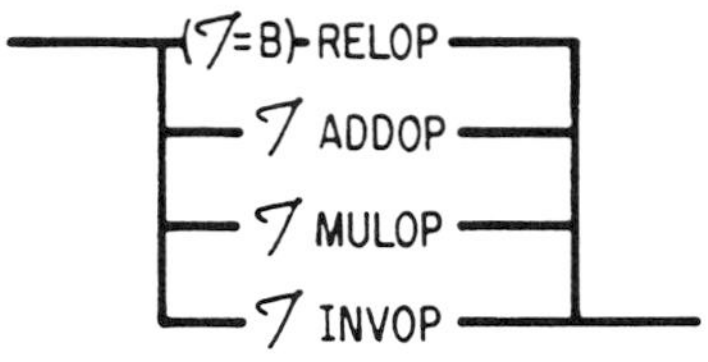

P12: FORMAL PARAMETER LIST (FORPARL)

$\theta_{DYN} = \{$ VEC, MAT, DIAG $\}$

$\theta = \{$ I, B, CH, CD, R, C, S, CS, P, ST, A, REC SET $\}$

$\theta' = \{$ I, B, CH, CD, R, C, S, CS, P, ST, A, REC SET $\}$

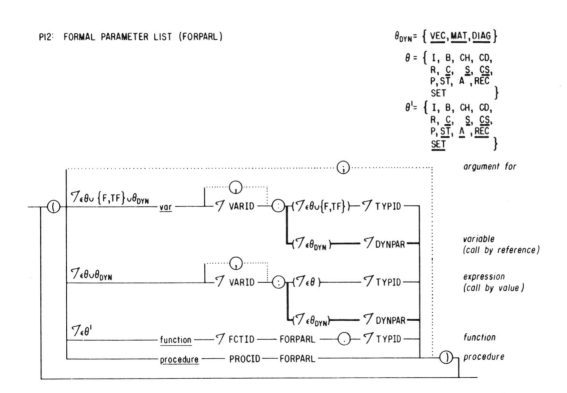

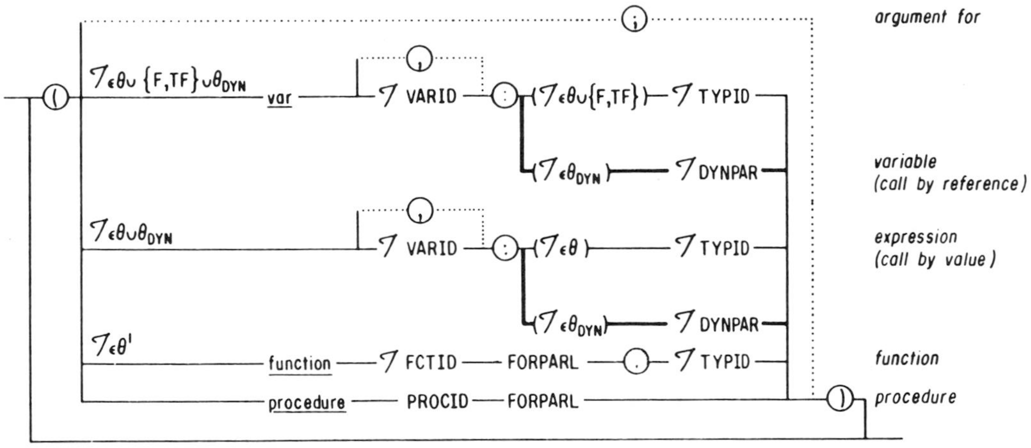

In an operator declaration at least one, at most two parameter.

PI3: τ DYNAMIC TYPE (τ DYNTYPE) $\tau \in \theta_{DYN} = \{$<u>VEC</u>, <u>MAT</u>, <u>DIAG</u>$\}$

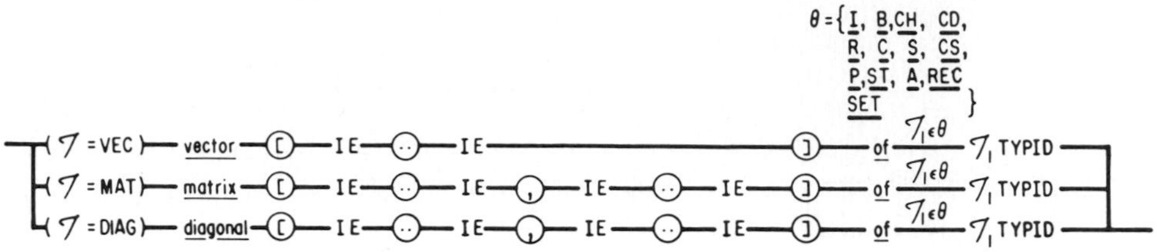

PI4: τ DYNAMIC PARAMETER (τ DYNPAR) $\tau \in \theta_{DYN} = \{$<u>VEC</u>, <u>MAT</u>, <u>DIAG</u>$\}$

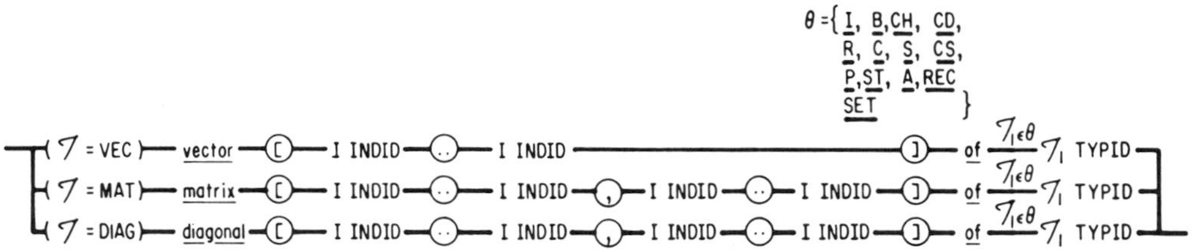

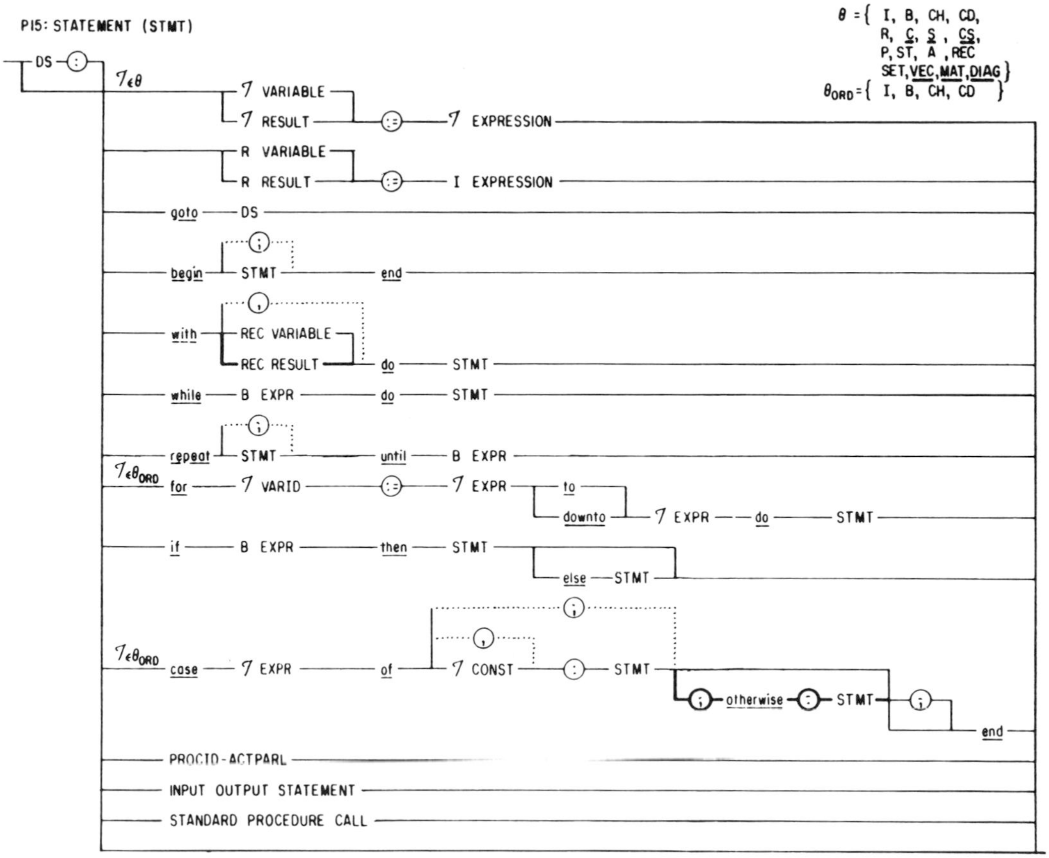
P15: STATEMENT (STMT)
θ = { I, B, CH, CD, R, C, S, CS, P, ST, A, REC SET, VEC, MAT, DIAG }
θORD = { I, B, CH, CD }
DS
:
τ∈θ
τ VARIABLE
τ RESULT
:=
τ EXPRESSION
R VARIABLE
R RESULT
I EXPRESSION
goto
begin
STMT
end
with
REC VARIABLE
REC RESULT
do
while
B EXPR
repeat
until
τ∈θORD
for
τ VARID
τ EXPR
to
downto
if
then
else
case
of
τ CONST
otherwise
PROCID-ACTPARL
INPUT OUTPUT STATEMENT
STANDARD PROCEDURE CALL

PI6: STANDARD PROCEDURE CALL $\theta_{ORD} = \{I, B, CH, CD\}$

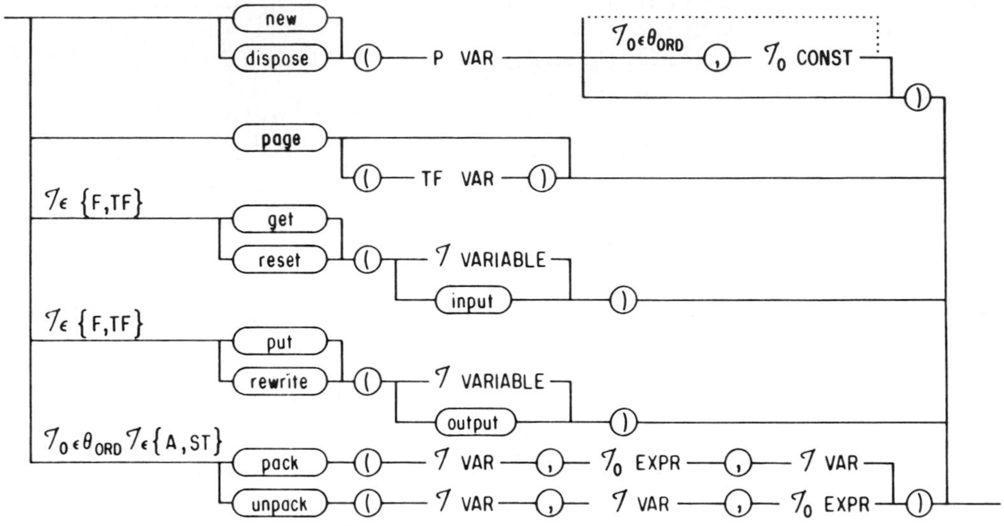

P17: INPUT OUTPUT STATEMENT

$\theta = \{$ I, B, CH, CD, R, C, S, CS, P, ST, A, REC, SET, $\}$

$\theta_{EMA} = \{$ R, C, VR, VC, MR, MC, $\}$

$\theta_{SEG} = \{$ S, CS, VS, VCS, MS, MCS, DS, DCS, $\}$

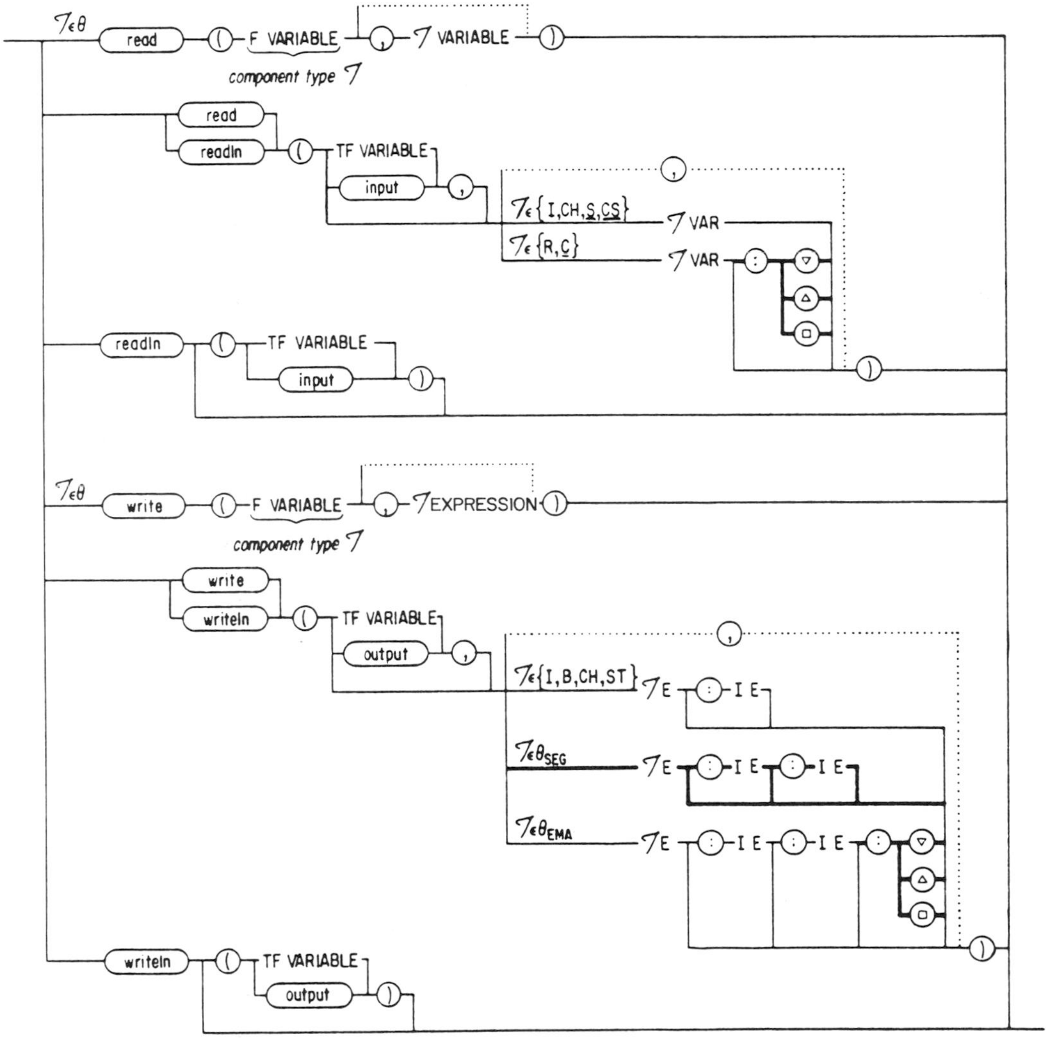

P18: $\mathcal{T}$ EXPRESSION ($\mathcal{T}$EXPR, $\mathcal{T}$E) $\mathcal{T}\epsilon\theta\cup\{F,TF\}$

$$\theta = \{ I, B, \underline{CH}, \underline{CD}, R, \underline{C}, \underline{S}, \underline{CS}, P, \underline{ST}, A, \underline{REC}, SET, \underline{VEC}, \underline{MAT}, \underline{DIAG}\}$$

```
                                          ─── 𝒯 SIMPEXP ───
  ┌──────────────────────────────────────────────────────────────────────┐
  └─(𝒯=B)── 𝒯₁, 𝒯₂ ε θ∪{F,TF} ── 𝒯₁ SIMPEXP ── RELOP ── 𝒯₂ SIMPEXP ──
```

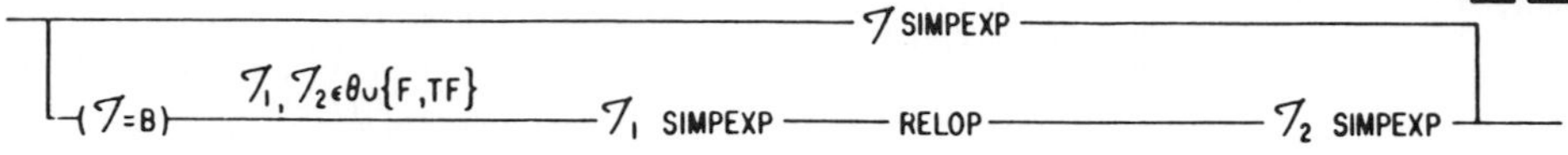

P19: $\mathcal{T}$ RELATIONAL OPERATOR ($\mathcal{T}$ RELOP) $\mathcal{T}\epsilon\theta$

$$\theta = \{ \begin{array}{llll} I, & B, & CH, & CD, \\ SI, & SB, & SCH, & SCD, \\ R, & \underline{C}, & \underline{S}, & \underline{CS}, \\ \underline{VR}, & \underline{VC}, & \underline{VS}, & \underline{VCS}, \\ \underline{MR}, & \underline{MC}, & \underline{MS}, & \underline{MCS}, \\ \underline{DR}, & \underline{DC}, & \underline{DS}, & \underline{DCS}, \\ P, & ST, & A, & REC\} \end{array}$$

declaration	result type	operator	types of operands (SIMPLE EXPRESSIONs)	meaning
nonstandard	B	$=$, $<>$, $<=$, $>=$, $<$, $>$, in, $><$	$\mathcal{T}_1\times\mathcal{T}_2$ with $\mathcal{T}_1,\mathcal{T}_2\epsilon\theta\cup\{F, TF\}$	according to the declaration
standard	B	$=$, $<>$	$\mathcal{T}_3\times\mathcal{T}_3$, $I\times R$, $R\times I$, $\underline{R\times C}$, $\underline{C\times R}$, $\underline{I\times C}$, $\underline{C\times I}$, ($\mathcal{T}_3\epsilon\theta$)	$=$, $\neq$
		$<=$, $>=$	$I\times I$, $I\times R$, $R\times I$, $R\times R$	$\leq$, $\geq$
			$B\times B$, $CH\times CH$, $CD\times CD$, $ST\times ST$	
			$\underline{S\times S}$, $\underline{VS\times VS}$, $\underline{MS\times MS}$, $\underline{DS\times DC}$	$\subseteq$, $\supseteq$
			$\underline{CS\times CS}$, $\underline{VCS\times VCS}$, $\underline{MCS\times MCS}$, $\underline{DCS\times DCS}$	
			$\underline{SI\times SI}$, $\underline{SB\times SB}$, $\underline{SCH\times SCH}$, $\underline{SCD\times SCD}$	
		$<$, $>$	$I\times I$, $I\times R$, $R\times I$, $R\times R$	$<$, $>$
			$B\times B$, $CH\times CH$, $CD\times CD$, $ST\times ST$	
		in	$I\times SI$, $B\times SB$, $CH\times SCH$, $CD\times SCD$	ϵ
			$\underline{I\times S}$, $\underline{R\times S}$, $\underline{VR\times VS}$, $\underline{MR\times MS}$, $\underline{DR\times DS}$	
			$\underline{I\times CS}$, $\underline{R\times CS}$, $\underline{C\times CS}$, $\underline{VC\times VCS}$, $\underline{MC\times MCS}$, $\underline{DC\times DCS}$	
		$><$	$\underline{S\times S}$, $\underline{VS\times VS}$, $\underline{MS\times MS}$, $\underline{DS\times DS}$	*disjoint*
			$\underline{CS\times CS}$, $\underline{VCS\times VCS}$, $\underline{MCS\times MCS}$, $\underline{DCS\times DCS}$	

P20: $\mathcal{T}$ SIMPLE EXPRESSION ($\mathcal{T}$ SIMPEXP) $\mathcal{T}\epsilon\theta\cup\{F, TF\}$

$$\theta = \{ I, B, \underline{CH}, \underline{CD}, R, \underline{C}, \underline{S}, \underline{CS}, P, \underline{ST}, A, \underline{REC}, SET, \underline{VEC}, \underline{MAT}, \underline{DIAG}\}$$

```
  ─ 𝒯 ε θ∪{F,TF} ──────────────────────────────── 𝒯 TERM ─────────────────────
  ┌──────────────────────────────────────────────────────────────────────────────┐
  │                     ┌······· 𝒯₂ ADDOP ·······┐                                 │
  └─(𝒯ε θ)── 𝒯₁ ε θ∪{F,TF}, 𝒯₂ ε θ ── 𝒯₁ TERM ─ 𝒯 ADDOP ─ 𝒯₃ ε θ∪{F,FT} ─ 𝒯₃ TERM ─
```

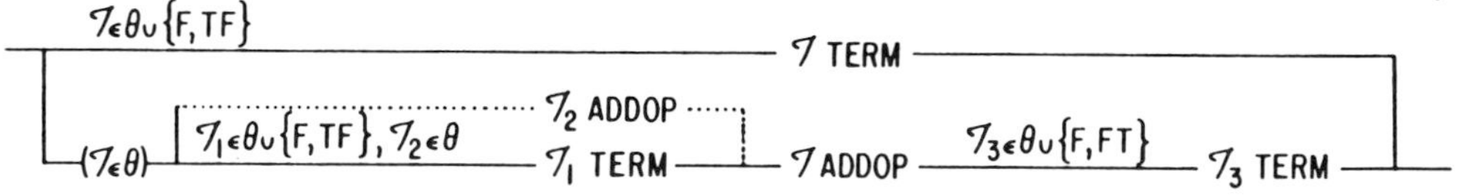

P21: $\mathcal{T}$ ADDING OPERATOR ($\mathcal{T}$ADDOP) $\mathcal{T} \in \theta$

$$\theta = \{ I, B, \underline{CH}, \underline{CD}, SI, SB, SCH, SCD, R, \underline{C}, \underline{S}, \underline{CS}, \underline{VR}, \underline{VC}, \underline{VS}, \underline{VCS}, \underline{MR}, \underline{MC}, \underline{MS}, \underline{MCS}, \underline{DR}, \underline{DC}, \underline{DS}, \underline{DCS}, \underline{P}, \underline{ST}, \underline{A}, \underline{REC} \}$$

declaration	result type	operator	types of operands (TERMs)	meaning
nonstandard		$+, -, \overset{\nabla}{+}, \nabla, \overset{\triangle}{+}, \triangle$, + *, $\underline{or}$	$\mathcal{T}_1 \times \mathcal{T}_2$ with $\mathcal{T}_1, \mathcal{T}_2 \in \theta \cup \{F, TF\}$	according to the declaration
standard	I	+, −	I × I	
	R	$+, -, \overset{\nabla}{+}, \nabla, \overset{\triangle}{+}, \triangle$	I × R, R × I, R × R	
	C	$+, -, \overset{\nabla}{+}, \nabla, \overset{\triangle}{+}, \triangle$	I × C, C × I, R × C, C × R, C × C	
	S	+, −	I × S, S × I, S × S	+, −
		+ *	S × S	*convex hull*
	CS	+, −	I × CS, CS × I, S × CS, CS × S, CS × CS	+, −
		+ *	CS × CS	*convex hull*
	VR	+, −	VR × VR	
	MR	+, −	MR × MR	
	DR	+, −	DR × DR	
	VC	+, −	VC × VC	
	MC	+,	MC × MC	
	DC	+, −	DC × DC	
	VS	+, −, + *	VS × VS	
	MS	+, −, + *	MS × MS	+, −, *convex hull*
	DS	+, −, + *	DS × DS	
	VCS	+, −, + *	VCS × VCS	
	MCS	+, −, + *	MCS × MCS	
	DCS	+, −, + *	DCS × DCS	
	B	$\underline{or}$	B × B	
	SI	+, −	SI × SI	
	SB	+, −	SB × SB	
	SCH	+, −	SCH × SCH	
	SCD	+, −	SCD × SCD	

P22: $\mathcal{T}$ TERM $\mathcal{T} \in \theta \cup \{F, TF\}$

$$\theta = \{ I, B, \underline{CH}, \underline{CD}, R, \underline{C}, \underline{S}, \underline{CS}, \underline{P}, \underline{ST}, \underline{A}, \underline{REC}, SET, \underline{VEC}, \underline{MAT}, \underline{DIAG} \}$$

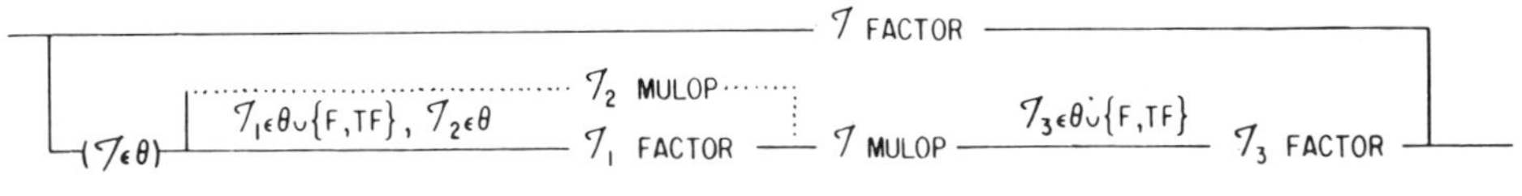

P23: $\mathcal{T}$ MULTIPLYING OPERATOR ($\mathcal{T}$ MULOP) $\mathcal{T} \in \theta$

$\theta = \{$ I, $\underline{\text{B}}$, $\underline{\text{CH}}$, $\underline{\text{CD}}$,
SI, SB, SCH, SCD,
R, $\underline{\text{C}}$, $\underline{\text{S}}$, $\underline{\text{CS}}$,
$\underline{\text{VR}}$, $\underline{\text{VC}}$, $\underline{\text{VS}}$, $\underline{\text{VCS}}$,
$\underline{\text{MR}}$, $\underline{\text{MC}}$, $\underline{\text{MS}}$, $\underline{\text{MCS}}$,
$\underline{\text{DR}}$, $\underline{\text{DC}}$, $\underline{\text{DS}}$, $\underline{\text{DCS}}$,
$\underline{\text{P}}$, $\underline{\text{ST}}$, $\underline{\text{A}}$, $\underline{\text{REC}}\}$

declaration	result type	operator	types of operands (FACTORs)	meaning
nonstandard	$\mathcal{T}$	$\underline{\text{mod}}$, $\underline{\text{div}}$ $*$, /, $\overset{*}{\nabla}$, ∇ $\overset{*}{\Delta}$, Δ, $\underline{\text{and}}$, $**$	$\mathcal{T}_1 \times \mathcal{T}_2$ with $\mathcal{T}_1, \mathcal{T}_2 \in \theta \cup \{F, TF\}$	according to the declaration
standard	I	$\underline{\text{mod}}$, $\underline{\text{div}}$, $*$	I×I	
	R	$*$, $\overset{*}{\nabla}$, $\overset{*}{\Delta}$	I×R, R×I, R×R,	
		/, ∇, Δ	I×I, I×R, R×I, R×R	
		$*$, $\overset{*}{\nabla}$, $\overset{*}{\Delta}$	VR×VR	*scalar product*
	C	$*$, $\overset{*}{\nabla}$, $\overset{*}{\Delta}$	I×C, C×I, R×C, C×R, C×C	
		/, ∇, Δ	I×C, C×I, R×C, C×R, C×C	
		$*$, $\overset{*}{\nabla}$, $\overset{*}{\Delta}$	VC×VC	*scalar product*
	S	$*$, /	I×S, S×I, S×S	
		$**$	S×S	*intersection*
		$*$	VS×VS	*scalar product*
	CS	$*$, /	I×CS, CS×I, S×CS, CS×S, CS×CS	
		$**$	CS×CS	*intersection*
		$*$	VCS×VCS	*scalar product*
	VR	$*$	I×VR, R×VR, MR×VR, DR×VR	
	MR	$*$	I×MR, R×MR, MR×MR, DR×MR, MR×DR	
	DR	$*$	I×DR, R×DR, DR×DR	
	VC	$*$	I×VC, R×VC, C×VC, MC×VC, DC×VC	
	MC	$*$	I×MC, R×MC, C×MC, MC×MC, DC×MC, MC×DC	
	DC	$*$	I×DC, R×DC, C×DC, DC×DC	
	VS	$*$	I×VS, S×VS, MS×VS, DS×VS	
		$**$	VS×VS	*intersection*
	MS	$*$	I×MS, S×MS, MS×MS, DS×MS, MS×DS	
		$**$	MS×MS	*intersection*
	DS	$*$	I×DS, S×DS, DS×DS	
		$**$	DS×DS	
	VCS	$*$	I×VCS, S×VCS, CS×VCS, MCS×VCS	
		$**$	VCS×VCS	*intersection*
	MCS	$*$	I×MCS, S×MCS, CS×MCS, MCS×MCS	
		$**$	MCS×MCS	*intersection*
	DCS	$*$	I×DCS, S×DCS, CS×DCS, DCS×DCS	
		$**$	DCS×DCS	
	B	$\underline{\text{and}}$	B×B	
	SI	$*$	SI×SI	
	SB	$*$	SB×SB	
	SCH	$*$	SCH×SCH	
	SCD	$*$	SCD×SCD	

P24: $\mathcal{T}$ FACTOR $\quad \mathcal{T} \in \theta \cup \{F, TF\}$

$\theta = \{$ I, B, CH, CD, R, C, S, CS, P, ST, A, REC, SET, VEC, MAT, DIAG $\}$

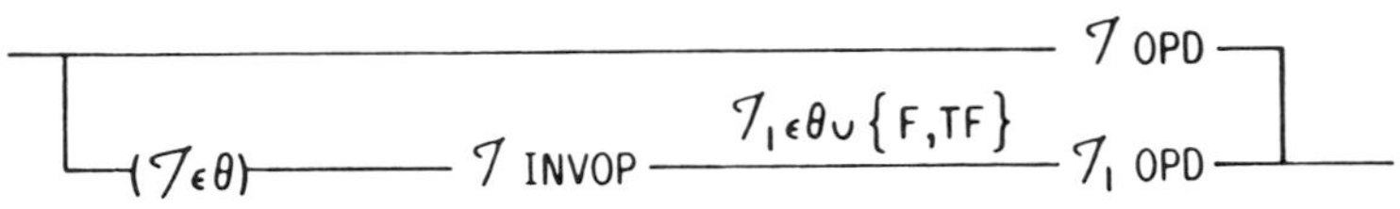

P25: $\mathcal{T}$ INVERTING OPERATOR ($\mathcal{T}$INVOP) $\quad \mathcal{T} \in \theta$

$\theta = \{$ I, B, CH, CD, SI, SB, SCH, SCD, R, C, S, CS, VR, VC, VS, VCS, MR, MC, MS, MCS, DR, DC, DS, DCS, P, ST, A, REC $\}$

declaration	result type	operator	type of (OPERAND)
nonstandard	$\mathcal{T}$	not, +, −	$\mathcal{T}_1 \in \theta \cup \{F, TF\}$
standard	B	not	B
	I	+, −	I
	R	+, −	R
	C	+, −	C
	S	+, −	S
	CS	+, −	CS
	VR	+, −	VR
	MR	+, −	MR
	DR	+, −	DR
	VC	+, −	VC
	MC	+, −	MC
	DC	+, −	DC
	VS	+, −	VS
	MS	+, −	MS
	DS	+, −	DS
	VCS	+, −	VCS
	MCS	+, −	MCS
	DCS	+, −	DCS

P26: $\mathcal{T}$ OPD ($\mathcal{T}$ OPERAND) $\qquad \mathcal{T} \in \theta \cup \{F, TF\}$

$\theta = \{$I, B, CH, CD, R, C, S, CS, P, ST, A, REC, SET, VEC, MAT, DIAG$\}$

$\theta_{EMA} = \{$R, VR, MR, C, VC, MC$\}$

$\theta_{CON} = \{$I, B, CH, CD, R, C, S, CS$\}$

$\theta_{SEG} = \{$S, VS, MS, DS, CS, VCS, MCS, DCS$\}$

$\theta_{SFCT} = \{$I, B, CH, CD, R, C, S, CS, VR, VC, VS, VCS, MR, MC, MS, MCS, DR, DC, DS, DCS$\}$

$\theta_{SET} = \{$SI, SB, SCH, SCD$\}$

$\theta_{ORD} = \{$I, B, CH, CD$\}$

β: $\theta_{SEG} \rightarrow \theta_{EMA}$

S	R
VS	VR
MS	MR
CS	C
VCS	VC
MCS	MC

i.e. $\beta\mathcal{T}$ denotes the basic type $\beta\mathcal{T} \in \theta_{EMA}$ of the segment type $\mathcal{T} \in \theta_{SEG}$

γ: $\theta_{SET} \rightarrow \theta_{ORD}$

SI	I
SB	B
SCH	CH
SCD	CD

i.e. $\gamma\mathcal{T}$ denotes the basic type $\gamma\mathcal{T} \in \theta_{ORD}$ of the set type $\mathcal{T} \in \theta_{SET}$

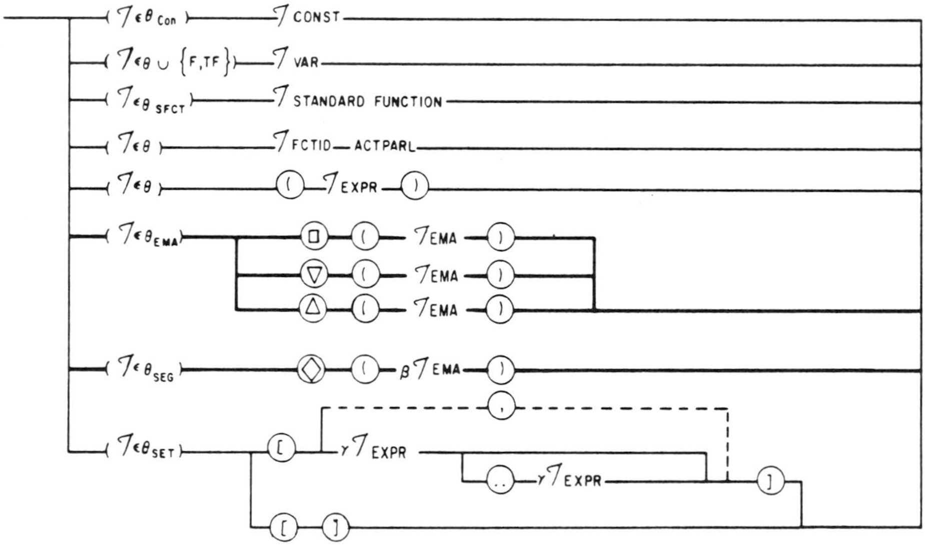

P27: R EMA (R EXPRESSION WITH MAXIMUM ACCURACY)

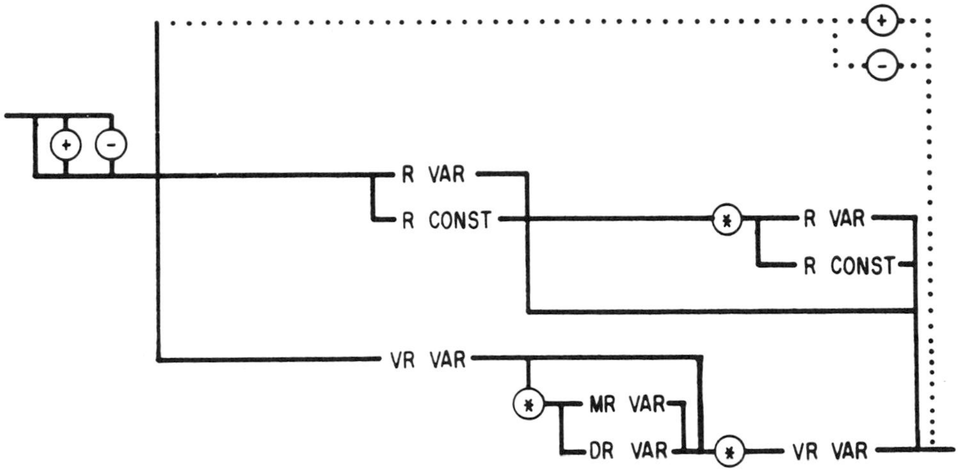

P28: C EMA (C EXPRESSION WITH MAXIMUM ACCURACY)

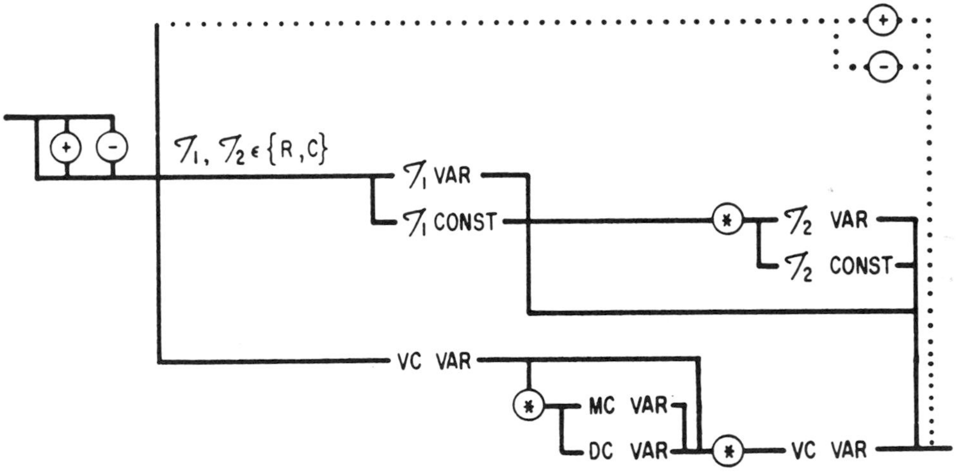

at least one C or VC VARIABLE or CONSTANT occurs

P29: VR EMA (VR EXPRESSION WITH MAXIMUM ACCURACY)

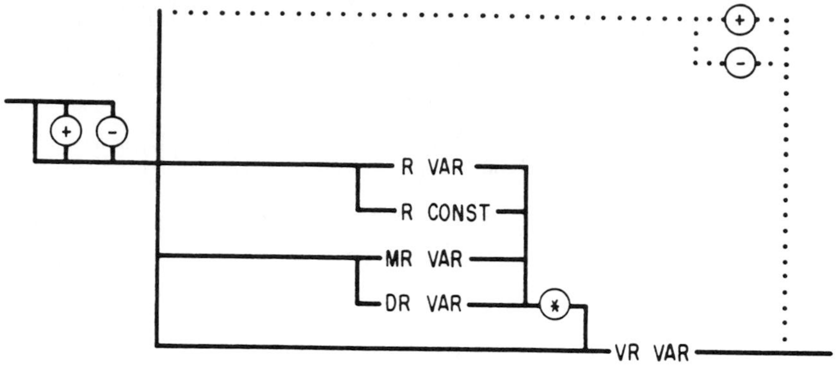

P30: MR EMA (MR EXPRESSION WITH MAXIMUM ACCURACY)

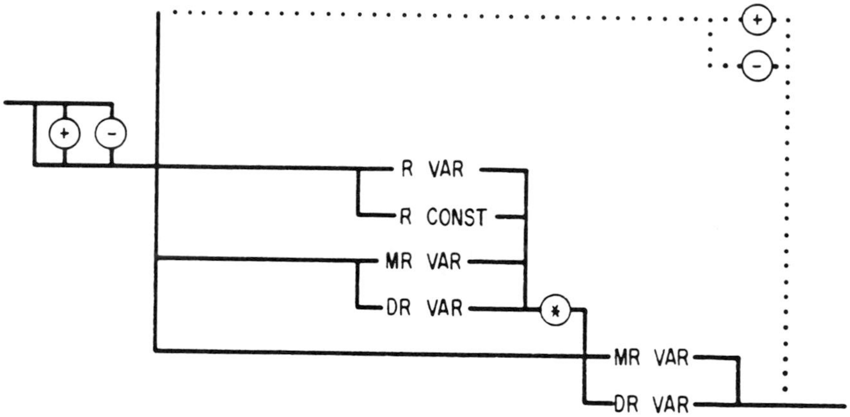

P31: VC EMA (VC EXPRESSION WITH MAXIMUM ACCURACY)

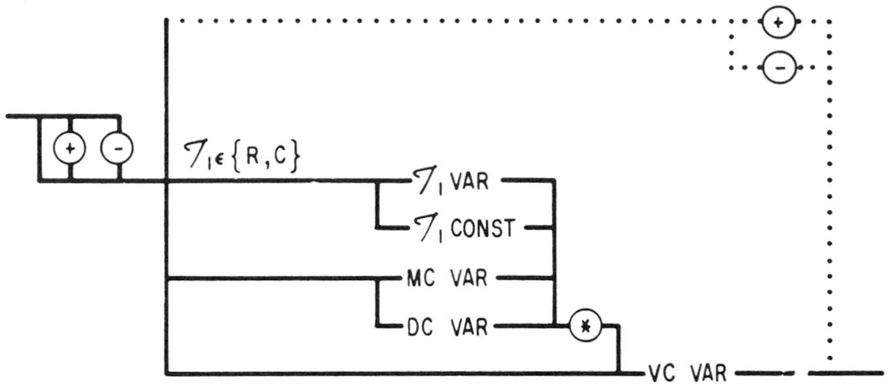

P32: MC EMA (MC EXPRESSION WITH MAXIMUM ACCURACY)

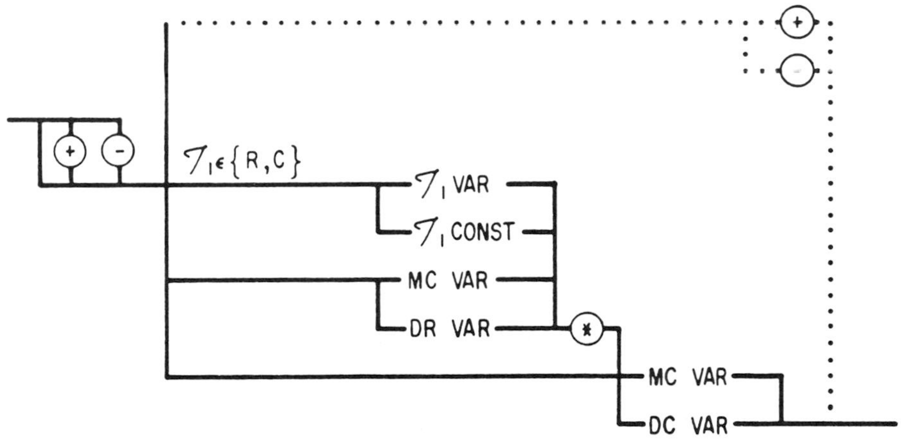

P33: ACTUAL PARAMETER LIST (ACTPARL)

$\theta = \{$ I, B, CH, CD, R, C, S, CS, P, ST, A, REC, SET, VEC, MAT, DIAG $\}$

$\theta' = \{$ I, B, CH, CD, R, C, S, CS, P, ST, A, REC, SET $\}$

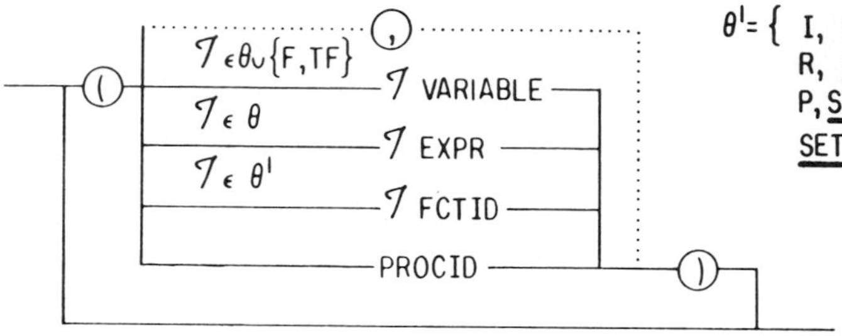

P34: τ VARIABLE (τ VAR)

$\tau \in \theta \cup \{F, TF\} \cup \theta_{DYN}$

$\theta = \{$ I, B, CH, CD, R, C, S, CS, P, ST, A, REC, SET $\}$

$\theta_{DYN} = \{$ VEC, MAT, DIAG $\}$

$\theta_{ORD} = \{$ I, B, CH, CD $\}$

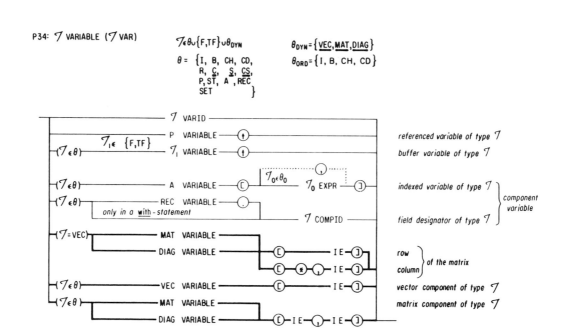

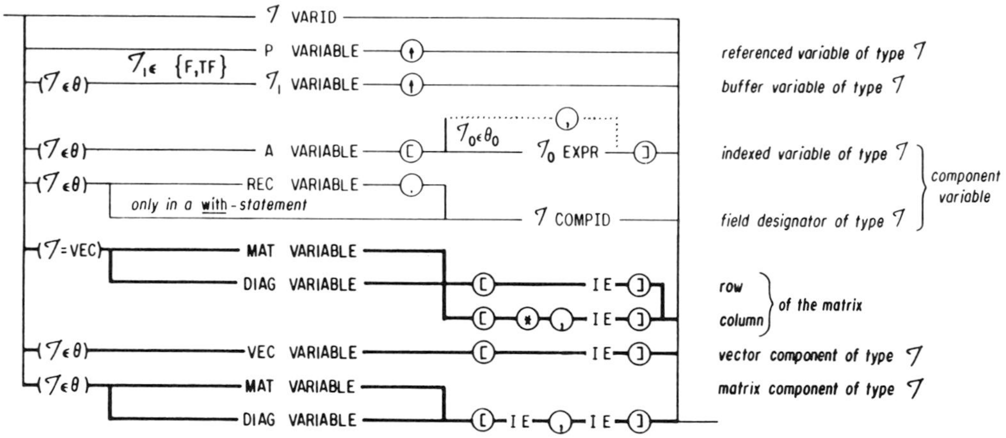

P35: $\mathcal{T}$ RESULT $\qquad \mathcal{T} \in \theta \cup \theta_{DYN}$

$\theta = \{$ I, B, CH, CD, R, C, S, CS, P, ST, A, REC, SET $\}$

$\theta_{ORD} = \{$ I, B, CH, CD $\}$

$\theta_{DYN} = \{$ VEC, MAT, DIAG $\}$

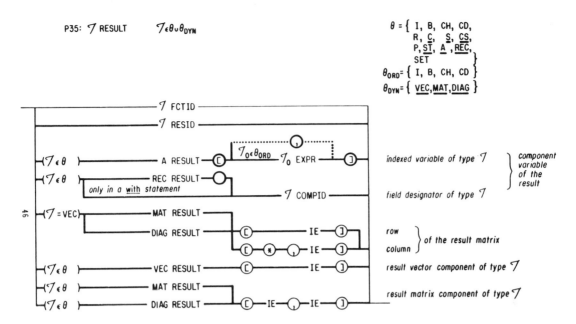

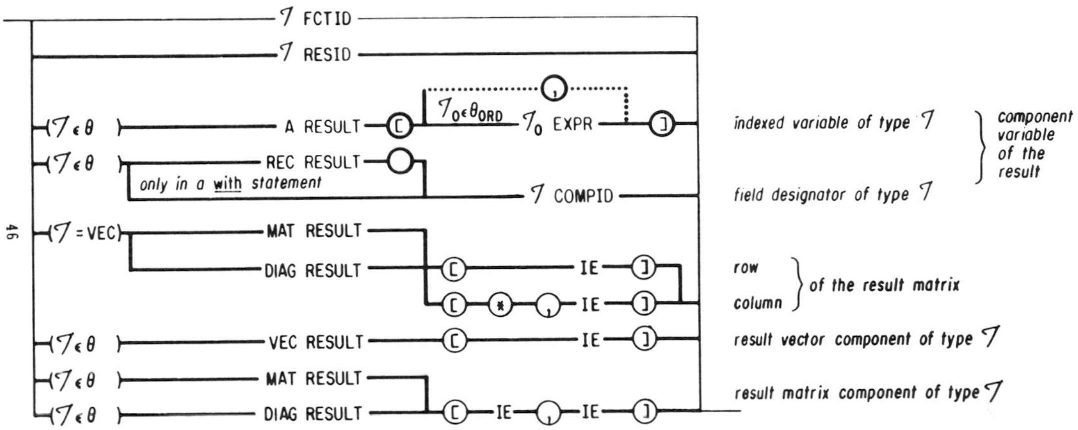

P36: 7 STANDARDFUNCTION

7∈{I, B, CH, CD, R, C, S, CS, VR, MR, VC, MC, VS, MS, VCS, MCS}

result type	function identifier	actual parameter list	meaning
I	ord	(I E), (B E), (CH E), (CD E)	
	succ, pred, abs, sqr	(I E)	
	trunc, round	(R E)	
B	eof	(input), (F VAR), (TF VAR) –[1]	
	eoln	(input), (TF VAR) –[1]	
	odd	(I E)	
	succ, pred	(B E)	
CH	chr	(I E)	
	succ, pred	(CH E)	
CD	succ, pred	(CD E)	
R	sqr, abs	(R E)	
	sqrt, exp, ln, arctan, sin, cos	(R E), (I E)	
	cabs	(C E)	*absolute value of the C expression*
	carg	(C E)	*argument of the C expression*
	cre, cim	(C E)	*real resp. imaginary part of the C expression*
	sabs	(S E)	*absolute value of the S expression*
	sinf, ssup	(S E)	*infimum resp. supremum of the S expression*
	csnorm	(CS E)	*sum of the absolute value of real part and the absolute value of imaginary part*
C	compl	(R E, R E)	*composes a complex value out of the two R expressions, where the first one becomes the real part and the second one the imaginary part*
	csinf, cssup,	(CS E)	*infimum resp. supremum of the CS expression*
	conj	(C E)	*conjugate of the C expression*
	csqr, csqrt, cexp, cln, carctan csin, ccos	(C E)	*complex versions of the corresponding real functions*
S	sgmt	(R E, R E)	*composes a segment out of the two R expressions, where the first one becomes the infimum the second one the supremum*
	csre, csim, csabs, csarg,	(CS E)	*segment versions of the corresponding real functions*
	ssqr, ssqrt sexp, sln, sarctan, ssin, scos	(S E)	

(1) *Actual Parameter list may be missing*

result type	function identifier	actual parameter list	meaning
CS	csgmt	(C E, C E)	*composes a complex segment out of the two* C *expressions*
	cscompl	(S E, S E)	*composes a complex segment out of the two* S *expressions*
	cssqr, cssqrt, csexp, csln csarctan, cssin, cscos, csconj	(CS E)	*complex segment version of the corresponding complex functions*
VR	vcre, vcim	(VC E)	*real resp. imaginary part of a* VC *expression*
	vsinf, vssup	(VS E)	*infimum resp. supremum of a* VS *expression*
	vrnull	(I E)	*produces a real vector of n zeros, where n is the value of the* I *expression*
	vrdiag	(DR E)	*takes the diagonal of the* DR *expression*
	vrdiagmat	(MR E)	*takes the diagonal of the* MR *expression*
MR	mcre, mcim	(MC E)	*real resp. imaginary part of the* MC *expression*
	msinf, mssup	(MS E)	*infimum resp. supremum of the* MS *expression*
	mrtransp	(MR E)	*transposes the* MR *expression*
DR	drnull, drid	(I E)	*produces the real zero matrix resp. the identity* $n \times n$ *matrix, where n it the value of the* I *expression*
	drvec	(VR E)	*produces a diagonal matrix with the* VR *expression as diagonal*
	drmat	(MR E)	*annihilates all non-diagonal elements of the* MR *expression*
VC	vcompl	(VR E, VR E)	*composes a complex vector out of two* VR *expressions*
	vcsinf, vcssup	(VCS E)	*infimum resp. supremum of the* VCS *expression*
	vcnull	(I E)	*produces a complex vector of n zeros, where n is the value of the* I *expression*
	vcdiag	(DC E)	*takes the diagonal of the* DC *expression*
	vcdiagmat	(MC E)	*takes the diagonal of the* MC *expression*
	vconj	(VR E)	*produces the conjugate of the* VR *expression*
MC	mcompl	(MR E, MR E)	*composes a complex matrix out of two* MR *expressions*
	mcsinf, mcssup	(MCS E)	*infimum resp. supremum of the* MCS *expression*
	mcherm, mconj	(MC E)	*produces the conjugate transpose resp. the conjugate of the* MC *expression*
DC	dcnull, dcid	(I E)	*produces the complex zero matrix resp. the identity* $n \times n$ *matrix, where n is the value of the* I *expression*
	dcvec	(VC E)	*produces a diagonal matrix with the* VC *expression as diagonal*
	dcmat	(MC E)	*annihilates all non-diagonal elements of the* MC *expression*

result type	function identifier	actual parameter list	meaning
VS	vsgmt	(VR E, VR E)	*composes a segment vector out of the two* VR *expressions*
	vcsre, vcsim	(VCS E)	*real resp. imaginary part of the* VCS *expression*
	vsnull	(I E)	*produces a segment vector of n zeros, where n is the value of the* I *expression*
	vsdiag	(DS E)	*takes the diagonal of the* DS *expression*
	vsdiagmat	(MC E)	*takes the diagonal of the* MC *expression*
MS	msgmt	(MR E, MR E)	*composes a segment matrix out of the two* MR *expressions*
	mcsre, mcsim	(MCS E)	*real resp. imaginary part of the* MCS *expression*
	mstransp	(MS E)	*transposes the* MS *expression*
DS	dsnull, dsid		*produces the segment zero matrix resp. the segment identity matrix*
	dsvec	(VS E)	*produces a diagonal matrix with the* VS *expression as diagonal*
	dsmat	(MS E)	*annihilates all non-diagonal elements of the* MS *expression*
VCS	vcscompl	(VS E, VS E)	*composes a complex segment vector out of the two* VS *expressions*
	vcsgmt	(VC E, VC E)	*composes a complex segment vector out of the two* VC *expressions*
	vcsnull	(I E)	*produces a complex segment vector of n zeros, where n is the value of the* I *expression*
	vcsdiag	(DCS E)	*takes the diagonal of the* DCS *expression*
	vcsdiagmat	(MCS E)	*takes the diagonal of the* MCS *expression*
	vcsconj	(VCS E)	*produces the conjugate of the* VCS *expression*
MCS	mcscompl	(MS E, MS E)	*composes a complex segment matrix out of two* MS *expressions*
	mcsgmt	(MC E, MC E)	*composes a complex segment matrix out of the two* MC *expressions*
	mcsherm, mcsconj	(MCS E)	*produces the conjugate transpose resp. the conjugate of the* MCS *expression*
DCS	dcsnull, dcsid	(I E)	*produces the complex segment zero resp. the identity matrix with dimension n × n, where n is the value of the* I *expression*
	dcsvec	(VCS E)	*produces a diagonal matrix with the* VCS *expression as diagonal*
	dcsmat	(MCS E)	*annihilates all non-diagonal elements of the* MCS *expression*

P37: PROGRAM IDENTIFIER (PID)

PROGRAM PARAMETER IDENTIFIER (PPID)

PROCEDURE IDENTIFIER (PROCID)

$\mathcal{T}_1$	TYPE IDENTIFIER	($\mathcal{T}_1$ TYPID)	$\mathcal{T}_1 \in \theta \cup \{F, TF\}$
$\mathcal{T}_1$	VARIABLE IDENTIFIER	($\mathcal{T}_1$ VARID)	$\mathcal{T}_1 \in \theta \cup \{F, TF\} \cup \theta_{DYN}$
$\mathcal{T}_1$	COMPONENT IDENTIFIER	($\mathcal{T}_1$ COMPID)	$\mathcal{T}_1 \in \theta \cup \{F, TF\}$
$\mathcal{T}_2$	CONSTANT IDENTIFIER	($\mathcal{T}_2$ CONID)	$\mathcal{T}_2 \in \theta_C = \{I, B, CH, CD, R, \underline{C}, \underline{S}, \underline{CS}\}$
$\mathcal{T}_3$	FUNCTION IDENTIFIER	($\mathcal{T}_3$ FCTID)	$\mathcal{T}_3 \in \theta^3 \cup \theta_{DYN}$
$\mathcal{T}_4$	RESULT IDENTIFIER	($\mathcal{T}_4$ RESID)	$\mathcal{T}_4 \in \theta^4 \cup \theta_{DYN}$
$\mathcal{T}_5$	INDEX IDENTIFIER	($\mathcal{T}_5$ INDID)	$\mathcal{T}_5 \in \theta_{ORD} = \{I, B, CH, CD\}$

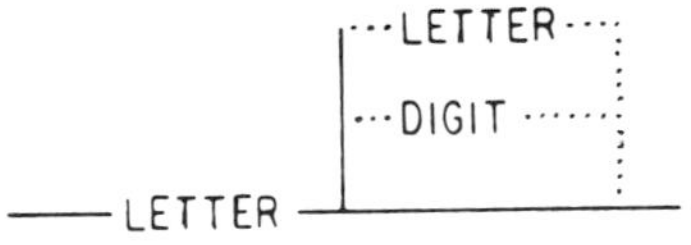

$$\theta = \{ I,\ B,\ CH,\ CD,\ R,\ \underline{C},\ \underline{S},\ \underline{CS},\ P,\ ST,\ A,\ REC,\ SET \}$$

$$\theta^3 = \{ I,\ B,\ CH,\ CD,\ R,\ \underline{C},\ \underline{S},\ \underline{CS},\ P,\ \underline{\underline{ST}},\ \underline{\underline{A}},\ \underline{\underline{REC}},\ \underline{\underline{SET}} \}$$

$$\theta^4 = \{ \underline{\underline{I}},\ \underline{\underline{B}},\ \underline{\underline{CH}},\ \underline{\underline{CD}},\ \underline{\underline{R}},\ \underline{C},\ \underline{S},\ \underline{CS},\ \underline{\underline{P}},\ \underline{\underline{ST}},\ \underline{\underline{A}},\ \underline{\underline{REC}},\ \underline{\underline{SET}} \}$$

P38: DIGIT SEQUENCE (DS)

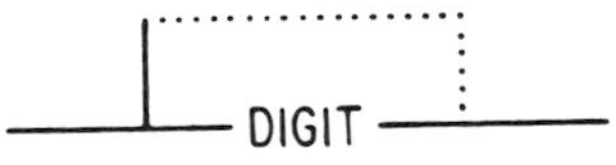

P39: STRING CHARACTER

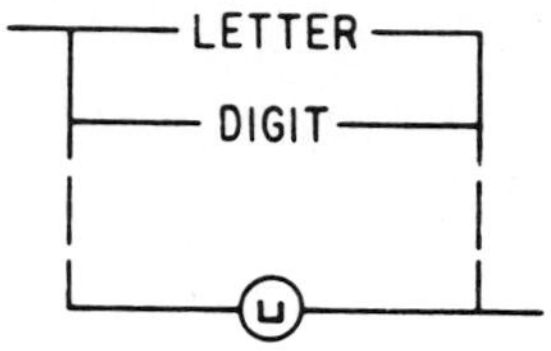

implementation dependent

P40: LETTER

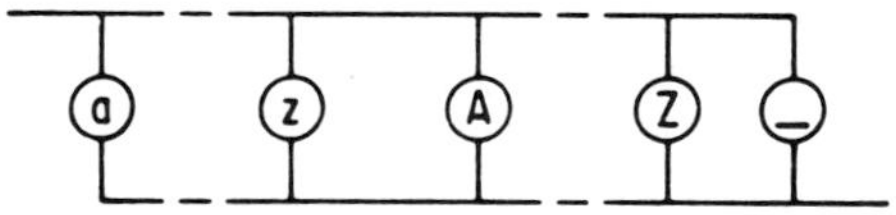

capital letters and underbar implementation dependent

P41: DIGIT

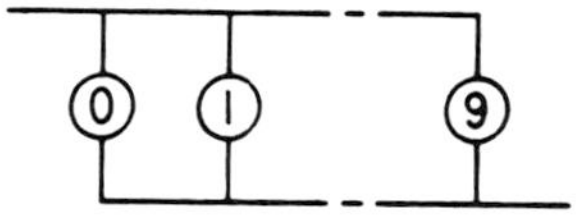

GLOSSARY OF SYNTAX VARIABLES

DENOTATION (ABBREVIATION)	*DEFINED IN DIAGRAM NO.*	*REFERENCED IN DIAGRAM NO.*
ACTUAL PARAMETER LIST (ACTPARL)	P33	P15,P26
7 ADDING OPERATOR (7 ADDOP)	P21	P11,P20
7 COMPONENT IDENTIFIER (7 COMPID)	P37	P8,P34,P35
7 CONSTANT (7 CONST)	P4	P3,P4,P7,P8,P15,P26,P27
CONSTANT DEFINITION (CONSTDEF)	P3	P2
7 CONSTANT IDENTIFIER (7 CONID)	P37	P3,P4
DECLARATION DEFINITION (DECDEF)	P2	P1,P10
DIGIT	P47	P37,P38,P39
DIGIT SEQUENCE (DS)	P38	P2,P4,P15
7 DYNAMIC PARAMETER (7 DYNPAR)	P14	P12
7 DYNAMIC TYPE (7 DYNTYPE)	P13	P9,P10
7 EXPRESSION (7 EXPR, 7 E)	P18	P13,P15,P16,P17,P26,P33,P34, P35
R EXPRESSION WITH MAXIMUM ACCURACY (R EMA)	P27	P26
C EXPRESSION WITH MAXIMUM ACCURACY (C EMA)	P28	P26
VR EXPRESSION WITH MAXIMUM ACCURACY (VR EMA)	P29	P26
MR EXPRESSION WITH MAXIMUM ACCURACY (MR EMA)	P30	P26
VC EXPRESSION WITH MAXIMUM ACCURACY (VC EMA)	P31	P26
MC EXPRESSION WITH MAXIMUM ACCURACY (MC EMA)	P32	P26
7 FACTOR	P24	P22
FIELD LIST (FL)	P8	P6
FORMAL PARAMETER LIST (FORPARL)	P12	P10,P12
7 FUNCTION IDENTIFIER (7 FCTID)	P37	P10,P12,P26,P33,P35
7 INDEX IDENTIFIER (7 INDID)	P37	P14
INPUT OUTPUT STATEMENT	P17	P15
7 INVERTING OPERATOR (7 INVOP)	P25	P11,P24
LETTER	P40	P37,P39
7 MULTIPLYING OPERATOR (7 MULOP)	P23	P11,P22
7 OPERAND (7 OPD)	P26	P24
7 OPERATOR (7 OP)	P11	P10
7 ORDINAL TYPE	P7	P6
ORDTYPE	P7	P6
PROCEDURE FUNCTION OPERATOR DECLARATION (PFODEC)	P10	P2
PROCEDURE IDENTIFIER (PROCID)	P37	P10,P12,P15,P33
PROGRAM	P1	
PROGRAM IDENTIFIER (PID)	P37	P1
PROGRAM PARAMETER IDENTIFIER (PPID)	P37	P1
RELATIONAL OPERATOR (RELOP)	P19	P11,P18
7 RESULT	P35	P15,P35
7 RESULT IDENTIFIER (7 RESID)	P37	P10,P35
7 SIMPLE EXPRESSION (7 SIMPEXP)	P20	P18
7 STANDARD FUNCTION	P36	P26
STANDARD PROCEDURE CALL	P16	P15
STATEMENT (STMT)	P15	P1,P10,P15
STRING CHARACTER	P39	P4
7 TERM	P22	P20
7 TYPE	P6	P5,P6,P8,P9
7 ,, TYPE		
7 TYPE DEFINITION (TYPEDEF)	P5	P2
7 TYPE IDENTIFIER (7 TYPID)	P37	P5,P6,P7,P8,P10,P12,P13,P14
7 VARIABLE (7 VAR)	P34	P15,P16,P17,P26,P27,P33,P34
VARIABLE DECLARATION (VARDEC)	P9	P2
7 VARIABLE IDENTIFIER (7 VARID)	P37	P9,P12,P15,P34

D2. Semantics

The semantics of the language extension, MATRIX PASCAL, is now presented. However in contrast to the syntax, we discuss only the semantics for the new aspects of the language extension. This discussion follows the format of the PASCAL ISO norm concept, see [1]. That is, the organization and numbering of the semantics of PASCAL in that norm is taken over here. (Note that because of this restriction to the new aspects, the numbering which appears while ordered is incomplete. For example the numbering here starts with 6.) The semantic remarks which are made are organized according to the section system of the norm including the use of section numbers and titles. Each such remark is either a replacement or modification of a corresponding remark in the norm or a new comment about the extension. Thus the semantics of the full extended language is that of the norm coupled with the modifications which we now present.

6. Requirements

6.1.5 Numbers

Rounding control causes rounding of the given constant $\square x$, $\bigtriangledown x$ or $\bigtriangleup x$ resp., to $\square(x)$, $\bigtriangledown(x)$ or $\bigtriangleup(x)$ resp. (see 6.7).

6.4.5 Compatible Types

In contrast to array types two dynamic types are compatible if they have identical number and length of their index ranges and their component types are compatible.

6.4.6 f)

Two dynamic types are assignment-compatible if they are compatible.

6.5.1 Variable Declaration

Vectors and matrices and diagonal matrices can be declared by special declarations containing dynamic types. In a variable declaration of the (main) program the index bounds of a dynamic type may only contain constants. When a procedure or a function of an operator containing a variable declaration with dynamic type is called the expressions of the index bounds must be evaluable. Diagonal matrices are special matrices. They are stored as vectors consisting of the diagonal elements and one additional non-diagonal element. The additional element is the internal representation of the equivalence class of all non-diagonal elements. In the case of diagonal matrices of numerical types (DR, DC, DS, DCS) all non-diagonal matrix elements are implicitly initialized with zero. The diagonal and the non-diagonal elements can be used like matrix components. An assignment to a non-diagonal element alters the value of all non-diagonal elements.

Example

```
        var  d: diagonal [1 . . 3, 1 . . 3] of real;
                i: integer;
        begin
        for  i := 1 to 3 do d[i, i] := i;
{the matrix now has the value:
        1       0       0
        0       2       0
        0       0       3}
d[1, 2] := 0.5
{the matrix now has the value:
        1       0.5     0.5
        0.5     2       0.5
        0.5     0.5     3 }
```

6.5.3.2 Indexed Variables

The action of selecting a component of a vector, matrix or diagonal matrix variable is the same as for a component of an array variable.

6.6.2 Function and Operator Declaration

6.6.2.1 Function Declaration

(Compare 6.6.2 in [1].) Notice that if the type of the function is an array type or record type, values also may be assigned to the components of the function result.

Example

```
type      string = array [1..10] of char;
function  vec: string;
var       i: 1..10;
begin     for i := 1 to 10 do
                vec [i] := 'a'
end
```

If the result is a dynamic type, then the expressions denoting the index ranges may contain formal parameters as well as global entities. These expressions are evaluated when the function is called.

6.6.2.2 Operator Declaration

The declaration of an operator may overload a standard operator or it may give an additional meaning to it. The type of an operator, i.e., the result type depends on the operator itself as well as on the ordering and the types of its parameters. The result and result type are subject to the same conventions as the result and result type of functions. The parameter list contains at least one and at most two parameters.

Example

```
type      string = array [1..10] of char;
operator  + (a, b: string) result: string;
var       i: 1..10;
begin     for i := 1 to 10 do
            if a[i] < b[i] then
```

```
            result [i] := a[i]
        else
            result [i] := b[i]
end
```

6.6.3 Parameters (P13)

6.6.3.1 General

There are no operator parameters. Instead of the conformant array schema dynamic parameters are introduced. In a formal parameter list the index identifiers of all dynamic parameters must be different; they define local constants for the procedure, function, or operator. When the procedure, function, or operator is called these constants are assigned the values of corresponding index bounds of the actual parameters.

6.6.3.6 Parameter List Congruity

The index bounds of the actual expression are assigned to the index type identifiers of a dynamic parameter.

6.6.3.7 Dynamic Type for the Result (P15)

The index expressions are to be evaluated prior to the beginning of the execution of the function block but after the evaluation of the parameter list. Therefore only constants, global entities, formal parameters or index type identifiers of dynamic parameters of the parameter list may occur in those expressions.

Example

```
operator + (a: vector [m .. n] of char;
            b: vector [p .. q] of char)
            res: vector [m .. n+q−p+1] of char;
var i:      integer;
begin       for i := m to n do
```

```
                res[i] := a[i];
            for i := 1 to q-p+1 do
                res[n+i] := b[p-1+i]
    end
```

6.7 Expressions (P19-P26)

Expressions are evaluated according to the rules of priority. The resultant type of an expression, simple expression, term or factor is the type of the operator last executed in its evaluation. The type of an operator depends on the types of the arguments. Type compatibility is displayed in the tables of P19, P21, P23 and P25. When directed roundings are to be employed the symbol "▽" (downwards) or "△" (upwards) is included in the operator symbol. The combinations "+*" and "**" and "><" denote additional operators.

In contrast to [1] the monadic operators "+" and "−" have the highest priority. This causes no change in the evaluation of an expression since all numerical types correspond to ringoids (see [12]) in which the property

$$\bigwedge_{\circ \in \{*,/\}} -(a \circ b) = (-a) \circ b = a \circ (-b)$$

holds.

6.7.1.1 Operands

If a rounding symbol occurs in front of an expression in parentheses (EMA), then this expression is evaluated with maximum accuracy. This means that the result is the exact value w of the expression rounded according to the given rounding control (see 6.7.2.2). In the case of $\Diamond$ the exact value is mapped into the segment space where the lower, resp. upper bound is computed by the rounding ∇w, resp. Δw. In such an expression (EMA) a constant which is not qualified by a rounding symbol must be representable in the floating-point system being employed.

In all the EMA diagrams the operators + , −, * are the standard operators in all cases. Thus a redefinition in the surrounding program has no influence.

6.7.2 Operators

6.7.2.1 General

Additional operators are

multiplying operators ▽*, △*, ▽, △, **

adding operators ▽+, △+, ▽, △, +*

relational operators ><

6.7.2.2 Arithmetic Operators and Rounding

6.7.2.2.1 Arithmetic Operators

According to P25 an inverting operator + or − can occur for all numeric types. The inverting operator + is the identity in each case while the operator − depends on the operand type as follows:

type of operand x	factor to be evaluated	Result
C	$-x$	$(-re(x), -im(x))$
S	$-x$	$[-\sup(x); -inf(x)]$
CS	$-x$	$([-\sup(re(x)); -inf(re(x))],$ $[-\sup(im(x)); -inf(im(x))])$

Vectors and matrices are inverted componentwise.

6.7.2.2.2 Rounding (compare Section B and Section E)

(a) Implicit rounding

The "implicit rounding" (symbolized by $\square$) is a mapping from the set $\mathbb{R}$ of the real numbers into the computer representable subset R with the following properties:

$(R1)$ $\quad \square(x) = x$ for all $x \in R$ $\quad$ (R denotes the type real)

$(R2)$ $\quad x \leq y$ implies $\square(x) \leq \square(y)$ for all $x, y \in \mathbb{R}$

$(R3)$ $\quad \square(-x) = -\square(x)$ for all $x \in \mathbb{R}$

For computer implementation of $\square$, see [12] Section 6.5 Figure 40.

(b) Directed roundings ∇, Δ

The rounding control ∇ resp. Δ is a mapping from $\mathbb{R}$ to R with the following properties:

$(R1)$ $\nabla(x) = x$ resp. $\Delta(x) = x$ for all $x \in R$ (R denotes the type real)

$(R2)$ $x \leq y$ implies $\nabla(x) \leq \nabla(y)$ *resp.* $\Delta(x) \leq \Delta(y)$ for all $x, y \in \mathbb{R}$

$(R4)$ $\nabla(x) \leq x$ resp. $x \leq \Delta(x)$ for all $x \in \mathbb{R}$

For computer implementation of ∇ and Δ, see [12] Section 6.5 Figure 41.

(c) Expression of type integer

The integer arithmetic is not extended.

(d) Expression of type real

Let $\bullet$ and $\circ$ be identical symbols from the set $\{+, -, *, /\}$, and let x, y be operands of admissible type. Let $x \circ y$ denote the correct result in $\mathbb{R}$ of the operation $\circ$. Then the result of the processor operation $\bullet$ may be read from the following table:

Expression to be evaluated	Result
$x \bullet y$	$\square(x \circ y)$
$x \overset{\bullet}{\nabla} y$	$\nabla(x \circ y)$
$x \overset{\bullet}{\Delta} y$	$\Delta(x \circ y)$

For computer implementation of the entries in the second column of this table see section E and [12].

The operators listed in P21, P23, P25 are to be implemented as described in Section E. The following properties must hold.

(e) Operators of type complex:

In each of the cases $C \times C$, $C \times I$, $I \times C$, $C \times R$, $R \times C$ the operands of type I and R are to be taken as being of type C. Let $\bullet$ and $\circ$ be identical symbols taken from the set $\{+, -, *, /\}$, and let x, y be operands of admissible type. Let $x \circ y$ denote the correct result in $\mathbb{C}$ of the operation $\circ$. Then the result of the processor operation $\bullet$ may be read from the following table:

Expression to be evaluated	Result
$x \bullet y$	$(\square(\mathrm{re}(x \circ y)), \square(\mathrm{im}(x \circ y)))$
$x \triangledown y$	$(\triangledown(\mathrm{re}(x \circ y)), \triangledown(\mathrm{im}(x \circ y)))$
$x \triangle y$	$(\triangle(\mathrm{re}(x \circ y))), \triangle(\mathrm{im}(x \circ y)))$

For computer implementation of the entries in the second column of this table see [12] and Section E.

(f) Operators of type segment:

In each of the cases $S \times I$, $I \times S$ the operand of type I is to be taken as being of type S.

Let $\bullet$ and $\circ$ be identical symbols taken from the set $\{+, -, *, /\}$, and let x, y be operands of admissible type. Let $x \circ y$ denote the correct result in $\mathbb{S}\mathbb{R}$ of the operation $\circ$. Then the result of the processor operation $\bullet$ may be read from the following table:

Expression to be evaluated	Result
$x \bullet y$	$[\triangledown(\inf(x \circ y)), \triangle(\sup(x \circ y))]$

For computer implementation of the entries in the second column of this table see [12] and Section E. The operators +* and ** resp., are defined as follows:

$x + * y$ represents $[\min(\inf(x), \inf(y)); \max(\sup(x), \sup(y))]$

and

$x{*}{*}y$ represents $[\max(\inf(x), \inf(y)); \min(\sup(x), \sup(y))]$

if the lower bound of the result is not greater than the upper bound. Otherwise an error shall be caused.

(g) Operators of type csegment:

In each of the cases $CS \times CS$, $CS \times S$, $S \times CS$, $CS \times I$, $I \times CS$ the operands of type I and S are to be taken as being of type CS. Let $\bullet$ and $\circ$ be identical symbols taken from the set $\{+, -, *, /\}$, and let x, y be operands of admissible type. Let $x \circ y$ denote the correct result in $\boldsymbol{S}$¢ of the operation $\circ$. Then the result of the processor operation $\bullet$ may be read from the following table:

Expression to be evaluated	Result
$x \bullet y$	$([\nabla \inf(\mathrm{re}(x \circ y)); \Delta \sup(\mathrm{re}(x \circ y))],$ $[\nabla \inf(\mathrm{im}(x \circ y)); \Delta \sup(\mathrm{im}(x \circ y))])$

For computer implementation of the entries in the second column of this table see [12] and Section E. The operators $+*$ and $**$, resp., are defined as follows:

$$(x_1, x_2) +* (y_1, y_2) \text{ represents } (x_1 +* y_1, x_2 +* y_2)$$

and

$$(x_1, x_2) ** (y_1, y_2) \text{ represents } (x_1 ** y_1, x_2 ** y_2)$$

if $x_1 ** y_1$ and $x_2 ** y_2$ are defined. Otherwise an error shall be caused.

(h) Operators of matrix and vector type:

All operators $\bullet \in \{+, -, *, +*, **\}$ are evaluated componentwise except for the multiplication of a vector or a matrix by a matrix.

For the multiplication of a vector or a matrix by a matrix the special scalar product is automatically applied. The scalar product is symbolized by the operator symbols $*$, $\overset{*}{\nabla}$, $\overset{*}{\Delta}$ in the cases of $VR \times VR$ and $VC \times VC$ and by the operator symbol $*$ in the cases $VS \times VS$ and $VCS \times VCS$.

Let x, y be vector operands of admissible type . Let $x \circ y$ denote the correct scalar product. Then the result of the processor operation $\ast$ may be read from the following table:

Expression to be evaluated	Result	
$x \ast y$	$\square(x \circ y)$	$\mathcal{T} \in \{VR, VC, VS, VCS\}$
$x \mathbin{\nabla\!\!\!\ast} y$	$\nabla(x \circ y)$	$\mathcal{T} \in \{VR, VC\}$
$x \mathbin{\triangle\!\!\!\ast} y$	$\triangle(x \circ y)$	

The number of the components of the two vectors have to be equal. Note that in case of a complex scalar product none of the operands is conjugated. For implementation see Section E.

Note that a diagonal matrix with non-zero non-diagonal elements is handled as a matrix.

The standard functions are listed in P31.

Remark: An essential simplification in the utilization of the functions listed in P36 may be made in the following way. The parameters specify the function and the result type. Then the function identifier need not repeat this information. Thus the number of identifiers can be considerably reduced.

Example:

*s*inf
*cs*inf
*vs*inf
*ms*inf
*vcs*inf
*mcs*inf

} are then replaced by the overloaded function identifier inf.

The following function computes the inverse of a given dynamic square matrix A according to the Schulz-Method. X1 is a given approximation of the inverse.

```
function  inv(var A,X1:  matrix [m1..n1,m2..n2] of real):
                         matrix [m1..n1,m1..n1] of segment;
          var R,Y,Z,Y0:  matrix [m1..n1,m1..n1] of segment;
                     I:  diagonal [m1..n1,m1..n1] of real;
          begin      I := drid;
                     R := ◊(2*I−X1*A);
                     Y0:= ◊(I−X1*A);
                     Z := Y0;
                     repeat   Y:= Z;
                              Z:= Y0+R*Y
                     until    Z in Y;
                     inv:= X1+Z
          end
```

The following part of a program computes a matrix expression $A*B + C*D$ with maximum accuracy.

```
var     A: matrix [1..2,1..4] of real;
        B: matrix [1..4,1..3] of real;
        C: matrix [1..2,1..1] of real;
        D: matrix [1..1,1..3] of real;
        Z: matrix [1..2,1..3] of segment;
begin      .   .   .
        Z:= ◊(A*B + C*D)
end
```

6.8.3.5 case-statement (P15)

If an <u>otherwise</u> clause occurs in the <u>case</u>-statement that clause is executed if the value of the case index is not equal to any case constant.

6.8.3.10 with-statement (P15)

If an REC RESULT occurs in the with-statement its components are subject to the same conventions and restrictions as function or result identifiers.

Example

```
type      polar = record rad, phi: real end;
operator * (a, b: polar) res: polar;
begin     with res do
          begin rad:= a.rad * b.rad;
                phi:= a.phi + b.phi
          end
end
```

6.9 Input/output

6.9.2 The Procedure Read

To read values for variables of type $\mathcal{T}$ from the input file there must be constants without rounding control on the input file. The type of constants must be assignment compatible with the type of the variables. The rounding control occurring in the input statement is executed as follows.

(a) In the case $\mathcal{T} = C$ the rounding operator pertains to the real and imaginary part.

(b) In the case $\mathcal{T} = S$ or $\mathcal{T} = CS$ the bounds of the segments are automatically rounded outwardly.

(c) In case of vectors and matrices ($\mathcal{T} \in \{VR, VC, VS, VCS, MR, MC, MS, MCS\}$) there must be a constant on the input file for every component. Matrices are read row-wise.

The rounding to the machine dependent floating point representation is modified by the optional rounding control in the following way (see (6.7.2.2.2).

Rounding control	Rounding executed	Value
□ , missing	implicit rounding	$\square(x)$
▽	downwardly directed rounding	$\triangledown(x)$
△	upwardly directed rounding	$\triangle(x)$

6.9.4 The Procedure Write

Values of expressions of type $\mathcal{T}$ are written as constants of type $\mathcal{T}$ on the output file. The rounding control occurring in the output statement is executed as follows.

In the case $\mathcal{T} = C$ all format specifications and rounding controls pertain to real and imaginary parts.

In the case $\mathcal{T} = S$ or $\mathcal{T} = CS$ the bounds of the segments are automatically rounded outwardly according to the format specifications.

In case of vectors and matrices ($\mathcal{T} \in \{VR, VC, VS, VCS, MR, MC, MS, MCS\}$) the components are written row-wise as constants of the type of the components.

6.9.4.2 Write Parameters

The values of TotalWidth and FracDigits shall be greater than or equal to zero. The value zero for FracDigits is ignored.

6.9.5.5 Real Type

The rounding is modified by the optional rounding control in the following way (see 6.7.2.2.2):

Rounding control	Rounding executed	Value
$\square$ / missing	implicit rounding	$\square(x)$
$\bigtriangledown$	downwardly directed rounding	$\bigtriangledown(x)$
$\bigtriangleup$	upwardly directed rounding	$\bigtriangleup(x)$

6.10 Hardware Representation

The additional symbols	may be represented by
$\square$	#=
$\bigtriangledown$	#<
$\bigtriangleup$	#>
$\Diamond$	##
$\bigtriangledown\!\!\!\!\!+$	+<
$\bigtriangledown\!\!\!\!\!-$	-<
$\bigtriangledown\!\!\!\!\!*$	*<
$\bigtriangledown\!\!\!\!\!/$	/<
$\bigtriangleup\!\!\!\!\!+$	+>
$\bigtriangleup\!\!\!\!\!-$	->
$\bigtriangleup\!\!\!\!\!*$	*>
$\bigtriangleup\!\!\!\!\!/$	/>

E. COMMENTS ON THE IMPLEMENTATION OF THE ARITHMETIC

In this section we comment on the arithmetic of MATRIX PASCAL. Theoretical aspects are reviewed, and details concerning reduction to practice are given also. Basic floating-point algorithms and roundings are discussed first, then the evaluation of expressions with maximum accuracy (EMA) and finally operations in the higher numerical spaces.

E1. Basic Floating-point Algorithms and Roundings

We begin our discussion by defining a floating-point system $R(b, \ell, \text{emin}, \text{emax})$ consisting of numbers of the form

$$\underbrace{\pm 0.m_1 m_2 \ldots m_\ell}_{\text{mantissa}} \bullet b^e$$

with emin $\leq e \leq$ emax, $0 \leq m_i \leq b-1$ for $i = 1(1)\ell$ and $m_1 \neq 0$ as well as a representer for $0 = 0.0\ldots0 \bullet b^{\text{emin}}$. We suppose that b and ℓ are natural numbers such that $b \geq 2$ and $\ell \geq 1$. We further suppose that emin, emax$\in\mathbb{Z}$. Unnormalized numbers with the special form

$$\pm 0.m_1 m_2 \ldots m_\ell \bullet b^{emin} \quad \text{with} \quad m_1 = 0$$

may be adjoined. Furthermore $+\infty$, $-\infty$, undefined numbers, etc., can be introduced (cf. [5]). All of these additional numbers affect the succeeding algorithms only in over/underflow recovery and in normalization. In the following the largest and smallest representable positive floating-point numbers are denoted by P and p, resp. The basic floating-point algorithms (level 2 routines) can be performed with an appropriate integer arithmetic (level 1 routines) which includes comparison, exact addition, multiplication and division with remainder (cf. [12], [18]). However, in principle only comparison and exact addition are necessary.

The result of a real operation applied to two floating-point numbers is a real number which is usually not exactly representable in the floating-point system. So rounding of the real result is critical.

We define roundings as mappings $\Box\colon D \to R$ with $D := \{x \in \mathbb{R} \mid -P \leq x \leq P\}$ where

$$\text{(R1)} \qquad \bigwedge_{x \in R} \Box x = x.$$

We require roundings only for the bounded domain D (compare [12], [18] and Section B above). When the argument x lies outside of D, we say that an overflow has occurred.

In the following development we are especially concerned with the monotone directed roundings, ∇ and $\triangle$, as well as the roundings which are both monotone and antisymmetric. Let $D^+ := \{x \in D \mid x \geq 0\}$ and $D^- = D \setminus D^+$. Consider the following roundings:

$$\bigwedge_{x \in D} \nabla x := \max \{y \in R \mid y \leq x\},$$ monotone downwardly directed rounding.

$$\bigwedge_{x \in D} \triangle x := \min \{y \in R \mid x \leq y\},$$ monotone upwardly directed rounding.

$$\bigwedge_{x \in D^+} \Box_b x := \nabla x \wedge \bigwedge_{x \in D^-} \Box_b x := -\Box_b (-x),$$ monotone rounding towards zero.

$$\bigwedge_{x \in D^+} \Box_0 x := \triangle x \wedge \bigwedge_{x \in D^-} \Box_0 x := -\Box_0 (-x),$$ monotone rounding away from zero.

Using the notation

$$\bigwedge_{x \in D^+} s_\mu(x) := \nabla x + (\triangle x - \nabla x)\frac{\mu}{b}, \ \mu = 1(1)b-1,$$

we define the following roundings $\Box_\mu \colon D \to R$, $\mu = 1(1)b-1$, which are in common use.

$$\bigwedge_{x \in [0,p)} \Box_\mu x := 0,$$

$$\bigwedge_{x \in [p,P]} \Box_\mu x := \begin{cases} \nabla x & \text{for } x \in [\nabla x, s_\mu(x)) \\ \triangle x & \text{for } x \in [s_\mu(x), \triangle x] \end{cases}$$

$$\bigwedge_{x \in D^-} \square_\mu x := -\square_\mu(-x).$$

If b is an even number, $\bigcirc := \square_{b/2}$ denotes the rounding to the nearest floating-point number.

The roundings listed above have many familiar properties. For instance we have

$$\nabla x = -\Delta(-x), \quad \Delta x = -\nabla(-x)$$

$$\square_b x = \mathrm{sign}(x)(\nabla |x|), \quad \square_0 x = \mathrm{sign}(x)(\Delta |x|).$$

All of these roundings are monotone $(R2)$ (see Section B). The roundings $\square_\mu$, $\mu = 0(1)b$, are also antisymmetric $(R3)$ (see Section B).

According to our theory the only way arithmetic can be defined for the first row of Figure 1 in Section B is by means of semimorphism. Thus we derive algorithms for the operations as defined by the following formula,

$$(RG) \quad \bigwedge_{x,y \in R} x \boxdot y := \square(x \circ y), \quad \circ \in \{+, -, *, /\}.$$

We implement *(RG)* for the roundings $\square_\mu$, $\mu = 0(1)b$, and for purposes of application to interval arithmetic, we implement it as well for ∇ and Δ.

In [12] it is shown by means of specific algorithms for $\circ \in \{+, -, *, /\}$, that whenever $x \circ y$ is not representable on the computer, it is sufficient to replace it by an appropriate and representable value $x \tilde{\circ} y$. The latter will have the property $\square(x \circ y) = \square(x \tilde{\circ} y)$ for all roundings $\square \in \{\nabla, \Delta, \square_\mu, \mu = 0(1)b\}$. Then $x \tilde{\circ} y$ can be used to define $x \boxdot y$ by means of the following relations.

$$(S) \quad \bigwedge_{x,y \in R} x \boxdot y = \square(x \circ y) = \square(x \tilde{\circ} y).$$

The algorithms which implement this relation can, in principle, be separated into the following five steps:

1. Decomposition of x and y, i.e., separation of x and y into mantissa and exponent. If floating-point numbers are not stored in a common word, this step is vacuous.

2. Determination of $x \tilde{\circ} y$. It may be that $x \tilde{\circ} y = x \circ y$.

3. Normalization of $x \tilde{\circ} y$. If the result of 2. is already normalized this step can be skipped.

4. Rounding of $x \tilde{\circ} y$ to determine $x \boxdot y = \square(x \circ y) = \square(x \tilde{\circ} y)$.

5. Composition, i.e., assembling of the mantissa and exponent of the result into a floating-point number. This step is vacuous if the floating-point numbers are not stored in words.

We proceed to formulate algorithms for the $\tilde{\circ}$, $\circ \in \{+, -, *, /\}$ satisfying condition (S). To do so one of the following registers may be used (cf. [12]):

i) long register: $2\ell + 1$ digits of base b and 1 carry bit

ii) short register: $\ell + 2$ digits of base b, 1 carry bit and 1 residual bit.

As expected, algorithms for the long register are simpler than algorithms for the short register. The condition *(RG)* and the rounding properties $(R1)$, $(R2)$, $(R3)$ and $(R4)$ must be strictly adhered to during every step of the implementation. We do this by an independent implementation of the five steps described above. Thus we insure that every rounding $\square \epsilon \{\triangle; \nabla; \square_\mu, \mu = 0(1)b\}$ applied to an intermediate result yields the correct rounded final result in the sense of (S).

E1.1 Addition and Subtraction

Without loss of generality we may assume that $ex \geq ey$.

i) long register:

We distinguish two cases:

a) $ex - ey \geq \ell + 2$: In case of the rounding $\square_\mu$, $1 \leq \mu \leq b-1$, y is too small in absolute value to influence the first ℓ digits of the sum, and we simply obtain $mz := mx$. However, in the case of the roundings ∇, $\triangle$, $\square_0$ or $\square_b$ an arbitrarily small $y \neq 0$ can change the mantissa mz. In every case we set $ez := ex$ and for the mantissa mz we set

$$mz := \begin{cases} mx, & \text{if } my = 0, \\ mx - b^{-(\ell+2)}, & \text{if } my < 0, \\ mx + b^{-(\ell+2)}, & \text{if } my > 0. \end{cases}$$

The latter two cases here, may be viewed as the subtraction resp. addition of 1 in the $(\ell + 2)$-nd digit of mx.

b) $ex - ey \leq \ell + 1$. After a shift of $ex - ey$ digits of the mantissa of y we compute the correct sum $mz := mx + my \bullet b^{(ex-ey)}$ in the long register.

ii) short register:

We distinguish three cases:

a) $ex - ey \geq \ell + 2$. This case can be treated as in case of the long register.

b) $ex - ey \leq 2$. In this case the two mantissas can be added correctly after shifting y.

c) $2 < ex - ey \leq \ell + 1$. Here we set the residual bit if a nonzero digit is shifted beyond the $(\ell + 2)$-nd register position. Shifting must not reset the residual bit. The residual bit is to be added depending on the sign of my, when the summation $mz := mx + my$ is performed.

Subtraction is reduced to addition upon changing the sign of y.

In the remarks concerning multiplication and division, both of which now follow, we do not distinguish, a priori, between the two registers.

E1.2 Multiplication

If $x = 0$ or $y = 0$ then $\square(x * y) = 0$ for every rounding $\square \in \{\nabla, \Delta, \square_\mu$ with $\mu = 0(1)b\}$. Otherwise we set $ez := ex + ey$ and $mz := mx * my$. The product of the mantissas can be represented exactly in the long register. In the short register we store the first $\ell + 2$ digits of the product and in the residual bit we note whether at least one of the $(\ell + 3)$-rd to the 2ℓ-th digits of mz is nonzero or not.

E1.3 Division

If $y = 0$ an error message is given. If $x = 0$ and $y \neq 0$ we have $\square(x/y) = 0$ for every $\square \epsilon \{\nabla, \Delta, \square_\mu$ with $\mu = 0(1)b\}$. Otherwise we set $ez := ex - ey$. If $|mx| \geq |my|$ we shift mx by 1 digit and correct ez by 1. Thus a normalized mz is produced. The cases in which x or y are unnormalized numbers have to be treated separately and in an obvious way. It suffices to compute $\ell + 1$ digits of mx/my. For the roundings ∇, Δ, $\square_0$ and $\square_b$ the remainder r must not be omitted. Indeed if $r \neq 0$, then

$$mz := \begin{cases} mz - b^{-(\ell+2)}, & \text{if } mz < 0, \\ mz + b^{-(\ell+2)}, & \text{if } mz > 0. \end{cases}$$

That is, the remainder r furnishes a correction in the $(\ell + 2)$-nd digit of mz; the correction being dependent upon the sign of z.

After performing addition, subtraction or multiplication a normalization has to be executed in a rather obvious way. If unnormalized numbers are allowed, slight changes must be made in both division and the normalization process which follows multiplication.

A more detailed description of these implementations can be found in [12] and [18]. In particular a collection of special critical cases are dealt with there in sufficient detail to show that all desired algorithmic objectives are attained.

E2. Expressions with Maximum Accuracy

We propose the availability of a long accumulator for the evaluation of expressions with maximum accuracy (EMA, see Section B, D). The long accumulator has the following capability:

(A) The accumulator can be set to zero.

(B) A floating-point number or a product of two or three such numbers can be added to or subtracted from the contents of the accumulator.

(C) The contents of the accumulator can be rounded to a floating-point number using one of the roundings described above.

E2.1 Real Expressions with Maximum Accuracy

Allowable real expressions with maximum accuracy (see diagram P27), in particular scalar products and bilinear forms, can be reduced to sums of the following form

$$\square\left(\sum_{i=1}^{n} p_i\right), \quad \text{where } p_i = \begin{cases} a_i & \text{or} \\ a_i * b_i & \text{or} \\ a_i * b_i * c_i, & \end{cases} \tag{1}$$

and a_i, b_i, $c_i \epsilon R(b, \ell, emin, emax)$ are floating-point numbers.*

We assume that these products can be executed exactly in the floating-point system $R(b, 3\ell, 3emin-2, 3emax)$. The sum described in (1) can be computed in a long accumulator A of the form:

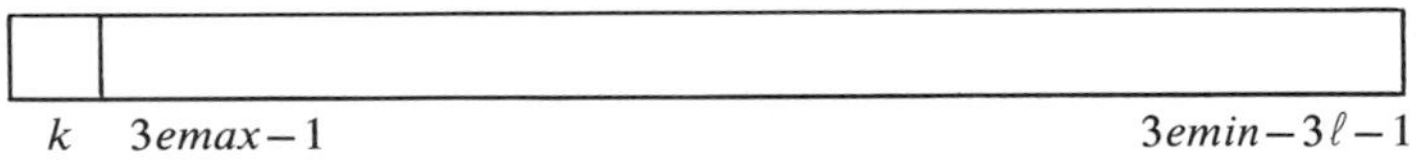

(2)

The long accumulator A may be interpreted as a fixed point number with $\{|3\ emin-3\ell-1| + (3emax-1) + k + 1\}$ digits of base b. The k leading digits of the accumulator A are provided to avoid overflow when adding another summand to the contents of A. Thus every sum (1), for $n \leq b^k$ summands can be computed exactly in the long accumulator A in (2).

The long accumulator A may be represented as a sequence of words (i.e., storage locations) in the computer. In a software procedure it might be simulated by an array of integers. The addition of $p_1 = 10^{50}$ and $p_2 = -10^{-50}$ shows that carries may need processing over long distances. This and other reasons motivate the use of two long accumulators: one to

---

* Let e.g. be $a \epsilon R$, $x = (x_i)_{i=1,\ldots,k}$, $y = (y_i)_{i=1,\ldots,k}$, $u = (u_i)_{i=1,\ldots,r}$, $v = (v_i)_{i=1,\ldots,s}$, $B = (b_{ij})_{i=1,\ldots,r, j=1,\ldots,s}$. Then

$$\square(a + x*y + u*B*v)$$

can be evaluated as follows:

$$\square\left(\sum_{i=1}^{n} p_i\right) = \square\left(a + \sum_{i=1}^{k} x_i * y_i + \sum_{i=1}^{r} \sum_{j=1}^{s} u_i * v_j * b_{ij}\right).$$

sum up the positive p_i, the other for the negative ones. This approach requires an algorithm to compute the difference of the numbers contained in the two long accumulators. Recall however that it is not the exact value of the difference which is needed but only its first $\ell + 1$ significant base b digits and a residual bit which indicates whether there is at least one non-zero digit following or not. Of course, in this subtraction some or all leading digits of the numbers represented by the long accumulator may be cancelled. Some special care has to be taken for this case. The $(\ell + 1)$-st digit of the difference is needed for the roundings $\square_\mu$, the residual bit for the roundings $\triangledown$ and $\triangle$.

The length required for the long accumulator (2) depends on the exponent range *emin..emax*. To simplify the final subtraction of the two accumulators we use two pointers which indicate the first and last non-zero digits in each accumulator, resp. The pointers to the first non-zero digits make for an easy computation of the sign of the difference, the pointers to the last non-zero digits make for a simple determination of the residual bit. The real number represented by the long accumulator can be converted to a floating-point number $x \in R(b, \ell, emin, emax)$ with the roundings $\triangledown$, $\triangle$ $\square_\mu$, $\mu = 0(1)b$. To realize $\diamond$ for EMA (see diagram P26) the contents of the long accumulator is rounded with $\triangledown$ and $\triangle$, resp., yielding the infimum and supremum of the resulting segment, resp.

There are other possibilities for the realization of a long accumulator. E.g., only one long accumulator will suffice when it is composed of partial words each having its own sign. Without going into details we point out that using this concept and possibly a redundant arithmetic most of the carry ripples vanish. Similar algorithms can be formulated using the methods described in [3], [12] and [18].

E.2.2 Complex Expressions with Maximum Accuracy

As in Section B.3 we name the basic data types I, R, ..., VC, ..., resp. Let $a, b \in VC$ be two complex floating-point vectors of the form $a = (a_i)_{i=1, \ldots, n}$, $b = (b_i)_{i=1,\ldots,n}$, $a_i, b_i \in C$. Then the scalar product is defined by

$$\square(a*b) = \square\Big(\sum_{i=1}^{n} a_i b_i\Big) := \begin{array}{l}\Big(\square\Big(\sum_{i=1}^{n} (Re(a_i)Re(b_i) - Im(a_i)Im(b_i))\Big),\\ \square\Big(\sum_{i=1}^{n} (Im(a_i)Re(b_i) + Re(a_i)Im(b_i))\Big)\Big).\end{array} \tag{3}$$

Complex bilinear forms xAy are defined in a similar manner. To avoid confusion, note that the complex vector x will not be conjugated automatically by the evaluation of this expression. The real and imaginary part of a complex expression with maximum accuracy have to be evaluated as two real expressions with maximum accuracy.

E2.3 Real and Complex Matrix and Vector Expressions with Maximum Accuracy

Such expressions are performed componentwise using the methods of E2.1 and E2.2.

E3. The Operations in the Higher Numerical Spaces

In analogy with the floating-point operations all operations in the higher numerical spaces are defined by semimorphism. The latter can be performed by componentwise execution of the basic floating-point operations in subsection E1 and of the expressions with maximum accuracy in subsection E2.

E3.1 The Arithmetic in C

The operations in C are defined by

$$(RG) \qquad \bigwedge_{(a,b),(c,d)\in C} (a, b) \boxdot (c, d) := \square((a, b) \circ (c, d)), \ \circ \in \{+, -, *, /\},$$

where the rounding $\square$: $\mathbb{C} \rightarrow C$ is defined by

$$\bigwedge_{(a,b)\in \mathbb{C}} \square(a, b) := (\square a, \square b) .$$

The rounding on the right-hand side denotes one of the roundings $\square_\mu$, $\mu = 0(1)b$. This leads to the following explicit representation of the operations for all couples of elements $(a, b), (c, d) \in C$.

$$(a, b) \boxplus (c, d) = (a \boxplus c, b \boxplus d),$$

$$(a, b) \boxminus (c, d) = (a \boxminus c, b \boxminus d),$$

$$(a, b) \boxast (c, d) = (\Box(ac - bd), \Box(ad + bc)),$$

$$(a, b) \boxslash (c, d) = \Big(\Box\Big(\frac{ac + bd}{c^2 + d^2}\Big), \Box\Big(\frac{bc - ad}{c^2 + d^2}\Big)\Big).$$

Division is not defined if $c = d = 0$.

Thus addition and subtraction can be executed in terms of the corresponding operations in R. The products can be evaluated by means of two real EMAs.

The quotient can be performed by another application of the methods of Section E2. In the quotient formula expressions of the form

$$q := (x_1 y_1 + x_2 y_2)/(u_1 v_1 + u_2 v_2)$$

occur with $x_i, y_i, u_i, v_i \in R$. First compute the products $x_i y_i$ and $u_i v_i$ of length 2ℓ. Then compute $\ell + 8$ digit approximations, $\tilde{n}$ and $\tilde{d}$, for the numerator and denominator of the quotient q. With these expressions, one gets an approximation $\tilde{q} := \tilde{n}/\tilde{d}$ of $\ell + 5$ digits for the exact quotient q. In most cases it turns out that $\Box\tilde{q} = \Box q$ for all roundings $\Box \in \{\nabla, \Delta, \Box_\mu, \mu = 0(1)b\}$. If $\Box\tilde{q} \neq \Box q$, the correct result $\Box q$ can be determined from $\tilde{q}$ and the residual

$$r := \Box(u_1 v_1 \tilde{q} + u_2 v_2 \tilde{q} - x_1 y_1 - x_2 y_2)$$

as a real EMA (see [6]).

The order relation $\leq$ in C is defined in terms of the order relation in R:

$$(a, b) \leq (c, d) :\Longleftrightarrow a \leq c \wedge b \leq d.$$

E3.2 The Arithmetic in S

The following formulas derived by semimorphism define the operations in S $(a, b, c, d \in R)$:

$$[a, b] \mathbin{\Diamond\hspace{-0.62em}+} [c, d] = [a \overset{+}{\triangledown} c,\ b \overset{+}{\triangle} d],$$

$$[a, b] \mathbin{\Diamond\hspace{-0.62em}-} [c, d] = [a \overset{-}{\triangledown} d,\ b \overset{-}{\triangle} c],$$

$$[a, b] \mathbin{\Diamond\hspace{-0.62em}*} [c, d] = [\min\{a \overset{*}{\triangledown} c,\ a \overset{*}{\triangledown} d,\ b \overset{*}{\triangledown} c,\ b \overset{*}{\triangledown} d\},\ \max\{a \overset{*}{\triangle} c,\ a \overset{*}{\triangle} d,\ b \overset{*}{\triangle} c,\ b \overset{*}{\triangle} d\}],$$

$$[a, b] \mathbin{\Diamond\hspace{-0.62em}/} [c, d] = [\min\{a \overset{/}{\triangledown} c,\ a \overset{/}{\triangledown} d,\ b \overset{/}{\triangledown} c,\ b \overset{/}{\triangledown} d\},\ \max\{a \overset{/}{\triangle} c,\ a \overset{/}{\triangle} d,\ b \overset{/}{\triangle} c,\ b \overset{/}{\triangle} d\}]\ .$$

The division is not defined if $0 \in [c, d]$. In the multiplication and division formulas, the minimum and maximum can be determined by considering the sign of the bounds of the interval operands.

In addition to the arithmetic operations in S, the intersection, $\cap$, and the convex hull, $\overline{\cup}$, have to be made available:

$$[a, b] \cap [c, d] := [\ \max\ \{a, c\},\ \ \min\ \{b, d\}],$$

$$[a, b] \mathbin{\overline{\cup}} [c, d] := [\ \min\ \{a, c\},\ \ \max\ \{b, d\}].$$

The intersection is not defined if $b < c$ or $d < a$.

The order relations $\leq$ and $\subseteq$ in S are defined by

$$[a, b] \leq [c, d] \ :\Longleftrightarrow\ a \leq c \wedge b \leq d,$$

$$[a, b] \subseteq [c, d] \ :\Longleftrightarrow\ c \leq a \wedge b \leq d.$$

Here the relation $\leq$ on the right-hand side denotes the order relation in R.

E3.3 The Arithmetic in CS

We obtain the following formulas for all couples of elements, $(A, B), (C, D) \in CS$ with $A = [a_1, a_2],\ B = [b_1, b_2],\ C = [c_1, c_2],\ D = [d_1, d_2] \in S$:

$$(A, B) \mathbin{\Diamond\hspace{-0.62em}+} (C, D) = (A \mathbin{\Diamond\hspace{-0.62em}+} C,\ B \mathbin{\Diamond\hspace{-0.62em}+} D) = ([a_1 \overset{+}{\triangledown} c_1,\ a_2 \overset{+}{\triangle} c_2],\ [b_1 \overset{+}{\triangledown} d_1,\ b_2 \overset{+}{\triangle} d_2]),$$

$$(A, B) \mathbin{\Diamond\hspace{-0.62em}-} (C, D) = (A \mathbin{\Diamond\hspace{-0.62em}-} C,\ B \mathbin{\Diamond\hspace{-0.62em}-} D) = ([a_1 \overset{-}{\triangledown} c_2,\ a_2 \overset{-}{\triangle} c_1],\ [b_1 \overset{-}{\triangledown} d_2,\ b_2 \overset{-}{\triangle} d_1]),$$

$$(A, B) \text{◈} (C, D) = \Big(\Diamond\,(\Box(AC-BD)),\ \Diamond(\Box(AD+BC))\Big)$$

$$= \Big([\nabla\{\min_{i,j=1,2}(a_ic_j) - \max_{i,j=1,2}(b_id_j)\},\ \Delta\{\max_{i,j=1,2}(a_ic_j) - \min_{i,j=1,2}(b_id_j)\}],$$

$$[\nabla\{\min_{i,j=1,2}(a_id_j) + \min_{i,j=1,2}(b_ic_j)\},\ \Delta\{\max_{i,j=1,2}(a_id_j) + \max_{i,j=1,2}(b_ic_j)\}]\Big),$$

$$(A, B) \Diamond\!\!/\, (C, D) = \Big(\Diamond(\Box(AC+BD)\ \boxed{/}\ \Box(CC+DD)),$$

$$\Diamond(\Box(BC-AD)\ \boxed{/}\ \Box(CC+DD))\Big).$$

The division is defined only if $0 \notin \Box(CC + DD)$. The operation $\boxed{/}$ occurring on the right-hand side of the last formula denotes the operation in S. The order relations $\leq$ and $\subseteq$, the intersection $\cap$ and the convex hull $\overline{\cup}$ in CS are defined componentwise.

The general rule for performing the operations upon operands of different types is to lift the operand of the simpler type into the type of the remaining operand by means of a transfer function. Then the operation can be executed as one of the inner operations all of which we have already dealt with.

E3.4 The Scalar Products for Floating-point Segments and Complex Segments

In order to realize the segment and complex segment matrix and vector products by semimorphism the following scalar products must be available. Let $a, b \in VS$ be two floating-point segment vectors of the form $a = (a_i)_{i=1,\ldots,n}$, $b = (b_i)_{i=1,\ldots,n}$; $a_i, b_i \in S$. Then the scalar product is defined by:

$$a \text{◈} b = \Diamond(a*b) = [\nabla \underline{s}, \Delta \overline{s}] =: \Diamond\, s$$

where

$$s := \left[\sum_{i=1}^{n} \min(\underline{a}_i\,\underline{b}_i,\ \underline{a}_i\,\overline{b}_i,\ \overline{a}_i\,\underline{b}_i,\ \overline{a}_i\,\overline{b}_i),\right. \qquad (4)$$

$$\left.\sum_{i=1}^{n} \max(\underline{a}_i\,\underline{b}_i,\ \underline{a}_i\,\overline{b}_i,\ \overline{a}_i\,\underline{b}_i,\ \overline{a}_i\,\overline{b}_i)\right].$$

Here $\underline{s}$ and $\overline{s}$ resp. denote the left and right bounds of the interval $s \in S$. Consider the computation of $\underline{s}$. By the definition (4) it suffices to determine two vectors v and w with the property:

$$v_i w_i := \min (\underline{a}_i\, \underline{b}_i, \underline{a}_i\, \overline{b}_i, \overline{a}_i\, \underline{b}_i, \overline{a}_i\, \overline{b}_i) \quad \text{for} \quad i = 1(1)n. \tag{5}$$

By distinguishing different cases for the multiplication of two intervals (cf. Table 1, p.99 in [12]) it is seen that only in the case $0 \in a_i$ and $0 \in b_i$ two products need to be computed in order to determine the minimum in (5). In all other cases one product suffices because of sign considerations. In the case $0 \in a_i$ and $0 \in b_i$ it has to be decided whether

$$\underline{a}_i\, \overline{b}_i \le \overline{a}_i\, \underline{b}_i \quad \text{or} \quad \underline{a}_i\, \overline{b}_i > \overline{a}_i\, \underline{b}_i.$$

For this purpose either the products $\underline{a}_i\, \overline{b}_i$ and $\overline{a}_i\, \underline{b}_i$ have to be computed precisely in double precision arithmetic or the difference of them is computed using the upper rounded scalar product as a special case of a real EMA.

$$\triangle(\underline{a}_i\, \overline{b}_i - \overline{a}_i\, \underline{b}_i) > 0 \iff \underline{a}_i\, \overline{b}_i > \overline{a}_i\, \underline{b}_i.$$

If $\triangle(\underline{a}_i\, \overline{b}_i - \overline{a}_i\, \underline{b}_i) \le 0$, then set $v_i = \underline{a}_i$ and $w_i = \overline{b}_i$, otherwise set $v_i = \overline{a}_i$ and $w_i = \underline{b}_i$. After determining the vectors v and w in the prescribed manner we have

$$\underline{s} = \triangledown(v * w).$$

Similar considerations lead to the computation of $\overline{s}$ using

$$\triangledown(\underline{a}_i\, \underline{b}_i - \overline{a}_i\, \overline{b}_i) < 0 \iff \underline{a}_i\, \underline{b}_i < \overline{a}_i\, \overline{b}_i.$$

Let $a, b \in VCS$ be two complex segment vectors of the form $a = (a_i)_{i=1,\ldots,n}$, $b = (b_i)_{i=1,\ldots,n}$, $a_i, b_i \in CS$. Then the scalar product is defined by

$$\lozenge(a*b) = \lozenge\left(\sum_{i=1}^{n} a_i b_i\right) := \begin{array}{l} \Big(\lozenge \sum_{i=1}^{n} (Re(a_i)Re(b_i) - Im(a_i)Im(b_i)), \\ \\ \lozenge \sum_{i=1}^{n} (Im(a_i)Re(b_i) + Re(a_i)Im(b_i))\Big). \end{array} \qquad (6)$$

The components of (6) are segment scalar products of $2n$ summands. These scalar products can be evaluated using the methods of E2.2.

E3.5 Matrix and Vector Operations

We are now going to define operations for the product sets (matrices and vectors) of all the sets with which we have dealt. All arithmetic operations are again defined by semimorphisms.

Let T be one of the sets (types) R, C, S, CS. We denote the set of $m \times n$ matrices with elements in T by

$$M_{mn}T := \{(a_{ij}) \mid a_{ij} \in T \text{ for } i = 1(1)m \wedge j = 1(1)n\}$$

and the set of n-dimensional vectors with elements in T by

$$V_nT := \{(a_i) \mid a_i \in T \text{ for } i = 1(1)n\}.$$

The operations defined below for matrices are directly extended to vectors if we identify the sets

$$V_nT \equiv M_{n1}T.$$

In $M_{mn}T$ we define operations $\circ : M_{mn}T \times M_{mn}T \rightarrow M_{mn}T$ by

$$\bigwedge_{(a_{ij}),(b_{ij}) \in M_{mn}T} (a_{ij}) \circ (b_{ij}) := (a_{ij} \circ b_{ij}).$$

Here the operation sign, $\circ$, on the right-hand side denotes certain operations, depending on the following two cases. In

Case $T = R$ or C: one of the rounded operations $\boxplus$, $\boxminus$,

Case $T = S$ or CS: one of the operations $\langle + \rangle$, $\langle - \rangle$, $\cap$, $\overline{\cup}$.

In addition to these operations, we define outer multiplications $*: T \times M_{mn}T \to M_{mn}T$ by

$$\bigwedge_{a \in T} \quad \bigwedge_{(b_{ij}) \in M_{mn}T} \quad a*(b_{ij}) := (a*b_{ij}).$$

Here the multiplication sign, $*$, on the right-hand side has two cases of realization also.

In

Case $T = R$ or C: rounded multiplication $\boxed{*}$,

Case $T = S$ or CS: multiplication $\langle * \rangle$.

Finally we define the products of matrices $*$: $M_{mn}T \times M_{np}T \to M_{mp}T$ by

$$\bigwedge_{(a_{ij}) \in M_{mn}T} \quad \bigwedge_{(b_{jk}) \in M_{np}T} \quad (a_{ij})*(b_{jk}) := \square\left(\sum_{j=1}^{n} a_{ij} b_{jk}\right).$$

The sums have to be calculated with the appropriate EMA of Section E2 or with the scalar product of Section E3.4.

For example, the multiplication of two complex matrices can be formulated as follows:

```
function cprod(a: matrix [n1..m1, n2..m2] of complex;
               b: matrix [n3..m3, n4..m4] of complex):
         matrix [n1..m1, n4..m4] of complex;

var row, col: integer;
begin
    for row:= n1 to m1                do
        for col:= n4 to m4 do
            cprod [row, col] := □ (a[row] * b[*, col])
end;
```

This function can be formulated as an operator similarly.

The order relation $\leq$ is defined componentwise in all spaces $M_{mn}T$ $(T \in \{R, C, S, CS\})$. For the spaces $M_{mn}T$ $(T \in \{S, CS\})$ the order relation $\subset$, the intersection $\cap$ and the convex hull $\overline{\cup}$ are defined componentwise also.

As before, the general rule for performing the matrix and vector operations for operands of different component types is to lift the operand with the simpler component type into the type of the remaining operand by means of a transfer function. Then the operation can be executed as one of the inner operations all of which we have already dealt with.

ACKNOWLEDGEMENTS

This paper was prepared during a stay at the University of Karlsruhe, in March 1981 and at the Mathematics Research Institute, Oberwolfach in March 1982.

We are grateful to U. Allendörfer, R. Kirchner, M. Neaga and H. W. Wippermann of the University of Kaiserslautern, whose implementation of the major part of the proposed language as well as whose experience contributed in an essential way to its well rounded form.

REFERENCES

[1] Addyman, A. M. (1980). A Draft Proposal for Pascal, SIGPLAN NOTICES, 15, S. 1-67.

[2] Alefeld, G., Herzberger, J. (1974). *Einführung in die Intervallrechnung,* Reihe Informatik, Band 12, Wissenschaftsverlag des Bibliographischen Instituts Mannheim.

[3] Bohlender, G. (1978). Genaue Berechnung mehrfacher Summen, Produkte und Wurzeln von Gleitkommazahlen und allgemeine Arithmetik in höheren Programmiersprachen, Dissertation, Universität Karlsruhe.

[4] Bohlender, G. et al. (1980). FORTRAN for Contemporary Numerical Computation, IBM Research Report RC 8348 and *Computing* 26, 277-314 (1981).

[5] Coonan, J. et al. (1979). A Proposed Standard for Floating-point Arithmetic, SIGNUM Newsletter.

[6] Haas, H. Ch. (1975). Implementierung der komplexen Gleitkommaarithmetik mit maximaler Genauigkeit, Diplomarbeit am Institut für Angewandte Mathematik, Universität Karlsruhe.

[7] Jensen, K, Wirth, N. (1976). *PASCAL User Manual and Report,* Lecture Notes in Computer Science, Vol. 18, Springer-Verlag, Berlin.

[8] Kaucher, E., Klatte, R., Ullrich, Ch. (1978). Benutzerfreundliche Darstellung der Syntax von *PASCAL* durch Syntaxdiagramme, Applied Computer Science, *Berichte zur praktischen Informatik* 11, 43-62, Hanser-Verlag, München.

[9] Kaucher, E., Klatte, R., Ullrich, Ch. (1978). Neuere Methoden zur Beschreibung von Programmiersprachen, *Jahrbuch Überblicke Mathematik,* Bibliographisches Institut, Mannheim.

[10] Kaucher, E., Klatte, R., Ullrich, Ch. (1978). *Höhere Programmiersprachen ALGOL, FORTRAN, PASCAL in einheitlicher und übersichtlicher Darstellung,* Reihe Informatik, Band 24, Wissenschaftsverlag des Bibliographischen Instituts Mannheim.

[11] Kulisch, U. (1976). *Grundlagen des Numerischen Rechnens-Mathematische Begründung der Rechnerarithmetik,* Reihe Informatik, Band 19, Wissenschaftsverlag des Bibliographischen Instituts Mannheim.

[12] Kulisch, U. W., Miranker, W. L. (1981). *Computer Arithmetic in Theory and Practice,* Academic Press.

[13] Moore, R. E. (1966). *Interval Analysis,* Englewood Cliffs, NJ, Prentice Hall Inc.

[14] Rutishauser, H. (1967). *Description of ALGOL-60,* Springer-Verlag, Berlin-Heidelberg.

[15] Rump, S. M. (1980). Kleine Fehlerschranken bei Matrixproblemen, Dissertation, Universität Karlsruhe.

[16] Wolff v. Gudenberg, J. (1980). Einbettung allgemeiner Rechnerarithmetik in *PASCAL* mittels eines Operatorkonzeptes und Implementierung der Standardfunktionen mit optimaler Genauigkeit, Dissertation, Universität Karlsruhe.

[17] Ullrich, Ch. P. (1982). A FORTRAN Extension for Scientific Computation. Proceedings of the 10th IMACS World Congress on Systems Simulation and Scientific Computation.

[18] Kulisch, U., Ullrich, Ch. (Hrsg.) (1982). Wissenschaftliches Rechnen und Progammiersprachen. Berichte des German Chapter of the ACM, 10, Teubner Stuttgart.

[19] Bohlender, G. et al. (1981). PASCAL-SC: A PASCAL for Contemporary Scientific Computation, IBM Research Report RC 9009.